Alan Howard, Pit-Mann Wong (Eds.)

Contributions to Several Complex Variables

Aspects of Mathematics
Aspekte der Mathematik

Editor: Klas Diederich

Vol. E1: G. Hector/U. Hirsch, Introduction to the Geometry of Foliations, Part A

Vol. E2: M. Knebusch/M. Kolster, Wittrings

Vol. E3: G. Hector/U. Hirsch, Introduction to the Geometry of Foliations, Part B

Vol. E4: M. Laska, Elliptic Curves over Number Fields with Prescribed Reduction Type

Vol. E5: P. Stiller, Automorphic Forms and the Picard Number of an Elliptic Surface

Vol. E6: G. Faltings/G. Wüstholz et. al., Rational Points
(A Publication of the Max-Planck-Institut für Mathematik, Bonn)

Vol. E7: W. Stoll, Value Distribution Theory for Meromorphic Maps

Vol. E8: W. von Wahl, The Equations of Navier-Stokes and Abstract Parabolic Equations

Vol. E9: A. Howard, P.-M. Wong (Eds.), Contributions to Several Complex Variables

Band D1: H. Kraft, Geometrische Methoden in der Invariantentheorie

The texts published in this series are intended for graduate students and all mathematicians who wish to broaden their research horizons or who simply want to get a better idea of what is going on in a given field. They are introductions to areas close to modern research at a high level and prepare the reader for a better understanding of research papers. Many of the books can also be used to supplement graduate course programs.
The series comprises two sub-series, one with English texts only and the other in German.

Alan Howard, Pit-Mann Wong (Eds.)

Contributions to Several Complex Variables

In Honour of Wilhelm Stoll

Springer Fachmedien Wiesbaden GmbH

Professors *Alan Howard* and *Pit-Mann Wong*,
Department of Mathematics, University of Notre Dame,
Post Office Box 398, Notre Dame, Indiana 46556, USA.

AMS Subject Classification: 32 06

Originally published by Friedr. Vieweg & Sohn Verlagsgesellschaft mbH, Braunschweig in 1986

Produced by W. Langelüddecke, Braunschweig

ISSN 0179-2156

ISBN 978-3-528-08964-1 ISBN 978-3-663-06816-7 (eBook)
DOI 10.1007/978-3-663-06816-7

Contents

Wilhelm Stoll
October 1984

Foreword

In 1960 Wilhelm Stoll joined the University of Notre Dame faculty as Professor of Mathematics, and in October, 1984 the university acknowledged his many years of distinguished service by holding a conference in complex analysis in his honour. This volume is the proceedings of that conference.

It was our priviledge to serve, along with Nancy K. Stanton, as conference organizers. We are grateful to the College of Science of the University of Notre Dame and to the National Science Foundation for their support.

In the course of a career that has included the publication of over sixty research articles and the supervision of eighteen doctoral students, Wilhelm Stoll has won the affection and respect of his colleagues for his diligence, integrity and humaneness. The influence of his ideas and insights and the subsequent investigations they have inspired is attested to by several of the articles in the volume.

On behalf of the conference partipants and contributors to this volume, we wish Wilhelm Stoll many more years of happy and devoted service to mathematics.

Alan Howard
Pit-Mann Wong

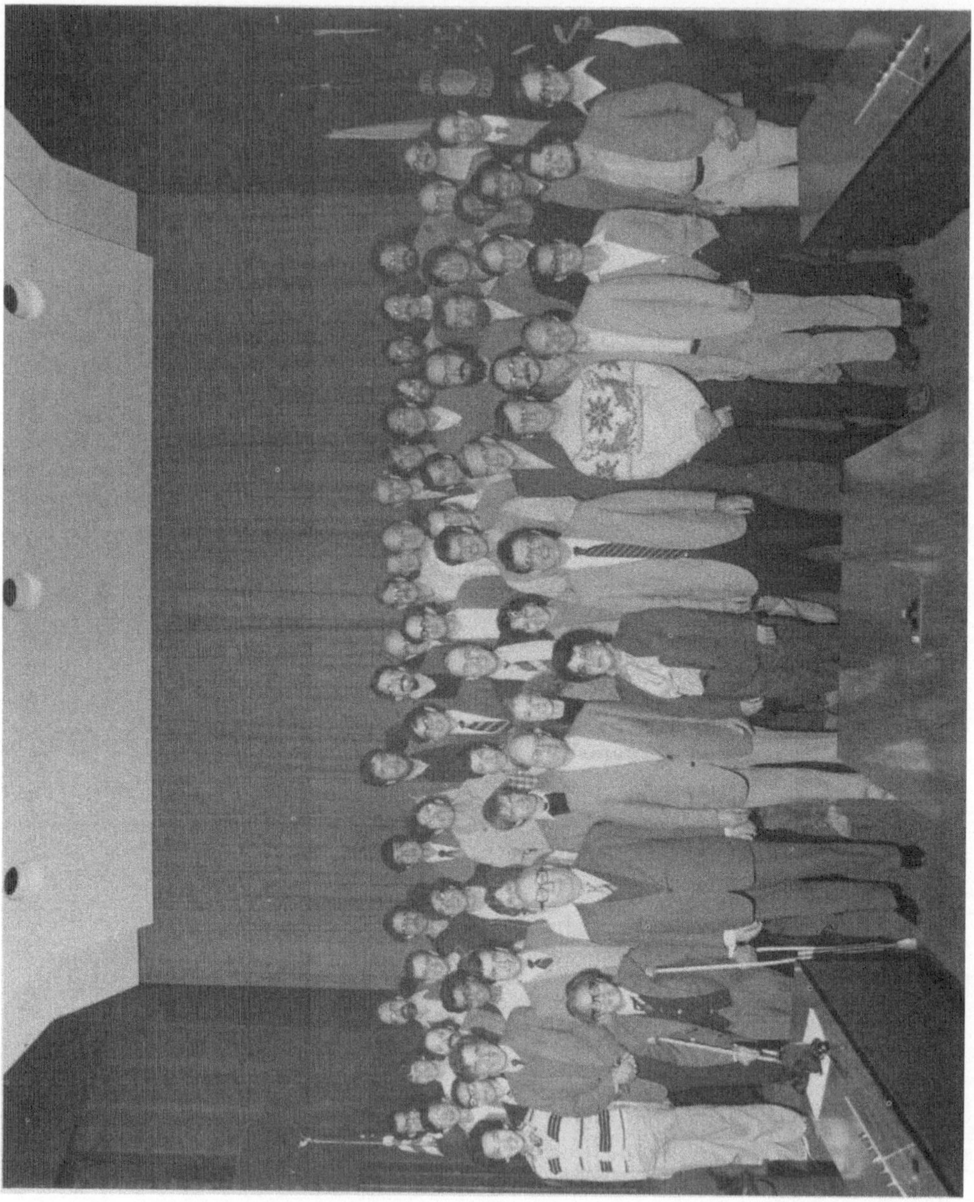

Group Picture of the Conference Participants

Scheme of the Group Picture (Names pages X, XI)

Participants on the Group Picture

1 Qi-keng LU, Professor, Chinese Academy of Science, Peking, China.

2 Pierre LELONG, Professor emeritus, University of Paris VI, France.

3 Wilhelm STOLL, Professor, University of Notre Dame.

4 Marilyn STOLL,

5 Timothy O'MEARA, Provost and Professor of Mathematics, Univ. of Notre Dame.

6 Mike SPURR, Professor, Rice University, Houston, Texas.

7 B. A. TAYLOR, Professor, University of Michigan, Ann Arbor, Mich.

8 Yi-Chuan PAN, Professor, Jackson State University, Jackson, Miss.

9 David BARRETT, Professor, Princeton University, Princeton, N.J.

10 Robert FOOTE, Professor, Texas Tech University, Lubbock, Texas.

11 Weiqi GAO, Student, University of Notre Dame.

12 Alan HOWARD, Professor, University of Notre Dame.

13 Dennis SNOW, Professor, University of Notre Dame.

14 Joanne SNOW, Professor, St. Mary's College, Notre Dame, Ind.

15 Robert MOLZON, Professor, University of Kentucky, Lexington, Ky.

16 Nancy STANTON, Professor, University of Notre Dame.

17 Mary Jo KREUZMAN, Student, University of Notre Dame.

18 Paula A. RUSSO, Professor, Michigan State Univ., East Lansing, Mich.

19 Eric BEDFORD, Professor, Indiana University, Bloomington, Ind.

20 Zbigniew SLODKOWSKI, Visiting Professor, Univ. of California, Los Angeles.

21 Hans GRAUERT, Professor, University of Göttingen, F.R.G.

22 Giorgio PATRIZIO, Professor, University of Rome II, Italy.

23 Alan HUCKLEBERRY, Professor, University of Bochum, F.R.G.

24 J. RAMANATHAN, Professor, University of Chicago, Illinois.

25 Henri SKODA, Professor, University of Paris VI, France.

26 Wanxi CHEN, Student, Univ. of Notre Dame (from Univ. of Science and Techn. of China).

27 Harry d'SOUZA, Professor, University of Michigan, Flint, Mich.

28 Chong-Kyu HAN, Professor, University of Alabama-Tuscaloosa, Univ. Alabama.

29 Stephen BELL, Professor, Princeton University, Princeton, N.J.

30 Yum-Tong SIU, Professor, Harvard University, Cambridge, Mass.

31 Klas DIEDERICH, Professor, University of Wuppertal, F.R.G.

32 Y. KIM, Student, University of Michigan, Ann Arbor, Mich.

33 Junjiro NOGUCHI, Professor, Tokyo Institute of Technology, Tokyo, Japan.

34 Daniel BURNS, Professor, University of Michigan, Ann Arbor, Mich.

35 Leonard SMILEY, Professor, University of Alaska, Anchorage, Alaska.

36 Yong In KIM, Professor, Michigan State Univ., East Lansing, Mich.

37 Pankaj TOPIWALA, Student, University of Michigan, Ann Arbor, Mich.

38 Walter BAILEY, Professor, University of Chicago, Chicago, Illinois.

39 Herbert ALEXANDER, Professor, University of Illinois at Chicago, Chicago, Illinois.

40 Pit-Mann WONG, Professor, University of Notre Dame.

41 Kam Wing LEUNG, Professor, Chinese University of Hong Kong.

42 Brian SMYTH, Professor, University of Notre Dame.

43 Andrew SOMMESE, Professor, University of Notre Dame.
44 Michael BUCHNER, Professor, University of New Mexico.
45 László LEMPERT, Professor, Eötvös Lóránd University, Budapest, Hungary.
46 Norman LEVENBERG, Professor, University of Kentucky, Lexington. Ky.
47 Chia-Chi TUNG, Professor, Mankato State University, Mankato, Minn.
48 Dan GROSS, Professor, Seton Hall University, South Orange, N.J.
49 Steven ZUCKER, Professor, Johns Hopkins University, Baltimore, Md.
50 Bernard SHIFFMAN, Professor, Johns Hopkins University, Baltimore, Md.
51 John WERMER, Professor, Brown University, Providence, R.I.
52 Gary HARRIS, Professor, Texas Tech University, Lubbock, Texas.
53 James FARAN, Professor, Johns Hopkins University, Baltimore, Md.
54 Thomas BLOOM, Professor, University of Toronto, Ontario, Canada.
55 Ian R. GRAHAM, Professor, University of Toronto, Ontario, Canada.
56 Mark ANEMA, Student, University of Michigan, Ann Arbor, Mich.
57 Patrick SMITH, Professor, University of Toronto, Ontario, Canada.
58 Harold BOAS, Professor, Texas A. & M. Univ., College Station, Texas.
59 Jay WOOD, Visiting Professor, University of Notre Dame.
60 Bruce GILLIGAN, Professor, University of Regina, Saskatchewan, Canada.

Arithmetic Hilbert Modular Functions III

Walter L. Baily, Jr.*

Department of Mathematics, University of Chicago
5734 S. University Ave., Chicago, IL 60637 U.S.A.

INTRODUCTION

The purpose of this paper, which is a continuation of [2,3,4,5], is to prove that the special values of Hilbert modular functions of level $\mathfrak{n}$ generate abelian extensions of certain CM fields, by using an essentially elementary theory of arithmetic Hilbert modular functions, based on the theory of congruence Eisenstein series. The main results are generalizations of the main results of Hecke's thesis [10]. They are also subsumed in more far-reaching results of Shimura and Taniyama [15,13,14]. But our methods are quite different from the latter's and stem directly from Hecke's original ideas.

We start from a totally real number field k of degree n over $\mathbb{Q}$ and consider a family of arithmetic subgroups of the projective group $PGL_{2+}(k)$ acting on the product of n upper half planes. Letting K be a purely imaginary quadratic extension of k, the various immersions of the multiplicative group $K^{\times}$ in $PGL_{2+}(k)$ determine fixed points in certain copies of H^n, which are called special points, each of which is associated to a certain order in K containing the ring $\mathfrak{o}$ of integers of k. Such a special point is realized as the image of an element τ of $K - k$ under those immersions of K in $\mathbb{C}$ for which the image of τ has positive imaginary parts. This collection of n immersions of K in $\mathbb{C}$ associated to τ also determines a CM-type for K and a reflex field $K^*(\tau)$. Defining arithmetic Hilbert modular functions as in [3,4], we show here that if f is an arithmetic Hilbert modular function of level $\mathfrak{n}$ and if (τ) is a special point associated to the principal order of K and if $K^*(\tau)$ is the corresponding reflex field, then $f((\tau))$ generates an abelian extension of $K^*(\tau)$; moreover, if $L_{\mathfrak{n}}$ is the abelian extension of $K^*(\tau)$ generated by <u>all</u> such special values of arithmetic Hilbert modular functions of level $\mathfrak{n}$, then the Galois group of $L_{\mathfrak{n}}/K^*(\tau)$ is isomorphic to a <u>sub-</u>

*Support from NSF Grant No. DMS-8401708

group of a certain class group of K. Cf. [10], pp. 55-57. The point of our developments here is to show much can be done without using the theory of families of abelian varieties and their moduli. Our treatment is based on [10, 3, 4, 5] and on a paper of Karel [11], in which the classical one variable case is similarly treated. Of course, we must emphasize the contribution of [7] in characterizing the more basic geometric and topological properties of the adelic space.

We postpone to a subsequent publication consideration of Shimura's reciprocity law for the abelian extension $L_{\mathfrak{n}}/K^*(\tilde{\Sigma})$ obtained from Kronecker type congruences for the special values.

0. NOTATION AND PRELIMINARY REMARKS.

In this article we use the notation of [5]. As a convenience to the reader we summarize some commonly used notation. Denote by k a totally real number field with $[k:\mathbb{Q}] = n > 1$, having ring of integers $\mathfrak{o}$. Let K be a purely imaginary quadratic extension of k. If $\alpha \in k$, $\alpha \gg 0$ will mean that α is totally positive. Let $\Sigma = (\sigma_1,\dots,\sigma_n)$ be the set of all immersions of k as a subfield of $\mathbb{R}$, arranged in a definite order; in general, $\tilde{\Sigma} = (\tilde{\sigma}_1,\dots,\tilde{\sigma}_n)$ will be a "lifting" or set of extensions of $\sigma_1,\dots,\sigma_n$ to immersions of K in $\mathbb{C}$. For a given lifting, define [15] the reflex field $K^*(\tilde{\Sigma})$ by $K^*(\tilde{\Sigma}) = \mathbb{Q}(\{\Sigma_{\tilde{\sigma}\in\tilde{\Sigma}}\, \eta^{\tilde{\sigma}} \mid \eta \in K\})$. Let A, $A(k)$, $A(K)$, I, $I(k)$ and $I(K)$ denote respectively the rings of adeles of $\mathbb{Q}$, of k, of K, and their groups of units, the ideles, each supplied with its customary topology. The subscripts ∞ and f will denote the projections of an adelic object to its archimedean and non-archimedean components respectively, and the subscript $+$ will indicate adelic objects with non-negative archimedean components. $\hat{\mathbb{Z}}$ resp. $\hat{\mathfrak{o}}$ will be the maximal compact subrings of A_f and of $A(k)_f$ respectively. On the other hand, $\mathbb{Q}$, k, and K are naturally imbedded in their rings of adeles and if α belongs to one of these fields, we use $\hat{\alpha}$ to denote α_f, the non-archimedean projection of the image of α in the adeles. Thus, if $\alpha \in \mathfrak{o}$, $\hat{\alpha} \in \hat{\mathfrak{o}}$, and $\hat{\mathfrak{o}}$ is the closure of the image of $\mathfrak{o}$ under this mapping.

We denote by H the upper half complex plane $\operatorname{Im} z > 0$, by H^n its n-th cartesian power, and by $i.e$, the point $(i,\ldots,i) \in H^n$, $i = \sqrt{-1}$. Moreover, G' is the algebraic group GL_2 defined over k and $G = R_{k/Q}G'$. We have a canonical isomorphism ϕ of $G'(A(k))$ onto $G(A)$ such that with respect to suitable integral structures we have $\phi(G'(\hat{\mathfrak{o}})) = G(\hat{\mathbb{Z}})$. P' is the group of upper triangular matrices in G' and $P = R_{k/Q}P'$.

If $\mathbb{K}$ is an open compact subgroup of $G(A_f)$, denote by $\Gamma(\mathbb{K})$ the corresponding arithmetic subgroup $G(Q) \cap G_+(R)\mathbb{K}$ of $G_+(\mathbb{Q})$, or its projection into $G_+(\mathbb{R})$, according to the context. We denote by $\mathbb{K}_\infty$ the isotropy group of $i.e \in H^n$ in $G_+(R)$.

For most other notation we refer to [5] . However, some modifications of our earlier conventions are needed.

The center of G is $Z = R_{k/Q}Z'$, where Z' is the center of G'. In [4] , the symbol $\mathfrak{A}(\mathbb{K},w)$ was used to denote the space of modular forms of weight w with respect to $\mathbb{K}$, and that space was defined as the complex vector space of functions Φ on $G_+(A)$ satisfying three conditions (i), (ii), and (iii) . In fact it turns out that (ii) is undesirable and unsuited for our purposes, and instead of replacing it with the condition that Φ transform according to some finite character on the center $Z(A)$ of $G_+(A)$, we drop the condition (ii) altogether. (In fact, condition (i) alone implies that $\Phi(zg) = \Phi(g)$ for all $g \in G_+(A)$ and for all z belonging to a subgroup of finite index in $Z(A)$. Indeed, the Eisenstein series (3), p. 596 of [3] does not satisfy (ii) unless the function s on double cosets does. Nevertheless, the further discussion of Eisenstein series, and in particular of their convergence, is not affected by this.)

Secondly, while most of the open compact subgroups $\mathbb{K}$ considered in [4] were principal congruence subgroups of $G'(\hat{\mathfrak{o}})$ of the form $\mathbb{K}(\pi) = \{k \in G'(\hat{\mathfrak{o}}) \mid k \equiv 1_2 \bmod \pi\}$, we want to consider here, as far as possible, an arbitrary open compact subgroup $\mathbb{K}$ of $G(A_f)$. In later sections we return to the principal congruence subgroups as a practical consequence of their relationship with ray class groups.

Returning to the purely imaginary quadratic extension K of k , let $B = (\beta_1, \beta_2)$ be a k-basis of K . If $\alpha \in K$, there is a unique two by two matrix $q_B(\alpha) \in M_2(k)$ such that

$$q_B(\alpha)\begin{pmatrix}\beta_1\\ \beta_2\end{pmatrix} = \begin{pmatrix}\alpha\beta_1\\ \alpha\beta_2\end{pmatrix}$$

The mapping $\alpha \longmapsto q_B(\alpha)$ is called a regular imbedding of K in $M_2(k)$, and may be extended in a natural way to an injection of $A(k)$-algebras of $A(K)$ into $M_2(A(k))$. If B is replaced by another k-basis B' of K , q_B is replaced by $q_{B'} = Sq_BS^{-1}$ for some $S \in GL_2(k)$. If $B = (\tau, 1)$, $\tau \in K - k$, we write q_τ in place of q_B .

1. SPECIAL POINTS AND RAY CLASS POLYNOMIALS

1.1 DEFINITIONS

We refer to §2 of [5]. Let $\textcircled{H} \subset G(A_f)$ be a set of representatives for the following decomposition of $G_+(A)$ into double cosets:

$$G_+(A) = \bigcup_{\theta \in \textcircled{H}} G_+(\mathbb{Q})\theta G_+(\mathbb{R})G(\hat{\mathbb{Z}})$$

Let K and $\tilde{\Sigma}$ be as in the introduction and $\tau \in K - k$ be such that $\tilde{\Sigma}(\tau) = \tilde{\Sigma}$, i.e., $\mathrm{Im}(\tau^{\tilde{\sigma}_j}) > 0$, $j = 1,\dots,n$. Then

$$(\tau) = (\tau^{\tilde{\sigma}_1},\dots,\tau^{\tilde{\sigma}_n}) \in H^n$$

and τ defines an imbedding q_τ of K in $M_2(k)$ which extends naturally to an imbedding of $A(K)$ in $M_2(A(k))$. We also associate to any $\tau \in K - k$ its reflex field $K^*(\tau) = K^*(\tilde{\Sigma}(\tau))$. We may assume that each $\theta \in \textcircled{H}$ is of the form $\theta = \begin{pmatrix}\theta' & 0\\ 0 & 1\end{pmatrix}$, $\theta' \in I(k)_f$, and let $v = v_\theta = \mathrm{id}.(\theta')^{-1}$. Let $\mathfrak{J}_{v,\tau}$ be the $\mathfrak{o}$-module $v\tau + \mathfrak{o}$ in K and let

$$\mathcal{R}_v(\tau) = \{x \in K \mid x\mathfrak{J}_{v,\tau} \subset \mathfrak{J}_{v,\tau}\}$$

be its order in K . If now we fix in K an order $\mathcal{R}$ containing the integers $\mathfrak{o}$ of k , we may define the subset $\Xi_v(\mathcal{R})$ of K by $\Xi_v(\mathcal{R}) = \{\tau \in K - k \mid \mathcal{R}_v(\tau) = \mathcal{R}\}$, and a subset $\Xi_v(\mathcal{R},\tilde{\Sigma})$ of $\Xi_v(\mathcal{R})$ by

$$\Xi_v(\mathcal{R}, \tilde{\Sigma}) = \{\tau \in \Xi_v(\mathcal{R}) \mid \tilde{\Sigma}(\tau) = \tilde{\Sigma}\} ,$$

and $\Xi_v(\mathcal{R}, \tilde{\Sigma})$ may be viewed as a subset of H^n by the mapping $\tau \longrightarrow (\tau) \cdot \varepsilon H^n$. We have

$$G(A_f) = \bigcup_{\theta \in \textcircled{H}} G_+(Q)_f \theta G(\hat{\mathbb{Z}}) .$$

For any $\theta \in \textcircled{H}$, define

$$X_\theta = \theta G_+(\mathbb{R}) G(\hat{\mathbb{Z}}) / \mathbb{K}_\infty G(\hat{\mathbb{Z}}) = \theta . H^n ,$$

$$\Xi_\theta(\mathcal{R}) = \{\theta.(\tau) \in X_\theta \mid \tau \in \Xi_{v_\theta}(\mathcal{R})\} ,$$

$$\Xi_{\theta,\infty}(\mathcal{R}) = \{g \in G_+(\mathbb{R}) \mid \theta . g(i.e) \in \Xi_\theta(\mathcal{R})\} ,$$

and

$$\Xi_{\theta,\infty}(\mathcal{R}, \tilde{\Sigma}) = \{g \in G_+(\mathbb{R}) \mid g(i.e) = (\tau) , \tau \in \Xi_{v_\theta}(\mathcal{R}, \tilde{\Sigma})\} ,$$

and then put $\Xi_A(\mathcal{R}) = G_+(Q) \Xi_\infty(\mathcal{R}) G(\hat{\mathbb{Z}})$, where

$$\Xi_\infty(\mathcal{R}) = \bigcup_{\theta \in \textcircled{H}} \Xi_{\theta,\infty}(\mathcal{R}) . \theta .$$

We define $\Xi_A(\mathcal{R}, \tilde{\Sigma})$ and $\Xi_\infty(\mathcal{R}, \tilde{\Sigma})$ analogously. (Cf. [5], §2.) We also carry over the notation $\Xi_{\mathbb{K}}(\mathcal{R})$, $\Xi_{\mathfrak{m}}(\mathcal{R})$, $C_{\mathfrak{m}}(\mathcal{R})$, etc., from §2 of loc. cit.

1.2 SPECIAL POINTS OF GIVEN TYPE

We now prove analogs of certain propositions of Karel [11] for application in our more general situation. Analogous to Lemma 1.2.6 of [11] we have

Proposition 1. Let $\theta \in \textcircled{H}$ and $\xi \in \Xi_{\theta,\infty}(\mathcal{R}, \tilde{\Sigma})$ (so that $\theta\xi = \xi\theta \in \Xi_\infty(\mathcal{R}, \tilde{\Sigma})$) , and let $\tau \in \Xi_{v_\theta}(\mathcal{R}, \tilde{\Sigma})$ be such that $(\tau) = \xi(i.e)$. Then each $x \in \Xi_A(\mathcal{R}, \tilde{\Sigma})$ can be written in the form

$$x = \gamma q_\tau(a^{-1}) \xi \theta \omega ,$$

for some $\gamma \in G_+(Q)$, $a \in I(K)_f$, and $\omega \in \mathbb{K}_\infty . G(\hat{\mathbb{Z}})$.

Proof. If $x \in \Xi_A(\mathcal{R}, \tilde{\Sigma})$, we have $x \in G_+(Q) \xi_1 \theta_1 G(\hat{\mathbb{Z}}) \mathbb{K}_\infty$ for some $\theta_1 \in \textcircled{H}$, where $\xi_1 \in \Xi_{\theta_1,\infty}(\mathcal{R}, \tilde{\Sigma})$. Put $v_1 = v_{\theta_1} = id.(\det(\theta_1))^{-1}$ and $v = id.(\det(\theta))^{-1}$, and let $\tau_1 \in K$ be such that $(\tau_1) = \xi_1(i.e)$. Then $\tau_1 \in \Xi_{v_1}(\mathcal{R}, \tilde{\Sigma})$, $\tau \in \Xi_v(\mathcal{R}, \tilde{\Sigma})$, so that $\tilde{\Sigma}(\tau) = \tilde{\Sigma}(\tau_1)$, and

$$\mathfrak{a}_\tau = v\tau + o \quad \text{and} \quad \mathfrak{a}_{\tau_1} = v_1\tau_1 + o$$

are proper $\mathcal{R}$-ideals of K , hence, so is $\mathfrak{a} = \mathfrak{a}_\tau^{-1} \mathfrak{a}_{\tau_1}$

since $\mathfrak{o} \subset \mathfrak{R}$ and $[K:k] = 2$. Since, therefore, $\mathfrak{a}$ is everywhere locally principal, there exists $a \in I(K)_f$ such that $\mathfrak{a} = K \cap A(K)_\infty a \hat{\mathfrak{R}}$. In the definition of $j(g)$ preceding Lemma 1 of [5], $j(g)$ <u>should have been defined</u> as the double coset $G_+(\mathbb{Q})g\mathbb{K}_\infty G(\hat{\mathbb{Z}})$ to which g belongs. Let us make this correction. Then, in that context, τ and τ_1 having already been chosen such that $\tilde{\Sigma}(\tau_1) = \tilde{\Sigma}(\tau) = \tilde{\Sigma}$, it follows from that Lemma 1 (<u>loc. cit.</u>) that $\theta_1\xi_1$ and $q_\tau(a^{-1})\theta\xi$ are in the same double coset with respect to $G_+(\mathbb{Q})$ on the left and $G(\hat{\mathbb{Z}})\mathbb{K}_\infty$ on the right, which completes the proof of Proposition 1.

1.3 ACTION OF THE GALOIS GROUP AND OF $G(A_f)$

As in [4] we introduce the graded algebra $\mathcal{A}(\mathbb{K},\mathbb{Q}_{ab})$ of $\mathbb{Q}_{ab}$-arithmetic modular forms with respect to $\mathbb{K}$ and, with respect to a certain action of the Galois group $\mathcal{G} = Gal(\mathbb{Q}_{ab}/\mathbb{Q})$, the graded $\mathbb{Q}$-subalgebra $\mathcal{A}(\mathbb{K},\mathbb{Q}_{ab})^{\mathbb{Q}}$ of $\mathbb{Q}$-rational modular forms with respect to $\mathbb{K}$. The respective subspaces of modular forms of weight w will be denoted by $\mathcal{A}(\mathbb{K},w,\mathbb{Q}_{ab})$ and $\mathcal{A}(\mathbb{K},w,\mathbb{Q}_{ab})^{\mathbb{Q}}$. Denote by $\mathcal{M}(\mathbb{K},\mathbb{Q}_{ab})$ resp. $\mathcal{M}(\mathbb{K},\mathbb{Q}_{ab})^{\mathbb{Q}}$ the algebra of $\mathbb{Q}_{ab}$-rational meromorphic modular functions resp. the algebra of $\mathbb{Q}$-rational meromorphic modular functions with respect to $\mathbb{K}$. As a convenient abbreviation, we also let $A(\mathbb{K}) = \mathcal{A}(\mathbb{K},\mathbb{Q}_{ab})^{\mathbb{Q}}$, $A(\mathbb{K},w) = \mathcal{A}(\mathbb{K},w,\mathbb{Q}_{ab})^{\mathbb{Q}}$, and $M(\mathbb{K}) = \mathcal{M}(\mathbb{K},\mathbb{Q}_{ab})^{\mathbb{Q}}$.

If $M \in G_+(A)$, we denote by $R(M)$ and $L(M)$ the right and left regular representations of M on the space of functions on $G_+(A)$, and for $\sigma \in \mathcal{G}$ we denote by $\beta(\sigma)$ the action defined in [5] of σ on elements of $\mathcal{A}(\mathbb{K},\mathbb{Q}_{ab})$, etc. According to §5.2 of [5], we have $R(M)\beta(\sigma) = \beta(\sigma)R(M)$ for $M \in G(A_f)$ and $\sigma \in \mathcal{G}$. If $\Phi \in \mathcal{A}(\mathbb{K},w)$ and $M \in G(A_f)$, we define $\Phi\|M = N\mathfrak{U}^w \Phi|M$ (as in §6.1 of [5]). Analogous to Lemma 3.2 of [11] we then have

<u>Proposition 2.</u> <u>If</u> $\Phi \in M(\mathbb{K})$ <u>and</u> $\Psi \in A(\mathbb{K})$, <u>and if</u> $M \in G(A_f)$, <u>then</u> $R(M)\Phi$ <u>and</u> $\Psi\|M/\Psi$ <u>belong to</u> $M(\mathbb{K}')$ <u>for some open compact subgroup</u> $\mathbb{K}'$ <u>of</u> $\mathbb{K}$. <u>If</u> $\mathbb{K}'$ <u>is an open compact subgroup of</u> $\mathbb{K}$ <u>and if</u> $\Phi \in M(\mathbb{K}')$ <u>and satisfies</u> $R(M)\Phi = \Phi$ <u>for every</u> $M \in \mathbb{K}$, <u>then</u> $\Phi \in M(\mathbb{K})$.

Proof. The proof is analogous to that given in §3.2 of [11] .

If $\mathbb{K}$ is an open compact subgroup of $G(A_f)$, denote by $pr_{\mathbb{K}}$ the canonical mapping of $G_+(A)$ onto its space of double cosets

$$V_{\mathbb{K}} = G_+(Q)\backslash G_+(A)/\mathbb{K}\mathbb{K}_\infty \quad .$$

1.4 OPERATION OF THE IDELES

According to Proposition 1 , given any element $\xi\theta \in \Xi_\infty(\mathcal{R}, \tilde{\Sigma})$ and $\tau \in \Xi_{v_\theta}(\mathcal{R}, \tilde{\Sigma})$ such that $\xi(i.e) = (\tau)$, each $x \in \Xi_A(\mathcal{R}, \tilde{\Sigma})$ can be written as $\gamma q_\tau(b)\xi\theta\omega$ in the notation there and with $b \in I(K)_f$. Then $q = q_\tau$ is a regular imbedding of K as a k-algebra into $M_2(k)$. We introduce for constant use the following conventions and notation: Suppose $\theta = \begin{pmatrix}\theta' & 0\\ 0 & 1\end{pmatrix}$, where $\theta' \in I(k)_f$, $v = v_\theta = id.(\theta')^{-1}$; then $\mathcal{R}$ is the order in K of the rank two o-module $v\tau + o$. Extend q to an imbedding of A(K) into $M_2(A(k))$ and define $q' = \theta^{-1}q\theta$. A simple calculation shows that $q'(\hat{\mathcal{R}}) \subset M_2(\hat{o})$, where $\hat{\mathcal{R}}$ is the closure of $\mathcal{R}$ in $A(K)_f$. Analogous to §3.3 of [11] , we have:

Proposition 3. If $a \in I(K)_f$ and $M \in G(A_f)$ are such that

(A) $$pr_{\mathbb{K}}(a*x) = pr_{\mathbb{K}}(xM) \quad ,$$

then for some $\beta \in K^\times$ we have

(B) $$M \in {}^{\omega^{-1}}q'(a^{-1}\hat{\beta}).\mathbb{K} \quad .$$

Conversely, (B) implies (A).

(Remark. The action $x \longmapsto a*x$ of $a \in I(K)_f$ on the special points $x \in \Xi_A(\mathcal{R}, \tilde{\Sigma})$ is that described in §2 of [5]. Cf. also [11], §1.2.5.)

Proof. (A) holds if and only if for some $\gamma \in G_+(Q)$,

(*) $$\xi q(a^{-1}b)\theta\omega \in \gamma\xi q(b)\theta\omega M\mathbb{K}\mathbb{K}_\infty \quad .$$

The archimedean component of this reads: $\xi \in \gamma_\infty\xi\mathbb{K}_\infty$, which is equivalent to having $(\tau) = \xi(i.e) = \gamma_\infty.(\tau)$, and since $q(K^\times)$ is the stabilizer of (τ) in $G_+(Q)$, this means $\gamma = q(\beta^{-1})$ for some $\beta \in K^\times$. Therefore the non-archimedean component of (*) reads

(**) $$q(a^{-1}b)\theta\omega \in q(b\hat{\beta}^{-1})\theta\omega M\mathbb{K}$$

which is equivalent to having $q(a^{-1})\theta\omega \in q(\hat{\beta}^{-1})\theta\omega M\mathbb{K}$, or $\omega^{-1}q'(\hat{\beta}a^{-1})\omega \in M\mathbb{K}$ or $M \in \omega^{-1}q'(\hat{\beta}a^{-1})\omega\mathbb{K}$, which is (B) . Conversely, (B) clearly implies (**) , and for all $\beta \in K^{\times}$, we have $(\tau) = q(\beta^{-1})_{\infty}(\tau)$, or if we put $\gamma = q(\beta^{-1})$, then $\xi \in \gamma_{\infty}\xi\mathbb{K}_{\infty}$, which, together with (**), gives (A). Q.E.D.

Henceforth we let the order $\mathcal{R}$ in K be the maximal order $\mathcal{O}$ of K . Let $\mathfrak{m}$ be a non-zero ideal in $\mathfrak{o}$ and let m be the smallest positive rational integer in $\mathfrak{m}$: $\mathbb{Z} \cap \mathfrak{m} = (m)$. Define the principal congruence subgroup $K(\mathfrak{m})$ of $G(\hat{\mathbb{Z}}) = G'(\hat{\mathfrak{o}})$ by

$$\mathbb{K}(\mathfrak{m}) = \{g \in G'(\hat{o}) \mid g \equiv 1_2 \mod \mathfrak{m}\}.$$

Then $G'(\hat{o}) = \mathbb{K}(1)$. We shall let $\mathbb{K}$ denote a normal open subgroup of $\mathbb{K}(1)$ containing some $\mathbb{K}(\mathfrak{m})$. The imbedding

$$q' : I(K)_f \longrightarrow G(A_f)$$

is a closed immersion which algebraically and topologically identifies $I(K)_f$ with the group $\theta^{-1}T(A_f)\theta$ for some $\mathbb{Q}$-torus T in G . Let $\tilde{H} = q'^{-1}(\mathbb{K})$. This is an open compact subgroup of $I(K)_f$. If $\mathbb{K} = \mathbb{K}(\mathfrak{m})$, $\tilde{H}$ is the principal congruence subgroup mod $\mathfrak{m}$ of the group $\hat{\mathcal{O}}^{\times}$ of non-archimedean idelic units of K . The class group

$$C(\mathcal{O}, \tilde{H}) = I(K)/I(K)_{\infty}.\tilde{H}.K^{\times}$$

operates on the set $\Xi_{\mathbb{K}}(\mathcal{O}, \tilde{\Sigma})$, the image of $\Xi_A(\mathcal{O}, \tilde{\Sigma})$ in $V_{\mathbb{K}}$ because $q(K^{\times}) \subset G_+(\mathbb{Q})$, $I(K)_{\infty}$ acts trivially on $\Xi_A(\mathcal{O})$ by the definition in §2 of [5] , and by assumption $q'(\tilde{H}) \subset {}^{\omega}\mathbb{K} = \mathbb{K}$ whenever $x = \gamma\xi\theta\omega \in \Xi_A(\mathcal{O})$, and $(\tau) = \xi$ (i.e) so that, according to the calculations of §2 of [5] , $\tilde{H}$ acts trivially on $\Xi_{\mathbb{K}}(\mathcal{O}, \tilde{\Sigma})$. Henceforth $\mathbb{K}$ will denote a subgroup of $G(A_f)$ satisfying these conditions, and such a subgroup will be called an n.o.c. subgroup of $G(A_f)$.

We let B be an orbit of $C(\mathcal{O}, \tilde{H})$ in $\Xi_{\mathbb{K}}(\mathcal{O}, \tilde{\Sigma})$. Following the ideas of [11] we wish to show that the 0-cycle B on $V_{\mathbb{K}}$ is defined over the reflex field $K^*(\tilde{\Sigma})$.

1.5 RATIONAL STRUCTURES. ARITHMETIC MODULAR FUNCTIONS.

In the statements and results of [5] regarding the field of rationality for algebras of modular forms, the norm $N\mathfrak{m}$ of $\mathfrak{m}$ may be replaced by m . For example, Prop. 2 of §5.2, loc. cit. may be strengthened to read: " $\mathcal{A}(\mathbb{K}(\mathfrak{m}))$ is the integral closure of $\tilde{\mathcal{E}}(\mathfrak{m})$ in its graded ring of homogeneous quotients and we have

$$\mathcal{A}(\mathbb{K}(\mathfrak{m})) = \mathcal{A}(\mathbb{K}(\mathfrak{m}), \mathbb{Q}_m) \otimes_{\mathbb{Q}_m} \mathbb{C} \text{ ,"}$$

and then Prop. 3 of loc. cit. applied to obtain immediately

Proposition 4. As a graded algebra over $\mathbb{Q}_m$, $\mathcal{A}(\mathbb{K}(\mathfrak{m}), \mathbb{Q}_m)$ is generated by $\mathcal{A}(\mathbb{K}(\mathfrak{m}))^{\mathbb{Q}} = A(\mathbb{K}(\mathfrak{m}))$.

For our given n.o.c. subgroup $\mathbb{K}$ of $G(A_f)$, which we assume to contain $\mathbb{K}(\mathfrak{m})$, we have that $\mathcal{A}(\mathbb{K})$ is a graded Q-subalgebra of $\mathcal{A}(\mathbb{K}(\mathfrak{m}))$ and, by Prop. 4, we have $\mathcal{A}(\mathbb{K}, w) \subset \mathcal{A}(K(\mathfrak{m}), w) = A(K(\mathfrak{m}), w) \otimes_{\mathbb{Q}} \mathbb{C}$. We may write $\mathbb{K}$ as a union of left cosets of $\mathbb{K}(\mathfrak{m})$, say

$$\mathbb{K} = \bigcup_{j=1}^{t} S_j \mathbb{K}(\mathfrak{m}) \quad .$$

Then if $\Phi \in A(\mathbb{K}(\mathfrak{m}), w)$, the average $\Psi = t^{-1} . \Sigma_{j=1}^{t} R(S_j)\Phi$ belongs to $\mathcal{A}(\mathbb{K}, w)$, and since $\beta(\sigma)R(S_j) = R(S_j)\beta(\sigma)$ for $j = 1,\ldots,t$ and all $\sigma \in \mathcal{G}$, we have $\Psi \in A(\mathbb{K}, w)$. Since the "average" of any $\Phi \in A(\mathbb{K}, w)$ is just Φ , we have, by linearization,

Proposition 5. $\mathcal{A}(\mathbb{K})$ is spanned by $A(\mathbb{K})$.

For a number field F , a modular function with respect to $\mathbb{K}$ is called F-arithmetic if it is the quotient of two F-arithmetic modular forms of the same weight. The F-algebra of such functions is denoted $\mathcal{M}(\mathbb{K}, F)$, and the subalgebra of those which are "holomorphic at the cusps" is denoted by $\mathcal{M}(\mathbb{K}, \{\kappa\}, F)$. The action β of $\mathcal{G}$ on $\mathcal{A}(\mathbb{K}, \mathbb{Q}_{ab})$ extends to an action on $\mathcal{M}(\mathbb{K}, \mathbb{Q}_{ab})$, and we denote the subalgebra of elements invariant under $\mathcal{G}$ by $M(\mathbb{K})$ resp. $M(\mathbb{K}, \{\kappa\})$.

1.6 THE $\mathbb{K}$-LEVEL CLASS POLYNOMIAL

Denote the algebra of quotients of $M(\mathbb{K}) \otimes_{\mathbb{Q}} K^*(\tilde{\Sigma})$ (with respect to its non-zero divisors) by $M(\mathbb{K}, K^*(\tilde{\Sigma}))$ or by $K^*(\tilde{\Sigma})M(\mathbb{K})$ and call this the compositum of $K^*(\tilde{\Sigma})$ and $M(\mathbb{K})$. Let $\Phi \in K^*(\tilde{\Sigma})M(\mathbb{K}) \cap \mathcal{M}(\mathbb{K}, \{\kappa\})$ be such that Φ is holomorphic on $\Xi_{\mathbb{K}}(\mathcal{O}, \tilde{\Sigma})$. (Cf. [11],§4.1.1.) Define the polynomial

$$H_B^{\Phi}(t) = \prod_{z \in B} (t - \Phi(z)) \quad .$$

This is analogous to the polynomial $T_{\mathfrak{m}}(t)$ introduced by Hasse [9 :p.78] and we shall prove here a result analogous to Satz

17, p. 84 of [9]; cf. also [11:§4.5.4].

We know from [5] that $\Xi_{(1)}(\mathcal{O}, \tilde{\Sigma})$ is a zero-cycle of $V_{(1)} = V_{\mathbb{K}(1)}$ rational over $K^*(\tilde{\Sigma})_0$. Now $\Xi_{\mathbb{K}}(\mathcal{O}, \tilde{\Sigma})$ is its pre-image under the natural (coset) mapping $\pi_{\mathbb{K}}$ of $V_{\mathbb{K}}$ onto $V_{(1)}$. Let $\mathcal{B}_{(1)} = (b_0,\dots,b_{N(1)})$ be a basis of $A(\mathbb{K}(1), w)$ over $\mathbb{Q}$ for some large w, and complement $\mathcal{B}_{(1)}$ to a basis $\mathcal{B}_{\mathbb{K}} = (b_0,\dots,b_{N(K)})$ of $A(\mathbb{K}, w)$ over $\mathbb{Q}$. If w is large enough, $\mathcal{B}_{(1)}$ resp. $\mathcal{B}_{\mathbb{K}}$ provides a set of projective coordinates for a projective imbedding of $V_{(1)}$ resp. of $V_{\mathbb{K}}$ as a reducible projective variety (of several components), rational over $\mathbb{Q}$ as a cycle in projective space. Then $\pi_{\mathbb{K}}$ is merely the projection on the first $N(1) + 1$ coordinates and therefore is defined over $\mathbb{Q}$. Hence, $\Xi_{\mathbb{K}}(\mathcal{O}, \tilde{\Sigma})$ is a zero cycle rational over $K^*(\tilde{\Sigma})_0$ also. (The components V_ω of $V_{\mathbb{K}}$ are rational over $\mathbb{Q}_m$, and $\mathcal{G}$ permutes them in a manner compatible with translation by the elements $\mu(\sigma) \in G(A_f)$ for $\sigma \in \mathcal{G} \simeq \hat{\mathbb{Z}}^{\times} \subset I(k)_f$. If $\Phi \in A(\mathbb{K},w)$, then by definition, $R(\mu(\sigma)^{-1})\Phi^{\sigma} = \Phi$ for all $\sigma \in \mathcal{G}$.)

What we wish to show is that B , as a subcycle of $\Xi_{\mathbb{K}}(\mathcal{O}, \tilde{\Sigma})$ is rational over $K^*(\tilde{\Sigma})$. Moreover, we can then see at the same time that for each of the narrow ideal class representatives v , the set $B_v = \pi_{\mathbb{K}}^{-1}(V_v) \cap B$, where V_v is one of the components of $V_{(1)}$, is also rational over $K^*(\tilde{\Sigma})$.

1.7 $\mathbb{K}$-LEVEL MODULAR CORRESPONDENCES AND THEIR FIXED POINTS

Let $C_{\mathbb{K}}(S_0) = \mathbb{K}S_0\mathbb{K}$ be the double coset determining a modular correspondence $\mathcal{C}_{\mathbb{K}}(S_0) \subset V_{\mathbb{K}} \times V_{\mathbb{K}}$, where S_0 belongs to $G(A_f)$, and $\mathbb{K}$ is an n.o.c. subgroup of $G(A_f)$. The projective coordinate ring of $V_{\mathbb{K}}$ has a $\mathbb{Q}$-structure $A(\mathbb{K})$ which was described in the last section. Let us assume that $S_0 \in M_2(\hat{\mathfrak{o}})$ (within our usual identifications). Let Φ be a homogeneous element of degree zero in the quotient algebra; that is, $\Phi = a/b$, where a , $b \in A(\mathbb{K})$ and are homogeneous of the same weight. Then we form the transformation polynomial of Φ with respect th $C_{\mathbb{K}}(S_0)$:

$$T_{\Phi,S_0,\mathbb{K}}(g)(X) = \prod_{j=1}^{N} (X - \Phi(gS_j)) , \quad C_{\mathbb{K}}(S_0) = \bigcup_{j=1}^{N} S_j\mathbb{K} .$$

By its choice, Φ satisfies $\beta(\sigma)\Phi = \Phi$ for every

$\sigma \in \mathcal{G} = \mathrm{Gal}(\mathbb{Q}_{ab}/\mathbb{Q})$. Moreover, $\beta(\sigma)R(S_j) = R(S_j)\beta(\sigma)$, $j = 1,\ldots,N$. Therefore, if

$$T_{\Phi,S_0,\mathbb{K}}(X) = X^N + \Sigma_{\nu \in N}\, a_\nu X^\nu \quad ,$$

then each coefficient is a $\mathbb{Q}_m$-arithmetic modular function with respect to $\mathbb{K}$ and belongs to $M(\mathbb{K})$. Moreover, Φ is also an element of $M(\mathbb{K})$.

Let $FP(C_{\mathbb{K}}(S_0))$ be the set of fixed or self-corresponding points of $V_{\mathbb{K}}$ with respect to $C_{\mathbb{K}}(S_0)$. Essentially as remarked in §2.5 of [4] , $FP(C_{\mathbb{K}}(S_0))$ (= $FP(S_0)$ in [4]) consists of a finite set of isolated points on $V_{\mathbb{K}}$ provided $C_{\mathbb{K}}(S_0)$ does not contain a scalar multiple of the identity (in [4], "if S_0 is not a scalar multiple of the identity," but this is not strong enough). In any case, if $\zeta \in FP(C_{\mathbb{K}}(S_0))$ and if $g \in G_+(A)$ is a pre-image of ζ , then for some $S \in C_{\mathbb{K}}(S_0)$ and for every $\Phi \in M(\mathbb{K})$ holomorphic at ζ and at the image ζ_S of gS we must have $\Phi(g) = \Phi(\zeta) = \Phi(\zeta_S) = \Phi(gS)$. Therefore, $\Phi(g) = \Phi(gS_j)$ for some $j = 1,\ldots,N$ and

(1) $\quad \Phi(g)^N + \Sigma_{\nu \in N}\, a_\nu(g)\Phi(g)^\nu = T_{\Phi,S_0,\mathbb{K}}(g)(\Phi(g)) = 0 \quad ,$

while if the image of $g' \in G_+(A)$ is not such a fixed point, there exists $\Phi \in M(K)$ holomorphic at g' and at the images of all the elements $g'S$, $S \in C_{\mathbb{K}}(S_0)$, such that $T_{\Phi,S_0,\mathbb{K}}(g')(\Phi(g')) \neq 0$. Since (1) is an algebraic relation among the values of elements of $M(\mathbb{K})$ at the pre-images g of points of $FP(C_{\mathbb{K}}(S_0))$, and each $a_\nu \in M(\mathbb{K})$, therefore is the quotient of homogeneous elements in the projective coordinates of the $\mathbb{Q}$-structure, it follows that $FP(C_{\mathbb{K}}(S_0))$ is a zero cycle on $V_{\mathbb{K}}$ and rational over $\mathbb{Q}$. By the same kind of argument as in [11,§3.5], which we shall not repeat here, the special points of $V_{\mathbb{K}}$ (that is, the images of the special points $K - k \longrightarrow H^n \simeq X_\omega \longrightarrow V_\omega$ in each component V_ω of $V_{\mathbb{K}}$) are the fixed points of all the modular correspondences $\mathcal{C}_{\mathbb{K}}(S_0)$, with $S_0 \in G_+(\mathbb{Q})_f$, such that $C_{\mathbb{K}}(S_0) = \mathbb{K}S_0\mathbb{K}$ does not contain a scalar multiple of the identity (in which case the set of fixed points of $C_{\mathbb{K}}(S_0)$ is finite).

As before, let $q = q_\tau$, θ , θ', and $v = v_\theta$ be such that the order of $v\tau + \mathfrak{o}$ is the maximal order $\mathcal{O}$ of K. Then, as remarked earlier, $q'(\hat{\mathcal{O}}) \subset M_2(\hat{\mathfrak{o}})$. Now define

$$V_{\mathbb{K}}^{q'} = \bigcap_{a \in K^{\times}} FP(C_{\mathbb{K}}(q'(\hat{a}))) \quad .$$

(Cf. [11:§4.1.3].) Since $a \in k^{\times}$ implies that $q(a)$ is a scalar multiple of the identity, the essential terms on the right hand side are those for $a \in K - k$. We have seen that each of the terms on the right hand side is a cycle rational over $\mathbb{Q}$. If $a \in K$ is a prime element for a principal prime ideal $\mathfrak{P} = (a) = a.\mathcal{O}$ of first degree in $\mathcal{O}$, then $C_{\mathbb{K}}(q'(\hat{a}))$ cannot contain a scalar multiple of the identity [11:§3.5]. Hence, for such a, $FP(C_{\mathbb{K}}(q'(\hat{a})))$ is a finite zero cycle rational over $\mathbb{Q}$. Therefore, in analogy with [11:§4.2.2], we have

<u>Proposition 5a.</u> <u>$V_{\mathbb{K}}^{q'}$ is a finite zero cycle on $V_{\mathbb{K}}$ rational over $\mathbb{Q}$.</u>

2. PROOF THAT $\tilde{B} = B \cup \iota * B$ IS A CYCLE RATIONAL OVER $K^*(\tilde{\Sigma})_0$

2.1 TOPOLOGICAL PROPERTIES OF CORRESPONDENCES AND THE ARTIN KERNEL

For the totally real field k, let $\mathfrak{z}'$ be the closure in $I(k)$ of the group $k^{\times}k^{\times}_{\infty+}$; in other words, $\mathfrak{z}'$ is the kernel of the Artin map [18]

$$\mathfrak{a}: \ I(k) \longrightarrow Gal(k_{ab}/k) \quad .$$

Let $\mathfrak{z}$ be the subgroup of $GL_2(A(k))$ consisting of the matrices $\zeta.1_2$, $\zeta \in \mathfrak{z}'$. Then $\mathfrak{z}$ is a subgroup of the center $Z(A(k))$ of $GL_{2+}(A(k))$ (whose center is the same as that of $GL_2(A(k))$).

<u>Proposition 6.</u> <u>For each</u> $g \in G_+(A)$, <u>the set</u> $G_+(\mathbb{Q})g\mathbb{K}_{\infty}\mathfrak{z}$ <u>is closed in</u> $G_+(A)$.

<u>Proof.</u> [11] We have the standard homomorphism Int: $G \longrightarrow Ad(G)$, and $Int(\mathbb{K}_{\infty})$ is a compact subgroup of $Ad(G)(A)$, while $Int(G_+(\mathbb{Q}))$ is discrete. Let $T = G_+(\mathbb{Q}).{}^{g}\mathbb{K}_{\infty}$; the image of this in $Ad(G)(A)$ is closed. If $x \in clos.(T.\mathfrak{z})$, then $Int(x) \in Int(T)$, so that $x \in c.T$ for some $c \in Z(A)$. We shall show that c may be chosen in $\mathfrak{z}$ and thus $T.\mathfrak{z}$ is closed, which implies that

$G_+(Q)g\mathbb{K}_\infty = T.\mathfrak{Z}.g$ is closed.

Of course, $x = c.t \in \text{clos.}(T.\mathfrak{Z})$, where $t \in T$ and may be written as $t = \gamma_0.k_0$, $\gamma_0 \in G_+(\mathbb{Q})$, $k_0 \in {}^g\mathbb{K}_\infty$. Since $x \in \text{clos.}(T.\mathfrak{Z})$, we may also write

$$x = \lim_{m \to \infty} \gamma_m k_m' c_m \ , \ \gamma_m \in G_+(Q) \ , \ k_m' \in {}^g\mathbb{K}'_\infty \ , \ c_m \in \mathfrak{Z}.E \ ,$$

where $\mathbb{K}_\infty'$ is the unique maximal compact subgroup of $\mathbb{K}_\infty$ (which is commutative), and E is the maximal finite subgroup of $Z(\mathbb{R})$, consisting of the n-tuples of matrices $(\pm 1_2, \ldots, \pm 1_2)$ associated to n (all real) archimedean places of k , because $\mathbb{K}_\infty = Z(\mathbb{R}).\mathbb{K}_\infty'$ and $Z(\mathbb{R}) \subset \mathfrak{Z}.E$. By taking a subsequence we may assume that $\lim k_m'$ exists and equals $k' \in {}^g\mathbb{K}_\infty$. Since c_m are all central, we obtain $ct = c\gamma_0 k_0 = \lim_{m \to \infty}(\gamma_m c_m)k'$, so that $\lim_{m \to \infty}(\gamma_m c_m) = ctk'^{-1}$. Now $\text{Int}(\gamma_m c_m) = \text{Int}(\gamma_m)$ for all m , therefore, $\text{Int}(\gamma_m)$ has a limit and since $\text{Int}(G_+(\mathbb{Q})) \subset \text{Int}(G)(\mathbb{Q})$ is discrete, we may suppose $\text{Int}(\gamma_m)$ is constant, say, $\text{Int}(\gamma_m) = \text{Int}(\gamma_1)$ for all m and $\gamma_m = \gamma_1 \zeta_m$, $\zeta_m \in Z(\mathbb{Q})$ for all m , $\gamma_1 \in G_+(\mathbb{Q})$. Thus, $z = \lim(\zeta_m c_m) = \gamma_1^{-1}ctk'^{-1} = c\gamma_1^{-1}\gamma_0 k_0 k'^{-1} \in Z(A)$, since $Z(A)$ is closed, and $\gamma_1^{-1}\gamma_0 = (c^{-1}z).(k_0k'^{-1})^{-1} \in (Z(A).{}^g\mathbb{K}_\infty) \cap G_+(\mathbb{Q})$. Looking at the non-archimedean part, this implies that $\gamma_1^{-1}\gamma_0 \in Z(\mathbb{Q}) = Z_+(\mathbb{Q})$. Also we then have

$$k_0k'^{-1} = (c^{-1}z).(\gamma_1^{-1}\gamma_0) \in Z(A) \cap {}^g K_\infty = Z(\mathbb{R}) \subset E.\mathfrak{Z} \ .$$

Now $c_m \in \mathfrak{Z}.E$ and $\zeta_m \in Z(\mathbb{Q})$ for all m . But $Z(\mathbb{Q})$ is contained in $\mathfrak{Z}.E$ and $\mathfrak{Z}$ is the closure of $k_{\infty+}^\times k^\times.1_2$, hence is closed, and therefore $\mathfrak{Z}.E$, being the union of at most 2^n cosets of $\mathfrak{Z}$, is also closed, so that

$$z = \lim_{m \to \infty}(c_m\zeta_m) \in \mathfrak{Z}.E \ .$$

Therefore

$$c = z.(k_0k'^{-1}).(\gamma_1^{-1}\gamma_0) \in \mathfrak{Z}.E \ ,$$

and $x \in c.T$. But since $E \subset Z(\mathbb{R}) \subset \mathbb{K}_\infty$, and $T = G_+(\mathbb{Q}).{}^g\mathbb{K}_\infty$, we see that $E.T = T$. Therefore, $x \in c'.T$, $c' \in \mathfrak{Z}$, and so $T.\mathfrak{Z}$ is closed, which is what we wanted.

2.2 FIXED POINTS OF MODULAR CORRESPONDENCES

<u>Definition.</u> <u>For a given open compact subgroup</u> $\mathbb{K}$ <u>of</u>

$G(A_f)$ <u>and for any element</u> Π <u>of</u> $G(A_f)$ <u>define</u>

$$FP_{\mathbb{K}}(\Pi) = \{z \in G_+(A) \mid z\Pi \in G_+(\mathbb{Q})z\mathbb{K}_\infty\mathbb{K} \quad\}.$$

<u>Define</u>

$$FP_\infty(\Pi) = \{z \in G_+(A) \mid z\Pi \in G_+(Q)z\mathbb{K}_\infty\mathfrak{z} \}.$$

Now we have

$$Z(\mathbb{R})Z(\mathbb{Q})\mathfrak{z} = E.\mathfrak{z} \subset Z(\mathbb{R})Z(\mathbb{Q})\mathcal{U}_{\mathfrak{n}}(\sigma) \subset Z(\mathbb{R})Z(\mathbb{Q})\mathbb{K}(\mathfrak{n})$$

for any non-zero (integral) ideal $\mathfrak{n} \subset \sigma$, where $\mathcal{U}_{\mathfrak{n}}(\sigma)$ $I(k)_f$ is the group of finite idelic units $\eta \equiv 1 \bmod \mathfrak{n}$, because $k^\times k^\times_{\infty+}\mathcal{U}_{\mathfrak{n}}(\sigma)$ is an open and hence closed neighborhood of $k^\times k^\times_{\infty+}$ for all such $\mathfrak{n}$. Therefore we have

$$FP_\infty(\Pi) \subset FP_{\mathfrak{n}}(\Pi) ,$$

for all $\mathfrak{n}$, where $FP_{\mathfrak{n}}(\Pi) = FP_{\mathbb{K}(\mathfrak{n})}(\Pi)$. In analogy with §4.4.3 of [11] we have

<u>Proposition 7.</u> <u>Let</u> Ω <u>be any compact open subgroup of</u> $G'(\hat{\sigma})$. <u>Then</u>

(A) $$FP_\infty(\Pi).\Omega = \bigcap_{\mathbb{K}} FP_{\mathbb{K}}(\Pi).\Omega ,$$

<u>where</u> $\mathbb{K}$ <u>runs over all n.o.c. subgroups of</u> $G(A_f)$.

<u>Remark.</u> The principal congruence subgroups of $K(\mathfrak{n})$ form a cofinal system among all open compact subgroups $\mathbb{K}$ of $G(A_f)$. Therefore it amounts to the same thing to formulate the statement (A) with this family of open compact subgroups.

<u>Proof.</u> We have just seen that the left side of the equality in (A) is contained in the right side. Therefore, it suffices to show that if $z \in FP_{\mathfrak{n}}(\Pi).\Omega$ for all $\mathfrak{n}$, then $z \in FP_\infty(\Pi)\Omega$. Let

$$T^{\mathfrak{n}} = z^{-1}G_+(Q)z\mathbb{K}_\infty\mathbb{K}(\mathfrak{n}) .$$

Then $zT^{\mathfrak{n}} = G_+(\mathbb{Q})z\mathbb{K}_\infty\mathfrak{z}\mathbb{K}(\mathfrak{n})$, and $G_+(\mathbb{Q})z\mathbb{K}_\infty\mathfrak{z}$ is closed by Prop. 6, while $\mathbb{K}(\mathfrak{n})$ is compact; hence, $T^{\mathfrak{n}}$ is closed. Let η be the continuous map

$$\eta : x \longmapsto x\Pi x^{-1} : \Omega \longrightarrow G_+(A) .$$

If $FP_{\mathfrak{n}'}(\Pi)$ is non-empty, and $\xi \in FP_{\mathfrak{n}'}(\Pi)$, then

$$\Pi \in \xi^{-1}G_+(Q)\xi\mathbb{K}_\infty\mathbb{K}(\mathfrak{n}') .$$

Thus $\eta^{-1}(T^{\mathfrak{n}'})$ is a closed subset of Ω for every $\mathfrak{n}'$, and since $z \in FP_{\mathfrak{n}'}(\Pi)\Omega$, this means that there exist ξ in $FP_{\mathfrak{n}'}(\Pi)$ and $x \in \Omega$ such that $z = \xi x^{-1}$, where ξ and $x \in \Omega$ depend on $\mathfrak{n}'$. Thus we have

$$\eta(x) = x\Pi x^{-1} \in x\xi^{-1}G_+(\mathbb{Q})\xi \mathbb{K}_\infty \mathbb{K}(\mathfrak{n}')x^{-1} = x\xi^{-1} G_+(\mathbb{Q}) \mathbb{K}_\infty \mathbb{K}(\mathfrak{n}')$$
$$= z^{-1}G_+(\mathbb{Q})z\mathbb{K}_\infty \mathbb{K}(\mathfrak{n}') = T^{\mathfrak{n}'} ,$$

so that $x \in \eta^{-1}(T^{\mathfrak{n}'})$. In deriving the last sequence of equalities, we have used the fact that $K(\mathfrak{n}')$ is a normal subgroup of $G'(\hat{o})$. Therefore $\eta^{-1}(T^{\mathfrak{n}'})$ is a closed non-empty subset of Ω for every $\mathfrak{n}'$, while $\mathbb{K}(\mathfrak{n}) \subset K(\mathfrak{n}')$ if $\mathfrak{n}'|\mathfrak{n}$, that is $T^{\mathfrak{n}} \subset T^{\mathfrak{n}'}$, so that the finite intersections of the closed subsets $\eta^{-1}(T^{\mathfrak{n}'})$ of the compact group Ω are non-empty. Hence there exists a point x_0 common to all of them and therefore for all $\mathfrak{n}'$:

$$x_0\Pi x_0^{-1} \in z^{-1}G_+(Q)z\mathbb{K}_\infty \mathbb{K}(\mathfrak{n}') ,$$

so that

$$zx_0\Pi x_0^{-1} \in \bigcap_{\mathfrak{n}} G_+(\mathbb{Q})z\mathbb{K}_\infty \mathbb{K}(\mathfrak{n}) .$$

The intersection on the right hand side is equal to $G_+(Q)z\mathbb{K}_\infty\mathfrak{Z} = D$, say, which is closed by Prop. 6. For if $\zeta = zx_0\Pi x_0^{-1} \in \bigcap_{\mathfrak{n}} D\,K(\mathfrak{n})$, say $\zeta = \theta_{\mathfrak{n}}k_{\mathfrak{n}}$, $\theta_{\mathfrak{n}} \in D$, $k_{\mathfrak{n}} \in K(\mathfrak{n})$ for each $\mathfrak{n}$, where $\mathfrak{n}$ tends "to ∞" through a sequence of ideals $\mathfrak{n}$ such that $\bigcap_{\mathfrak{n}} \mathbb{K}(\mathfrak{n}) = \{1\}$, then $\theta_{\mathfrak{n}} = \zeta k_{\mathfrak{n}}^{-1}$ and $\lim \theta_{\mathfrak{n}} = \zeta \in \mathrm{clos.}(D) = D$. Therefore $zx_0\Pi x_0^{-1} \in D$, $x_0 \in \Omega$, and $zx_0\Pi \in G_+(Q)zx_0\mathbb{K}_\infty\mathfrak{Z}$, so that $zx_0 \in FP_\infty(\Pi)$ and $z \in FP_\infty(\Pi).\Omega$. Q.E.D.

2.3 RATIONALITY OF $\tilde{\mathfrak{B}}$ over $K^*(\tilde{\Sigma})_0$

Proposition 8. (§4.4.4 of [11].) Let ι be the involutory element defined in §3 of [5] and put $\tilde{\mathfrak{B}} = \mathfrak{B} \cup \iota^*\mathfrak{B}$. Then for any n.o.c. subgroup $\mathbb{K}$ of $G(A_f)$ the 0-cycle $\tilde{\mathfrak{B}}$ on $V_{\mathbb{K}}$ is rational over $K^*(\tilde{\Sigma})_0$.

(Remark. To avoid complicating notation, we omit any label referring to $\mathbb{K}$ from $\mathfrak{B}$.)

Proof. If $\Pi \in G(A_f)$ and $\det(\Pi) \in \det(G_+(Q)_f).\det(\mathbb{K})$, the set of fixed points of the modular correspondence $\mathbb{K}\Pi\mathbb{K} = C_{\mathbb{K}}(\Pi)$ is the projection in $V_{\mathbb{K}}$ of the set $FP_{\mathbb{K}}(\Pi)$; we have seen that this cycle, $FP_{\mathbb{K}}(C_{\mathbb{K}}(\Pi))$, is $\mathbb{Q}$-rational. Moreover, if $\mathbb{K}'$ is another open compact subgroup of $G(A_f)$ and $\mathbb{K}' \subset \mathbb{K}$, then $FP_{\mathbb{K}'}(\Pi) \subset FP_{\mathbb{K}}(\Pi)$, hence

$$FP_{\mathbb{K}'}(\Pi).\Omega \subset FP_{\mathbb{K}}(\Pi).\Omega$$

and

$$\bigcap_{\mathbb{K}':\text{n.o.c.}} FP_{\mathbb{K}'}(\Pi)\Omega = \bigcap_{\mathbb{K}'\subset\mathbb{K}} FP_{\mathbb{K}'}(\Pi)\Omega$$

so that for any open compact subgroup Ω of $\mathbb{K}$ we have

$$FP_{\infty}(\Pi)\Omega = \bigcap_{\mathbb{K}'\subset\mathbb{K}} FP_{\mathbb{K}'}(\Pi)\Omega \quad .$$

Therefore,

$$pr_{\mathbb{K}}(FP_{\infty}(\Pi)\Omega) = pr_{\mathbb{K}}(\bigcap_{\mathbb{K}'\subset\mathbb{K}} FP_{\mathbb{K}'}(\Pi)\mathbb{K}) \quad ,$$

and if we let $\Omega = \mathbb{K}$, this shows that

$$pr_{\mathbb{K}}(FP_{\infty}(\Pi)) = pr_{\mathbb{K}}(\bigcap_{\mathbb{K}'\subset\mathbb{K}} FP_{\mathbb{K}'}(\Pi)\mathbb{K})$$

$$= pr_{\mathbb{K}}(\bigcap_{\mathbb{K}'\subset\mathbb{K}} G_{+}(Q)FP_{\mathbb{K}'}(\Pi)\mathbb{K}_{\infty}\mathbb{K})$$

(since obviously $G_{+}(Q)FP_{\mathbb{K}'}(\Pi)\mathbb{K}_{\infty} = FP_{\mathbb{K}'}(\Pi)$)

$$= \bigcap_{\mathbb{K}'\subset\mathbb{K}} pr_{\mathbb{K}}(G_{+}(Q)FP_{\mathbb{K}'}(\Pi)\mathbb{K}_{\infty}\mathbb{K})$$

(because each double coset $G_{+}(Q)a\mathbb{K}_{\infty}\mathbb{K}$ is the complete inverse image under $pr_{\mathbb{K}}$ of $pr_{\mathbb{K}}(a)$, so that "the intersection of the projections is the projection of the intersection"), and the last expression is finally equal to the intersection over $\mathbb{K}'\subset\mathbb{K}$ of the sets $pr_{\mathbb{K}}(FP_{\mathbb{K}'}(\Pi))$, which intersection is defined over $\mathbb{Q}$. That is, we have:

(*) $\quad pr_{\mathbb{K}}(FP_{\infty}(\Pi))$ is a zero cycle rational over $\mathbb{Q}$.

Now we assume B to be the orbit under $C_{\mathbb{K}}(\mathcal{O})$ (by way of its action on $\Xi_{\mathbb{K}}(\mathcal{O},\tilde{\Sigma})$, the projection of $\Xi_{A}(\mathcal{O},\tilde{\Sigma})$ into $V_{\mathbb{K}}$) of a point $\theta.(\tau)$ in the image of $\Xi_{\infty}(\mathcal{O},\tilde{\Sigma})$. This may be done without loss of generality because $\Xi_{A}(\mathcal{O},\tilde{\Sigma}) = G_{+}(Q)\Xi_{\infty}(\mathcal{O},\tilde{\Sigma})G(\hat{\mathbb{Z}})\mathbb{K}_{\infty}$, so that each pre-image in $G_{+}(A)$ of a point of $\Xi_{\mathbb{K}}(\mathcal{O},\tilde{\Sigma})$ contains an element of the form $\xi\theta\omega$, where $\omega \in G(\hat{\mathbb{Z}})$, and $\xi \in \Xi_{\theta,\infty}(\mathcal{O},\tilde{\Sigma})$, and if Φ is an F-arithmetic modular form with respect to $\mathbb{K}$, then $R(\omega)\Phi$ is, too, because ω normalizes $\mathbb{K}$; thus, as remarked in [11] the <u>family</u> of "ray class" polynomials with respect to $\mathbb{K}$ on an orbit B of $C_{K}(\mathcal{O})$ in $\Xi_{\mathbb{K}}(\mathcal{O},\tilde{\Sigma})$ is invariant under right translation by $\omega \in G(\hat{\mathbb{Z}})$; hence, we may take a point of the orbit B to be the image $\theta.(\tau)$ of a point $\xi\theta \in \Xi_{\infty}(\mathcal{O},\tilde{\Sigma})$. Suppose now that $\Pi = q'(\pi)$ with $\pi = \Lambda_{f}$ for some $\Lambda \in K$ such that $K = k(\Lambda)$.

We first show that $\tilde{B} \subset pr_{\mathbb{K}}(FP_{\infty}(q'(\pi)))$.

On account of the results of §§2,3 of [5] , $\Xi_{A}(\mathcal{O},\tilde{\Sigma})$ is stable under the action (described there) of $I(K)_{f}$ and of ι on $\Xi_A(\mathcal{O}, \tilde{\Sigma})$. Now $B = \{a*\theta_\cdot(\tau) \mid a \in I(K)_f\}$, and and $\tilde{B} = B \cup \iota * B$; hence $\tilde{B} \subset \Xi_{\mathbb{K}}(\mathcal{O}, \tilde{\Sigma})$, anyway. If $a \in I(K)_f$, $a*\theta_\cdot(\tau)$ is the image in $V_{\mathbb{K}}$ of $G_+(Q)q(a^{-1})\xi\theta\,\mathbb{K}\mathbb{K}_\infty$ and $\iota * a * \theta_\cdot(\tau)$, that of $G_-(Q)q(\iota)_\infty q(a^{-1})\xi\theta\,\mathbb{K}\mathbb{K}_\infty$. Now

$$G_+(Q)q(a^{-1})\xi\theta q'(\pi)\mathbb{K}_\infty = G_+(Q)q(\Lambda)_\infty^{-1}q(a^{-1})\xi\theta\,\mathbb{K}_\infty = $$
$$= G_+(Q)q(a^{-1})\xi\theta\,\mathbb{K}_\infty ,$$

because $q(\Lambda)_\infty^{-1}\xi\,\mathbb{K}_\infty = \xi\,\mathbb{K}_\infty$, since $q = q_\tau$, with τ as above, while

$$G_-(Q)q(\iota)_\infty q(a^{-1})\xi\theta q'(\pi)\mathbb{K}_\infty = G_-(Q)q(\pi)q(\iota)_\infty q(a^{-1})\xi\theta\,\mathbb{K}_\infty =$$
$$= G_-(Q)q(\Lambda)_\infty^{-1}q(\iota)_\infty q(a^{-1})\xi\theta\,\mathbb{K}_\infty =$$
$$= G_-(Q)q(\iota)_\infty q(a^{-1})q(\bar{\Lambda})_\infty^{-1}\xi\theta\,\mathbb{K}_\infty = G_-(Q)q(\iota)_\infty q(a^{-1})\xi\theta\,\mathbb{K}_\infty$$

for the same reasons. Hence, both $a*\theta(\tau)$ and $a*\iota*\theta(\tau) = \iota*\bar{a}*\theta(\tau)$ belong to the fixed points of the correspondence $C_{\mathbb{K}}(q'(\pi))$, i.e., $\tilde{B} \subset FP_\infty(q'(\pi))$. (N.B. For any $y \in G_-(Q)$, y normalizes $G_+(Q)$; therefore, if $x \in G_+(Q)$, $G_-(Q) = G_+(Q)y = G_+(Q).{}^y x.y = G_+(Q)yx$.)

Suppose, conversely, that z belongs to the intersection of

$$\Xi_A(\mathcal{O}, \tilde{\Sigma}) = G_+(Q)\Xi_\infty(\mathcal{O}, \tilde{\Sigma})\mathbb{K}_\infty G(\hat{\mathbb{Z}})$$

and of $FP_\infty(q'(\pi))$. If we show that $pr_{\mathbb{K}}(z) \in \tilde{B}$ for every such z , it will follow that

$$\tilde{B} \supset \Xi_{\mathbb{K}}(\mathcal{O}, \tilde{\Sigma}) \cap pr_{\mathbb{K}}(FP_\infty(q'(\pi))) ,$$

because each element on the right hand side is the image under $pr_{\mathbb{K}}$ of an element of $\Xi_A(\mathcal{O}, \tilde{\Sigma}) \cap FP_\infty(q'(\pi))$, and thus it will follow that

$$\tilde{B} = \Xi_{\mathbb{K}}(\mathcal{O}, \tilde{\Sigma}) \cap pr_{\mathbb{K}}(FP_\infty(q'(\pi))) ,$$

and the right hand side is a zero cycle rational over $K^*(\tilde{\Sigma})_0$. (In fact this will show that $\tilde{B} = V_{\mathbb{K}}^{q'} \cap \Xi_{\mathbb{K}}(\mathcal{O}, \tilde{\Sigma})$.)

But for a given such z , we may, by Prop. 1, write

$$z = \gamma\xi q(a)\theta\eta , \gamma \in G_+(Q) , \theta \in \textcircled{H} , a \in I(K)_f, \eta \in G(\hat{\mathbb{Z}})\mathbb{K}_\infty ,$$

and where $\xi(i.e) = (\tau)$. Write $\eta = \eta_1 u_1$, $\eta_1 \in G(\hat{\mathbb{Z}})$, $u_1 \in \mathbb{K}_\infty$. We now show that $\theta\eta\theta^{-1}$ belongs either to $q(I(K)_f)\mathbb{K}_\infty$ or to $q(\iota_f I(K)_f)\mathbb{K}_\infty$, and this says that $z \in \gamma q(c)\xi\theta\,\mathbb{K}_\infty$, where c is either in $I(K)_f$ or in $I(K)_f\iota_f$, so that $z \in B \cup \iota * B = \tilde{B}$, which is what we want.

In fact we have

$$z = \gamma\xi q(a)\theta\eta \in FP_\infty(\Pi) \subset FP_{\mathbb{K}'}(\Pi)$$

for every n.o.c. subgroup $\mathbb{K}'$ of $G(A_f)$. Hence, $zq'(\pi) \in G_+(Q)z\mathbb{K}_\infty\mathbb{K}'$ and $pr_{\mathbb{K}'}(zq'(\pi)) = pr_{\mathbb{K}'}(z)$; therefore, by Prop. 3 (with $a = 1$), there exists $b_{\mathbb{K}',f} \in \hat{K}^\times$ such that

$$q'(\pi) \in \eta^{-1}\theta^{-1}q(b_{\mathbb{K}',f})\theta\eta\mathbb{K}' \quad ,$$

so that $q'(\pi) = \eta^{-1}\theta^{-1}q(b_{\mathbb{K}',f})\theta\eta\kappa'$ for some $\kappa' \in \mathbb{K}'$. For the rest of this proof, we may assume $\mathbb{K}' = \mathbb{K}(\mathfrak{n})$ for some integral ideal $\mathfrak{n}$ of $\mathfrak{o}$. If we let $\mathfrak{n}$ "tend to infinity" in such a way that $\mathbb{K}(\mathfrak{n})$ tends to $\{1\}$, then $\kappa' = \kappa_{\mathfrak{n}} \longrightarrow 1$, so that $b_{\mathbb{K}',f} = b_{\mathfrak{n},f}$ must also tend to a limit b_0 which belongs to the closure of $\hat{K}^\times$ in $I(K)_f$ (because the extension of q to $I(K)$ is a closed immersion of $I(K)$ into $G_+(A)$, since the image $q(I(K))$ is equal to $T(A(k))$ for a certain maximal torus T of GL_2, defined over k). Let $\theta\eta = \omega = \omega' u_1$, so that $\omega' = \theta\eta_1 \in G(A_f)$, and put $\omega'' = \theta\eta_1\theta^{-1}$. Then $q'(\pi) = \omega'^{-1}q(b_0)\omega'$ for some b_0 in the closure of $\hat{K}^\times$ in $I(K)_f$, or

$$\omega' q'(\pi)\omega'^{-1} = \omega'' q(\pi)\omega''^{-1} = q(b_0) \quad .$$

Now $K = k(\Lambda)$, $\pi = \Lambda_f$, and $q = q_\tau$. We want to prove that $b_0 \in \hat{K}^\times$, too. For this purpose we use

2.3.1 A LEMMA BASED ON A RESULT OF E. ARTIN [1]

Lemma 1. Let q be a regular imbedding of K in $M_2(k)$, extended to an imbedding of $A(K)$ in $M_2(A(k))$. Suppose that b_0 belongs to the closure of $\hat{K}^\times$ in $I(K)$, that $\hat{c}$ belongs to $\hat{K}^\times$, $c \in K^\times$, such that $K = k(c)$, and that for some $\omega \in G(A_f)$ we have $\omega^{-1}q(b_0)\omega = q(\hat{c})$. Then ω normalizes the subalgebra $q(A(K))$ of $M_2(A(k))$ and $b_0 = \hat{b}$ for some $b \in K^\times$.

Proof. According to [18: p.62] , if β_1 , β_2 are a k-basis of K , there is a canonical identification of $A(K)$ with $A(k)\beta_1 + A(k)\beta_2$. Since K is a CM-extension of the totally real field k , there exists $\delta \in K$ such that $K = k(\delta)$, $\delta^2 \in k$, and $\delta^2 \ll 0$. Hence

$$A(K) \simeq A(k) + A(k)\delta$$

which is an isomorphism of $A(k)$-algebras. Hence we may write uniquely

$$\hat{c} = \hat{\gamma} + \hat{\varepsilon}\hat{\delta} \text{ , with } \gamma,\varepsilon \in k \text{ ,}$$

$$b_0 = \alpha + \beta\hat{\delta} \text{ , with } \alpha,\beta \in A(k) \text{ ,}$$

and $q(A(k))$ is contained in the center of $M_2(A(k))$. Therefore from the relation $\omega^{-1}q(b_0)\omega = q(\hat{c})$ we have (since q is an algebra homomorphism)

$$q(\hat{\gamma}) + q(\hat{\varepsilon})^{\omega}q(\hat{\delta}) = q(\alpha) + q(\beta)q(\hat{\delta})$$

and $\varepsilon \neq 0$ since $K = k(\hat{c})$. Therefore we have

(2) $$^{\omega}q(\hat{\delta}) = q(\hat{\varepsilon})^{-1}\{q(\alpha) + q(\beta)q(\hat{\delta}) - q(\hat{\gamma})\} \in q(A(K)) \text{ .}$$

That is, since $A(K)_f = A(k)_f + A(k)_f\hat{\delta}$,

(3) $$^{\omega}q(A(K)_f) \subset q(A(K)_f) \text{ .}$$

Let $\mathfrak{g}$ be any finite place of k , and let $K_{\mathfrak{g}}$ be the tensor product $k_{\mathfrak{g}} \otimes_k K = k_{\mathfrak{g}} + k_{\mathfrak{g}}\delta$. Then (3) implies that

(4) $$^{\omega}q(K_{\mathfrak{g}}) \subset q(K_{\mathfrak{g}}) = q(k_{\mathfrak{g}}) + q(k_{\mathfrak{g}})q(\hat{\delta}) \text{ .}$$

Now $q(K_{\mathfrak{g}}) = q(k_{\mathfrak{g}}) + q(k_{\mathfrak{g}})q(\delta)$ contains the $k_{\mathfrak{g}}$-rational elements $T(k_{\mathfrak{g}})$ of a maximal torus in $GL_2(k_{\mathfrak{g}})$. Since $\omega_{\mathfrak{g}} \in GL_2(k_{\mathfrak{g}})$, $\mathrm{Inn}(\omega_{\mathfrak{g}})$ is an automorphism of $M_2(k_{\mathfrak{g}})$ and (4) implies that $^{\omega_{\mathfrak{g}}}q(K_{\mathfrak{g}}) = q(K_{\mathfrak{g}})$, and hence $\omega_{\mathfrak{g}}$ belongs to the normalizer of $T(k_{\mathfrak{g}}) = q(K_{\mathfrak{g}}) \cap GL(k_{\mathfrak{g}})$. The normalizer of $T(k_{\mathfrak{g}})$ in $GL_2(k_{\mathfrak{g}})$ contains its centralizer, namely $T(k_{\mathfrak{g}})$, as a subgroup of index at most two. The $\mathfrak{g}$-component of the (relatively) involutory element ι also normalizes $T(k_{\mathfrak{g}})$, but $^{q(\iota)}q(\delta) = -q(\delta)$. Hence, the normalizer of $T(k_{\mathfrak{g}})$ in $GL_2(k_{\mathfrak{g}})$ is $T(k_{\mathfrak{g}}) \cup q(\iota)T(k_{\mathfrak{g}})$. Therefore, $\omega_{\mathfrak{g}} \in T(k_{\mathfrak{g}}) \cup q(\iota)T(k_{\mathfrak{g}})$ for all finite places $\mathfrak{g}$ of k . Thus, $(^{\omega}q(\delta))_{\mathfrak{g}} = \pm q(\delta)_{\mathfrak{g}} = \xi_{\mathfrak{g}}q(\delta)_{\mathfrak{g}}$, say, where $\xi_{\mathfrak{g}} = \pm 1$ for each $\mathfrak{g}$; therefore, since 1 and $\delta = \delta_{\mathfrak{g}}$ are linearly independent over $k_{\mathfrak{g}}$ for each $\mathfrak{g}$, we have $\alpha = \hat{\gamma}$ and $q(\hat{\varepsilon})_{\mathfrak{g}} = \pm q(\beta)_{\mathfrak{g}} = q(\xi_{\mathfrak{g}})q(\beta)_{\mathfrak{g}}$. Since $\varepsilon \neq 0$, we have $\beta_{\mathfrak{g}} \neq 0$ for all $\mathfrak{g}$. Call ξ the idelic unit $(\xi_{\mathfrak{g}})_{\mathfrak{g}\text{:finite}}$ in $I(K)_f$. Clearly $\xi^2 = 1$.

By hypothesis, there is a sequence $\{b_n\} \subset K^{\times}$ such that $b_0 = \lim_{n \to \infty} \hat{b}_n$ in $I(K)_f$. We may write each $b_n = \alpha_n + \beta_n\delta$, with $\alpha_n , \beta_n \in k$, and $\hat{b}_n \longrightarrow b_0$ in the <u>idelic</u> topology (which is finer than the adelic). Since $b_0 \in I(K)$, b_0^{-1} exists and $\hat{b}_n^{-1} \longrightarrow b_0^{-1}$, thus $b_n \neq 0$ for almost all n , and if m , n are very large and $m > n$, then

$b_m = \varepsilon_{m,n} b_n$, where $\varepsilon_{m,n}$ belongs to a small congruence subgroup of the units of K . Since K is a pure imaginary quadratic extension of k , the totally positive units of k are a subgroup of finite index in the units of K , so by Chevalley's theorem [6] on the units, we may assume each $\varepsilon_{m,n}$ is a totally positive unit of k . Let $\eta_1, \ldots, \eta_t$ be an independent set of generators of the totally positive units of k $(t = [k:Q] - 1)$. Since ${}^{\omega}q(\hat{\delta}) = q(\xi\hat{\delta})$, we have $q(\hat{\gamma}) + q(\hat{\varepsilon}\xi)q(\hat{\delta}) = q(\alpha) + q(\beta)q(\hat{\delta})$, and if $\hat{b}_n = \hat{\alpha}_n + \hat{\beta}_n\hat{\delta}$, we have in the limit

$$q(\hat{\alpha}_n) + q(\hat{\beta}_n)q(\hat{\delta}) \longrightarrow q(\alpha) + q(\beta)q(\hat{\delta})$$

in the idelic topology. By discarding the early terms in the sequence, we have for some large enough n_0 :

$$\hat{b}_m = \hat{\varepsilon}_{m,n_0}(\hat{\alpha}_{n_0} + \hat{\beta}_{n_0}\hat{\delta}) \quad , \quad \beta_{n_0} \neq 0 \ ,$$

(since $\beta \neq 0$ because 1 and δ are linearly independent over k : cf.(2)), where $\hat{\varepsilon}_m = \hat{\varepsilon}_{m,n_0}$ is a Cauchy sequence in $I(K)_f$, and each $\varepsilon_m \in \mathfrak{o}^{\times}_{+}$, the totally positive units of k . Hence, if $\hat{\beta}_m = \hat{\varepsilon}_m\beta_{n_0}$, we have $\hat{\beta}_m \longrightarrow \beta = \hat{\varepsilon}\xi$, in the idelic topology. Therefore, $\hat{\varepsilon}^{-1}\hat{\beta}_m \longrightarrow \xi = (\xi_{\mathfrak{g}})$, in the idelic topology. Since $\xi_{\mathfrak{g}} = \pm 1$ for every finite place, $\hat{\varepsilon}^{-1}\hat{\beta}_m = \hat{\theta}_m$, where we may assume $\hat{\theta}_m$ is a unit of K , hence (since ε , $\beta_m \in k$), a unit of k for all m . Say $\theta_m = u_m\theta_0$, where $\theta_0 \in \mathfrak{o}^{\times}$ and $u_m \in \mathfrak{o}^{\times}_{+}$ for all m , and

$$\hat{u}_m = \hat{\eta}_1^{x_{1,m}} \cdot \ldots \cdot \hat{\eta}_t^{x_{t,m}} \quad , \quad x_{j,m} \in \mathbb{Z} \quad .$$

We may take a subsequence such that for each $j = 1,\ldots,t$, $x_{j,m}$ converges to $\bar{x}_j \in \hat{\mathbb{Z}}$ (the standard compactification of $\mathbb{Z}$). Then

$$\xi = \lim(\hat{\theta}_m) = \hat{\theta}_0 \lim \hat{u}_m = \hat{\theta}_0 \eta_1^{\bar{x}_1} \cdot \ldots \cdot \eta_t^{\bar{x}_t} \quad .$$

Since $\xi^2 = 1$, this gives

$$\hat{\eta}_1^{2\bar{x}_1} \cdot \ldots \cdot \hat{\eta}_t^{2\bar{x}_t} = \hat{\theta}_0^{-2} = \hat{\eta}_1^{a_1} \cdot \ldots \cdot \hat{\eta}_t^{a_t} \quad , \quad a_j \in \mathbb{Z} \ ,$$

since $\theta_0^2 \in o^{\times}_{+}$. By a lemma of Artin [1] , this implies that $2\bar{x}_j = a_j$, $j = 1,\ldots,t$, hence $\bar{x}_j = a_j/2 \in \hat{\mathbb{Z}} \cap \frac{1}{2}\mathbb{Z} = \mathbb{Z}$. That is,

$$\xi = \hat{\theta}_0 \hat{\eta}_1^{x_1} \cdot \ldots \cdot \hat{\eta}_t^{x_t} \quad , \quad x_j \in \mathbb{Z} \quad , \; j = 1, \ldots, t \quad ,$$

so that ξ is in fact equal to $\hat{\eta}$ for a unit $\eta = \pm 1 \varepsilon k$. Thus $\beta = \hat{\varepsilon}\hat{\eta} = \pm \hat{\varepsilon} \varepsilon \hat{k}$. Since $\alpha = \hat{\gamma}$, we obtain $b_0 = \hat{\gamma} \pm \hat{\varepsilon}\hat{\delta} \varepsilon \hat{K}^{\times}$, which proves the lemma.

2.3.2 COMPLETION OF THE PROOF OF PROPOSITION 8

Returning to the proof of Proposition 8, this shows that $b_0 \varepsilon \hat{K}^{\times}$. But

$$\omega'' q(\pi) \omega''^{-1} = q(b_0) \quad ,$$

thus $\mathrm{Int}(\omega'')$ gives an automorphism of $q(\hat{K})$ over $q(\hat{k})$, and since $A(K) = A(k) + A(k)\Lambda$, it follows that this automorphism extends to one of $A(K)$. Clearly we must have $b_0 = \pi$ or $b_0 = \overline{\pi}$. $T(k_{\mathfrak{g}}) = q(K_{\mathfrak{g}}^{\times})$ is its own centralizer in $GL_2(k_{\mathfrak{g}})$ and $N(k_{\mathfrak{g}}) = T(k_{\mathfrak{g}}) \cup \iota T(k_{\mathfrak{g}})$ is the normalizer of $T(k_{\mathfrak{g}})$. It follows that either $\mathrm{Int}(\omega'')$ is the identity or $\mathrm{Int}(\iota\omega'')$ is the identity, and therefore we have either $\omega'' \varepsilon q(A(K)^{\times})$ or $\omega'' \varepsilon q(A(K)^{\times}\iota_f$, as in the case described in [11:§4.4.4]. Thus $z = \gamma\xi q(a)\omega' u_1$ belongs to

$$G_+(\mathbb{Q})\xi[q(I(K)_f) \cup q(\iota_f I(K)_f)]\,\theta \mathbb{K}_\infty \quad ,$$

the full pre-image of $\tilde{B}$. Hence

$$\tilde{B} = \Xi_{\mathbb{K}}(\mathcal{O}, \tilde{\Sigma}) \cap \mathrm{pr}_{\mathbb{K}}(FP_\infty(q'(\pi))) \quad ,$$

which shows that $\tilde{B}$ is a zero-cycle rational over $K^*(\tilde{\Sigma})_0$.

3 THE RATIONALITY OF B OVER $K^*(\tilde{\Sigma})$

<u>Proposition 9. The orbit</u> B <u>of</u> $C_{\mathbb{K}}(\mathcal{O})$ <u>in</u> $\Xi_{\mathbb{K}}(\mathcal{O}, \tilde{\Sigma}) \subset V_{\mathbb{K}}$ <u>is rational over the reflex field</u> $K^*(\tilde{\Sigma})$.

In the statement and proof of this proposition, the notation retains the same meanings as in the statement and proof of Proposition 8.

<u>Proof.</u> We already know that $\tilde{B} = B \cup \iota * B$ is rational over the totally real subfield $K^*(\tilde{\Sigma})_0$ of $K^*(\tilde{\Sigma})$. Either B and $\iota * B$ are equal or they are disjoint; for if $a * \theta(\tau) = \iota * a' * \theta(\tau) = \overline{a'} * \iota * \theta(\tau)$ for some a, $a' \varepsilon I(K)_f$, then clearly $B = \iota * B$, so that $B = \tilde{B}$, and B is rational even over

$K^*(\tilde{\Sigma})_0$.

Now suppose B and $\iota * B$ are disjoint. We want to construct an arithmetic modular function Φ in the compositum $M(\mathbb{K}, K^*(\tilde{\Sigma}))$ of $K^*(\tilde{\Sigma})$ and $M(\mathbb{K})$, such that Φ is constant on each of B and $\iota * B$ and such that $\Phi(B) \neq \Phi(\iota * B)$.

3.1 STABILIZER SUBGROUPS AND FIXED POINTS OF CORRESPONDENCES

We start from the chosen point $pr_{\mathbb{K}}(\xi\theta)$ of the orbit B, where θ is an element of $\textcircled{H}$ and $\xi \in \Xi_{\theta,\infty}(\mathcal{O}, \tilde{\Sigma})$. If $z \in \Xi_A(\mathcal{O}, \tilde{\Sigma})$, then by Prop. 1, z belongs to the set $G_+(Q)q(I(K)_f)\xi\theta\omega\mathbb{K}_\infty$ for some $\omega \in G(\hat{\mathbb{Z}})$. Let $(\tau) = \xi(i.e)$, so that $\theta.(\tau) \in \Xi_\theta(\mathcal{O}, \tilde{\Sigma})$. For the present, z is any element of $\Xi_A(\mathcal{O}, \tilde{\Sigma})$, and not necessarily in B . Put $\eta = \theta\omega$ and define

$$\mathcal{G}(z) = z^{-1}G_+(Q)z\mathbb{K}_\infty \cap G(A_f) \quad .$$

Suppose that $g = z^{-1}\gamma' z k'_\infty \in \mathcal{G}(z)$, $k'_\infty \in K_\infty$ and that $z = \gamma q(a)\xi\theta\omega k_\infty$, $\gamma \in G_+(Q)$, $a \in I(K)_f$, and $k_\infty \in \mathbb{K}_\infty$. Then the requirement $g \in \mathcal{G}(z)$ reads, for the archimedean part,

$$\xi^{-1}\gamma_\infty{}^{-1}\gamma'_\infty\gamma_\infty\xi \in \mathbb{K}_\infty \text{ , for some } \gamma' \in G_+(\mathbb{Q}) \quad ,$$

or $\gamma_\infty^{-1}\gamma'_\infty\gamma_\infty \in {}^{\xi}\mathbb{K}_\infty \cap G_+(Q)_\infty = q_\tau(K^\times)_\infty$, where $q = q_\tau$ and q', etc., will have their usual meanings. Thus, $\gamma^{-1}\gamma'\gamma = q(c)$, where $c \in K^\times$. Since $g \in G(A_f)$, the conditions on the non-archimedean part read

$$g = \eta^{-1}q(a)^{-1}\gamma_f{}^{-1}\gamma_f'\gamma_f q(a)\eta = \eta^{-1}q(a)^{-1}q(\hat{c})q(a)\eta = \eta^{-1}q(\hat{c})\eta \quad ,$$

where $\hat{c} = c_f$. Thus

$$\mathcal{G}(z) = \eta^{-1}q(\hat{K}^\times)\eta \quad .$$

If $\mathbb{K}'$ is an n.o.c. subgroup of $\mathbb{K}$, let $\mathcal{G}_{\mathbb{K}'}(z) = \mathcal{G}(z) \cap \mathbb{K}'$.

On the other hand, for a fixed element b of $K^\times$ let

$$X_{q'} = \mathrm{Int}\,\mathbb{K}\,(q'(\hat{b})) = \{kq'(\hat{b})k^{-1} \mid k \in \mathbb{K}\},$$

$$X_{\mathbb{K}'}(z) = \mathcal{G}(z) \cap (X_{q'}\mathbb{K}') \text{ , and } X^\infty_{q'}(z) = \mathcal{G}(z) \cap X_{q'} \quad .$$

Let $Y_{\mathbb{K}'}(z)$ resp. $Y^\infty_{\mathbb{K}'}(z)$ denote $X_{\mathbb{K}'}(z)\mathbb{K}'$ resp. $X^\infty_{q'}(z)\mathbb{K}'$, and $Z_{\mathbb{K}'}(z)$ resp. $Z^\infty_{\mathbb{K}'}(z)$ denote the images of these in $V_{\mathbb{K}'}$. Clearly $Y^\infty_{\mathbb{K}'}(z) \subset Y_{\mathbb{K}'}(z)$ and $Z^\infty_{\mathbb{K}'}(z) \subset Z_{\mathbb{K}'}(z)$ for all n.o.c. subgroups $\mathbb{K}'$ of $\mathbb{K}$. We claim that $Z_{\mathbb{K}'}(z)$ is always finite and that for sufficiently small $\mathbb{K}'$, we have

$Z^{\infty}_{\mathbb{K}'}(z) = Z_{\mathbb{K}'}(z)$. First of all, trivially,

$$\mathrm{Int}\,\mathbb{K}(q'(\hat{b})) \subset \mathbb{K}q'(\hat{b})\mathbb{K} = C_{\mathbb{K}}(q'(\hat{b})) ,$$

which is compact and open in $G(A_f)$; since $\mathbb{K}'$ is open and contained in $\mathbb{K}$, this implies that $X_{q'}$ contains representatives of only finitely many left cosets of $\mathbb{K}'$ in $G(A_f)$, hence $X_{\mathbb{K}'}(z)$ modulo $\mathbb{K}'$ on the right is finite (possibly empty); therefore, $Z_{\mathbb{K}'}(z)$ and $Z^{\infty}_{\mathbb{K}'}(z)$ are finite. Suppose $X_{\mathbb{K}'}(z)$ is non-empty for all such $\mathbb{K}'$, that is

$$X_{q'}\mathbb{K}' \cap \mathscr{G}(z) = X_{q'}\mathbb{K}' \cap \omega^{-1}q'(\hat{K}^{\times})\omega \neq \emptyset , \qquad \omega = \theta^{-1}\eta ,$$

for all n.o.c. subgroups $\mathbb{K}'$ of $\mathbb{K}$, where $\hat{K} = K_f$. Then for each $\mathbb{K}'$ there exist $c_{\mathbb{K}'} \in K^{\times}$, $k_{\mathbb{K}'} \in \mathbb{K}$, and $k'_{\mathbb{K}'} \in \mathbb{K}'$ such that

$$(*) \qquad \omega^{-1}q'(\hat{c}_{\mathbb{K}'})\omega = k_{\mathbb{K}'}q'(\hat{b})k_{\mathbb{K}'}^{-1}k'_{\mathbb{K}'} .$$

Let $\mathbb{K}'$ "shrink to $\{1\}$" through a linearly ordered sequence of n.o.c. subgroups constituting a neighborhood basis for the identity in $G(A_f)$, and for each $\mathbb{K}'$, let $c_{\mathbb{K}'}$, $k_{\mathbb{K}'}$, and $k'_{\mathbb{K}'}$ be chosen to satisfy (*) . Automatically $k'_{\mathbb{K}'}$ tends to the identity in the usual topology on $G(A_f)$, and since $\mathbb{K}$ is compact, by going to a subsequence we may assume

$$\lim_{\mathbb{K}' \longrightarrow \{1\}} k_{\mathbb{K}'} = k_0 \in \mathbb{K} .$$

Then

$$\lim_{\mathbb{K}' \longrightarrow \{1\}} \hat{c}_{\mathbb{K}'} = c_0$$

exists and belongs to the closure of $\hat{K}^{\times}$ in $I(K)_f$. We claim that c_0 belongs to $\hat{K}^{\times}$. In fact, we have from taking the limit in (*) that $\omega^{-1}q'(c_0)\omega = k_0q'(\hat{b})k_0^{-1} \in X_{q'}$, or $(\eta k_0)^{-1}q(c_0)\eta k_0 = q'(\hat{b})$, and applying Lemma 1 with $\omega = \eta k_0\theta^{-1}$, we see that $c_0 \in \hat{K}^{\times}$, as claimed.

Let $z \in \Xi_A(\mathcal{O},\hat{\Sigma})$, fix the n.o.c. subgroup $\mathbb{K}$ of $G(A_f)$, and let $\mathbb{K}'$ be an n.o.c. subgroup of K .

<u>Proposition 10. For all sufficiently small</u> $\mathbb{K}'$ <u>we have</u>

$$X_{\mathbb{K}'}(z) \subset X_{q'}(z)\mathbb{K} ,$$

<u>where</u> $X_{q'}(z) = X^{\infty}_{q'}(z) = X_{q'} \cap \mathscr{G}(z)$.

<u>Proof.</u> For any $\mathbb{K}' \subset \mathbb{K}$ we have

$$X_{\mathbb{K}'}(z) = \mathcal{G}(z) \cap X_{q'}\mathbb{K}' \subset X_{q'}\mathbb{K} = \mathbb{K}q'(\hat{b})\mathbb{K} \quad ,$$

so that $X_{\mathbb{K}'}(z)$ contains representatives from only a fixed finite set of left cosets of $\mathbb{K}$ for all $\mathbb{K}'$. If $\mathbb{K}' \subset \mathbb{K}$, write

$$X_{\mathbb{K}'}(z)\mathbb{K} = \bigcup_{\alpha \in A(\mathbb{K}')} \alpha\mathbb{K} \quad ,$$

where $A(\mathbb{K}')$ is a finite set. If $\mathbb{K}'' \subset \mathbb{K}'$, then $X_{\mathbb{K}''}(z) \subset X_{\mathbb{K}'}(z)$, hence we may assume that $A(\mathbb{K}'') \subset A(\mathbb{K}')$, and if we put $A(\infty) = \bigcap_{\mathbb{K}'} A(\mathbb{K}')$, then there exists $\mathbb{K}_0'$ such that $A(\mathbb{K}_0') = A(\infty)$. It suffices to show that $A(\infty)\mathbb{K} = X_{q'}(z)\mathbb{K}$. If $\alpha \in A(\infty)$, then for every $\mathbb{K}' \subset \mathbb{K}$ there exist $q_{\mathbb{K}'} \in K^{\times}$, $k_{\mathbb{K}'} \in \mathbb{K}$, and $k'_{\mathbb{K}'} \in \mathbb{K}'$ satisfying (*), i.e., such that

$$g_{\mathbb{K}'} = \eta^{-1}q(\hat{c}_{\mathbb{K}'})\eta = k_{\mathbb{K}'}q(\hat{b})k_{\mathbb{K}'}^{-1}k'_{\mathbb{K}'} \in X_{\mathbb{K}'}(z) \quad ,$$

and also such that $\alpha \in g_{\mathbb{K}'}\mathbb{K}$. As before, passing to the limit as $\mathbb{K}'$ shrinks to $\{1\}$ we obtain $k_0 \in \mathbb{K}$, $c_0 \in K^{\times}$ such that $\hat{c}_0 = \lim \hat{c}_{\mathbb{K}'}$ and $g_0 = \eta^{-1}q(\hat{c}_0)\eta = k_0 q(\hat{b})k_0^{-1} \in X_{q'}(z)$, and therefore for small enough $\mathbb{K}'$ we have $\eta^{-1}q(\hat{c}_{\mathbb{K}'})\eta \in g_0\mathbb{K}$; then $\alpha \in g_{\mathbb{K}'}\mathbb{K} = g_0\mathbb{K}$ for all sufficiently small $\mathbb{K}'$, hence $\alpha\mathbb{K} = g_0\mathbb{K} \subset X_{q'}(z)\mathbb{K}$ for all $\alpha \in A(\mathbb{K}_0')$, proving the result.

3.2 MULTIPLIER POLYNOMIALS

The coset space $X_{\mathbb{K}'}(z)\mathbb{K}'/\mathbb{K}'$ is finite, because $[\mathbb{K}:\mathbb{K}']$ is finite, and may be identified with $X_{\mathbb{K}'}(z)/\mathcal{G}_{\mathbb{K}'}(z)$, where $\mathcal{G}_{\mathbb{K}'}(z) = \mathcal{G}(z) \cap \mathbb{K}'$, by means of the mapping of cosets

$$x\mathcal{G}_{\mathbb{K}'}(z) \longrightarrow x\mathbb{K}' \quad ;$$

for if $x, y \in X_{\mathbb{K}'}(z)$ and if $y\mathcal{G}_{\mathbb{K}'}(z) \subset x\mathbb{K}'$, then $x^{-1}y \in \mathbb{K}' \cap \mathcal{G}(z) = \mathcal{G}_{\mathbb{K}'}(z)$.

Let $pr_{\mathbb{K}'/\mathbb{K}}$ be the canonical projection of $V_{\mathbb{K}'}$ onto $V_{\mathbb{K}}$, $pr^{\infty}_{\mathbb{K}}$ be the canonical projection of $V_{\infty} = G_+(Q)\backslash G_+(A)/\mathbb{K}_{\infty}$ onto $V_{\mathbb{K}}$, and $pr_{\mathbb{K}}$ be that of $G_+(A)$ onto $V_{\mathbb{K}}$. Then

$$pr_{\mathbb{K}}(X_{\mathbb{K}'}(z)) = pr_{\mathbb{K}'/\mathbb{K}}(X_{\mathbb{K}'}(z)/\mathcal{G}_{\mathbb{K}'}(z))$$

is a (finite) zero cycle on $V_{\mathbb{K}}$, as is also $pr_{\mathbb{K}}(X_{q'}(z))$ which it contains.

Fix $\psi \in A(\mathbb{K}, w)$ for some $w > 0$ such that ψ is non-zero at every point of $\Xi_A(\mathcal{O}, \tilde{\Sigma})$. If $M \in G(A_f)$, define for any $g \in G_+(A)$ such that $\psi(g) \neq 0$,

$$\psi_M(g) = (\psi \| M)(g)/\psi(g) \quad ,$$

where $\psi \| M$ is defined as in §6.1 of [5] . Define a polynomial in the indeterminate t, for given $z \in \Xi_A(\mathcal{O}, \tilde{\Sigma})$, by

$$P_{\psi,\mathbb{K}',b}(t ; z) = \prod_{M \in X_{\mathbb{K}'}(z)\mathbb{K}'/\mathbb{K}'} (t - \psi_M(z)) .$$

Similarly define $P_{\psi,\infty,b}(t ; z)$ with $X^{\infty}_{q'}(z)$ in place of $X_{\mathbb{K}'}(z)$. For the time being, we fix $b \in K$, such that $K = k(b)$, and ψ, and omit the indices ψ and b. It is easy to see that

$$X_{\mathbb{K}'}(z\omega) = \omega^{-1} X_{\mathbb{K}'}(z)\omega \quad , \qquad \omega \in \mathbb{K} .$$

As M runs over $X_{\mathbb{K}'}(z)\mathbb{K}'/\mathbb{K}'$, $M' = \omega^{-1}M\omega$ runs over $X_{\mathbb{K}'}(z\omega)\mathbb{K}'/\mathbb{K}'$. If we define the ideal $\mathfrak{U}(M)$ and the function $\psi \| M$ as in the reference cited above, then $\mathfrak{U}(M) = \mathfrak{U}(M')$ is the same and equal to $\mathfrak{U}(q'(\hat{b}))$ for every $M \in X_{\mathbb{K}'}(z)\mathbb{K}$ and we have for each such M, $\omega \in \mathbb{K}$, and $\kappa \in \mathbb{K}_\infty$:

$$\begin{aligned}\psi_{M'}(z\omega\kappa) &= N(\mathfrak{U})^w \psi(z\omega\kappa M')/\psi(z\omega\kappa) = N(\mathfrak{U})^w \psi(zM\omega\kappa)/\psi(z\omega\kappa) \\ &= N(\mathfrak{U})^w \psi(zM)/\psi(z) = \psi_M(z) \quad ,\end{aligned}$$

since $\psi(g\omega) = \psi(g)$, $\psi(g\kappa) = \chi_w(\kappa)\psi(g)$ for all $g \in G_+(A)$, $\omega \in \mathbb{K}$, $\kappa \in \mathbb{K}_\infty$. Therefore,

$$P_{\mathbb{K}'}(t ; z\omega\kappa) = P_{\mathbb{K}'}(t ; z) \quad \text{and} \quad P_\infty(t ; z\omega\kappa) = P_\infty(t ; z) \quad ,$$

so that the (coefficients of the) polynomials $P_{\mathbb{K}'}(t ; \ )$ and $P_\infty(t ; \ )$ may be viewed as functions on the subset $\tilde{B}$ of $V_{\mathbb{K}}$ (practically as in [11]) .

3.3 RATIONALITY OF THE MULTIPLIER POLYNOMIAL

<u>Definition.</u> <u>Let</u> $\mathcal{F}$ <u>be a</u> $\mathbb{Q}$-<u>algebra of modular functions on</u> $V_{\mathbb{K}}$. <u>If</u> A <u>is a subset of</u> $V_{\mathbb{K}}$ <u>and</u> $\tilde{A}$ <u>is its complete inverse image in</u> $G_+(A)$ <u>and if</u>

$$v \longmapsto P(t ; v)$$

<u>(with indeterminate</u> t) <u>is a polynomial-valued function on</u> A, <u>we say that</u> P <u>is of type</u> $\mathcal{F}[t]$ <u>on</u> A <u>or on</u> $\tilde{A}$ <u>if there exist functions</u> $\Phi_N, \ldots, \Phi_0 \in \mathcal{F}$ <u>such that for every</u> $v \in A$

we have

$$P(t\,;\,v) \;=\; \Sigma_{\nu=0}^{N}\,\Phi_\nu(v)\,t^\nu\,.$$

In many cases, we shall take $\mathcal{F} = M(\mathbb{K})$, the $\mathbb{Q}$-structure on the algebra of modular functions on $V_{\mathbb{K}}$.

Lemma 2. If $\mathbb{K}'$ is an n.o.c. subgroup of $\mathbb{K}$, then each of the polynomial-valued functions

$$v \longrightarrow P_{\mathbb{K}'}(t\,;\,v)$$

is of type $M(\mathbb{K})[t]$ on $\tilde{\mathcal{B}}$.

Proof. (Cf.[11].) One first shows that $P_{\mathbb{K}'}(t\,;\,v)$ is the monic g.c.d. of two monic polynomials of type $M(\mathbb{K})[t]$ on $\tilde{\mathcal{B}}$.

The double coset $C_{\mathbb{K}'}(q'(\hat{b})) = \mathbb{K}'q'(\hat{b})\mathbb{K}'$, as in [4] , §2.4, determines a modular correspondence $\mathcal{C}_{\mathbb{K}'}(q'(\hat{b})) \subset V_{\mathbb{K}'}^{*} \times V_{\mathbb{K}'}^{*}$. If A is any subset of $V_{\mathbb{K}'}^{*}$, let $A.\mathcal{C}_{\mathbb{K}'}(q'(\hat{b}))$ denote the set of points of $V_{\mathbb{K}'}^{*}$ associated in the usual way to A by this correspondence. Let S be the finite set

$$\Xi_{\mathbb{K}'}(\mathcal{O},\tilde{\Sigma}) \cup \Xi_{\mathbb{K}'}(\mathcal{O},\tilde{\Sigma})\,.\,\mathcal{C}^{*}_{\mathbb{K}'}(q'(\hat{b}))\qquad ,$$

where $\mathcal{C}^{*}_{\mathbb{K}'}(q'(\hat{b}))$ is the union of all iterates $\mathcal{C}^{i}_{\mathbb{K}'}(q'(\hat{b}))$ of $\mathcal{C}_{\mathbb{K}'}(q'(\hat{b}))$, $i \le t$, for some suitably large t .

Let $f_1,\ldots,f_\ell \in M(\mathbb{K}')$ be such that $M(\mathbb{K}')$ is equal to the quotient algebra of the $\mathbb{Q}$-algebra generated by $f_1,\ldots,f_\ell$. We may assume that each is regular on S and that $f_1,\ldots,f_\ell$ separate the distinct points of S . Denote by $u = (u_1,\ldots,u_\ell)$ a variable point of $\mathbb{C}^\ell$ and let

$$F_u(x) \;=\; u_1 f_1(x) + \;.\;.\;.\; + u_\ell f_\ell(x)\quad ,\quad x \in V_{\mathbb{K}'}\quad .$$

If $x\,,x' \in S$, $x \neq x'$, then $F_u(x) \neq F_u(x')$ for some $u \in \mathbb{Q}^\ell$. Now if $M\,,M' \in X_{q'} = \mathrm{Int}(\mathbb{K})(q'(\hat{b}))$, and $z \in G_+(A)$, let

$$T(z\,;\,M,M') = \{u \in \mathbb{C}^\ell \mid \psi_M(z) + F_u(zM) - F_u(z) \neq \psi_{M'}(z)\}\,.$$

With $\mathbb{K}' \subset \mathbb{K}$, the set $T(z;M,M')$ is unchanged if M or M' is replaced by $M\omega$ or $M'\omega$ for any $\omega \in \mathbb{K}'$. Hence, there are only finitely many such sets for each z , and for each $z \in \Xi_A(\mathcal{O},\tilde{\Sigma})$, the family of such sets depends only on $pr_{\mathbb{K}'}(z)$: If $\omega \in \mathbb{K}'$, $\kappa \in \mathbb{K}_\infty$, then

$$\psi_M(z\kappa\omega) + F_u(z\kappa\omega M) - F_u(z\kappa\omega) - \psi_{M'}(z\kappa\omega)$$
$$= \psi_{\omega_M}(z) + F_u(z.{}^{\omega}M) - F_u(z) - \psi_{\omega_{M'}}(z) \quad .$$

If $v \in S$, let $z \in G_+(A)$ be such that $pr_{\mathbb{K}'}(z) = v$, $T(v)$ be the intersection of those sets $T(z;M,M')$ that are non-empty, and $T = \bigcap_{v \in S} T(v)$. If we have

(#) $pr_{\mathbb{K}'}(zM) = pr_{\mathbb{K}'}(z)$ and $\psi_M(z) = \psi_{M'}(z)$,

then $T(z;M,M')$ is empty; otherwise, $T(z;M,M')$ is a Zariski open and dense subset of $\mathbf{C}$. Therefore, T is Zariski open and dense, too. Let $u_0 \in \mathbf{Q}^{\ell} \cap T$. If $z \in G_+(A)$ and $pr_{\mathbb{K}'}(z) \in S$, then for $M,M' \in X_{q'}$, we have

(##) $\psi_M(z) + F_{u_0}(zM) - F_{u_0}(z) = \psi_{M'}(z)$

only if $T(z;M,M')$ is empty, i.e., only if (#) holds.

If $u \in \mathbf{C}^{\ell}$ and $z \in G_+(A)$, set

$$Q(t;u,z) = \prod_{M \in X_{q'}\mathbb{K}'/\mathbb{K}'} (t - \psi_M(z) - F_u(z) + F_u(zM)).$$

Since this depends only on $pr_{\mathbb{K}'}(z)$, we may replace z by a variable $v' \in V_{\mathbb{K}'}$. If $u \in \mathbf{Q}^{\ell}$, $Q(t;u,v')$ is of type $M(\mathbb{K}')[t]$ by the usual arguments on the symmetry of its coefficients, and is regular on $\Xi_{\mathbb{K}}(\mathcal{O}, \tilde{\Sigma})$. Now

$$Q(t;0,v') = \prod_{M \in X_{q'}\mathbb{K}'/\mathbb{K}'} (t - \psi_M(v')) ,$$

while for given $M \in X_{q'}\mathbb{K}'$, we have (##) for some M' in $X_{q'}\mathbb{K}'$ only if (#) holds because of the choice of u_0 . If $z \in \Xi_A(\mathcal{O}, \tilde{\Sigma})$, then $pr_{\mathbb{K}'}(zM) = pr_{\mathbb{K}'}(z)$ is equivalent to having $M \in \mathcal{G}(z)\mathbb{K}'$ by Prop. 3, and the monic polynomial g.c.d. of $Q(t;0,z)$ and of $Q(t;u_0,z)$ is, since $X_{q'}\mathbb{K}' \cap \mathcal{G}(z) = X_{\mathbb{K}'}(z)$, $\prod_{M \in X_{\mathbb{K}'}(z)/\mathbb{K}'} (t - \psi_M(z)) = P_{\mathbb{K}'}(t;z)$.

As remarked in [11] in the classical case, it is easy also here to see that for every $a \in I(K) \cup \iota I(K)$ one has

$$X_{\mathbb{K}'}(a*z) = X_{\mathbb{K}'}(z) \quad \text{and} \quad X_{q'}(a*z) = X_{q'}(z) \quad .$$

Therefore, the functions $v \longmapsto P_{\mathbb{K}'}(t;v) = P_{\mathbb{K}'}(t;z)$ (for every z such that $pr_{\mathbb{K}}(z) = v \in \tilde{B}$) are of constant degree on $\tilde{B}$. Hence we may use the theorem of Weber-Perron [9,17] to calculate the polynomial $P_{\mathbb{K}'}(t\,;\,z)$ as the g.c.d. of $Q(t\,;\,0,z)$ and $Q(t\,;\,u_0,z)$, so the coefficients of $P_{\mathbb{K}'}(t\,;\,z)$ are indeed elements of $M(\mathbb{K}')$ on the inverse image under $pr_{\mathbb{K}'/\mathbb{K}}$ of $\tilde{B}$ in $V_{\mathbb{K}'}$. Since they are right invariant under $\mathbb{K}$, the polynomials are in fact of type $M(\mathbb{K})[t]$ on $\tilde{B}$ itself by Prop. 2. This proves the lemma.

3.4 PROOF THAT B IS RATIONAL OVER $K^*(\tilde{\Sigma})$ (PROP.9)

We now suppose that $pr_{\mathbb{K}}(z) \in \tilde{B}$. We have chosen the orbit B to contain the image $\theta.(\tau)$ of $\xi\theta$, where $\xi \in \Xi_{\theta,\infty}(\mathcal{O}, \tilde{\Sigma})$ and $q = q_\tau$. Then $z \in G_+(Q)q(c)\xi\theta\mathbb{K}\mathbb{K}_\infty$ for some c belonging to $I(K)_f$ or to $I(K)_f\iota_f$, so that we may assume either

(a) $\quad z = \gamma q(a)\xi\theta\eta u$, if $pr_{\mathbb{K}}(z) \in B$, or

(b) $\quad z = \gamma q(a\iota_f)\xi\theta\eta u$, if $pr_{\mathbb{K}}(z) \in \iota^*B$,

where $\gamma \in G_+(Q)$, $a \in I(K)_f$, $\eta \in \mathbb{K}$, $u \in \mathbb{K}_\infty$. To describe $X_{q'}(z) = \mathcal{J}(z) \cap X_{q'}$ in these two cases: If z is as in (a), and $x \in X_{q'}(z)$, then we have

(c) $\quad x = q'(\hat{c}) = \omega q'(\hat{b})\omega^{-1}$

for some $c \in K^\times$ and $\omega \in \mathbb{K}$. If we put $\omega' = {}^\theta\omega$, this reads $q(\hat{c}) = {}^{\omega'}q(\hat{b})$, and by Lemma 1 this implies either $c = b$ or $c = \bar{b}$, depending on whether $\omega' \in q(I(K)_f)$ or $\omega' \in q(\iota_f I(K)_f)$. In case (b), the relation (c) is replaced by

(c') $\quad x = q'(\hat{\bar{c}}) = \omega q'(\hat{b})\omega^{-1}$,

or $q(\hat{\bar{c}}) = {}^{\omega'}q(\hat{b})$, leading to the same conclusion with the alternatives reversed. In either case, $X_{q'}(z)$ consists of one or both elements in the set

$$\{q'(\hat{b}),\ q'(\hat{\bar{b}})\}\ .$$

But if for a single z, it contains both of them, a simple calculation shows that $B = \iota^*B$, contrary to our assumptions. Of course $X_{q'}(z)$ is not empty because in case (a) or in case (b) we may take $c = b$ respectively $c = \bar{b}$ and in either case, ω to be the identity, to obtain an element of $X_{q'}(z)$. So let us suppose $pr_{\mathbb{K}}(z) \in B$ as in case (a) and let $z' = \iota^*z \in \iota^*B$.

When $\mathrm{card}(X_{q'}(z)) = 1$, the proposition implies that for sufficiently small $\mathbb{K}'$, the polynomial $P_{\mathbb{K}'}(t\,;\,z)$ has a single root $\psi_{M_0}(z)$, where $X_{q'}(z) = \{M_0\}$. The polynomial function

$$v \longrightarrow P_{\mathbb{K}'}(t\,;\,\bar{z}) = (t - \psi_{M_0}(\bar{z}))^N , \qquad v = pr_{\mathbb{K}}(\bar{z}) ,$$

for some integer $N > 0$, is of type $M(\mathbb{K})[t]$ on $\tilde{B} = B \cup \iota * B$, and this implies that the polynomial function

$$v \longmapsto t - \psi_{M_0}(\bar{z})$$

is also of type $M(\mathbb{K})[t]$ on $\tilde{B}$. In other words, ψ_{M_0} coincides on $\tilde{B}$ with the restriction of some function in $M(\mathbb{K})$. We now calculate $\psi_{M_0}(z)$ and $\psi_{M_0}(z')$.

We have $z = \gamma q(a^{-1})\xi\eta$, and without loss of generality may assume $M_0 = \eta^{-1} q(\hat{b})$ (instead of $\eta^{-1} q(\hat{\bar{b}})$: the second possibility is treated similarly). Now

$$\psi_{M_0}(z) = (\psi\|M_0)(z)/\psi(z)$$

and $(\psi\|M_0)(z) = N\mathfrak{A}^w \psi(zM_0)$, where w is the weight of ψ. $\mathfrak{A}$ is the ideal $\mathfrak{a}^{-2}\nu$, where $\nu = \det(M_0) = N_{K/k}(b)$ and $\mathfrak{a}^{-1}$ is the $\mathfrak{o}$-ideal of all $\lambda \in k$ such that λM_0 belongs to $M_2(\hat{\mathfrak{o}})$. We have

$$\psi(zM_0) = \psi(\xi q(a^{-1})\eta\eta^{-1} q(\hat{b})) = \psi(\xi q(a^{-1})q(\hat{b})\eta) =$$
$$= \psi(q(b)_\infty^{-1}\xi q(a^{-1})\eta) .$$

Since $q = q_\tau$, where $(\tau) = \xi(i.e)$, we further have $q(b)_\infty^{-1}\xi = \xi k_\infty$ for some $k_\infty \in \mathbb{K}_\infty$. Therefore

$$\psi(q(b)_\infty^{-1}\xi q(a^{-1})\eta) = \psi(\xi k_\infty q(a^{-1})\eta) =$$
$$= \chi_w(k_\infty)\psi(\xi q(a^{-1})\eta) = \chi_w(k_\infty)\psi(z) .$$

So the problem is to evaluate $\chi_w(k_\infty)$:

$$\chi_w(k_\infty) = \chi_w(\xi^{-1}q(b)_\infty^{-1}\xi) = j(\xi^{-1}q(b)_\infty^{-1}\xi, i.e)^w =$$
$$= j(q(b)_\infty^{-1}, (\tau))^w ,$$

by the cocycle relation.

The point (τ) of H^n is to be identified with the element $\tau \in K$,

$$(\tau) = \tau^{\tilde{\Sigma}} = (\tau^{\tilde{\sigma}_1}, \ldots, \tau^{\tilde{\sigma}_n}) ,$$

and $q(b)$ is the matrix $\begin{pmatrix} \alpha & \beta \\ \gamma & \delta \end{pmatrix} \in M_2(k)$ defined by the equations

$$b.\tau = \alpha\tau + \beta \quad ,$$
$$b = \gamma\tau + \delta \quad .$$

Define the type norm $N_{\Sigma}^{\sim}$ as before and let $\begin{pmatrix}\alpha' & \beta' \\ \gamma' & \delta'\end{pmatrix} = \begin{pmatrix}\alpha & \beta \\ \gamma & \delta\end{pmatrix}^{-1}$. Then we have

$$j(q(b)_{\infty}^{-1}, (\tau)) = N_{\Sigma}^{\sim}((\alpha'\delta' - \beta'\gamma')(\gamma'\tau + \delta')^{-2}) =$$
$$= N_{\Sigma}^{\sim}((b\bar{b})^{-1}b^2) = N_{\Sigma}^{\sim}(b\bar{b}^{-1}) \quad .$$

Moreover, $\mathfrak{V} = \mathfrak{a}^{-2}\nu = \mathfrak{a}^{-2}b\bar{b}$, and $\psi(zM_0)$, by these calculations is equal to $N_{\Sigma}^{\sim}(b\bar{b}^{-1})^w\psi(z)$. Therefore,

$$(\psi\|M_0)(z) = N\mathfrak{V}^w N_{\Sigma}^{\sim}(b\bar{b}^{-1})^w\psi(z) = N\mathfrak{a}^{-2w}N_{\Sigma}^{\sim}b^{2w}\psi(z) \quad ,$$

and hence $\psi_{M_0}(z) = N\mathfrak{a}^{-2w}N_{\Sigma}^{\sim}b^{2w}$.

In the same way we may calculate $\psi_{M_0}(z')$. Since

$$z' = \iota * z \in G_-(\mathbb{Q})q(\iota)_{\infty}\xi q(a^{-1})\eta = G_+(\mathbb{Q})q(\iota)_f\xi q(a^{-1})\eta \quad ,$$

while $M_0 = \eta^{-1}q(b)\eta$, we have

$$z'.M_0 \in G_+(\mathbb{Q})q(\iota)_f q(\hat{b})\xi q(a^{-1}) = G_+(\mathbb{Q})q(\hat{\bar{b}})q(\iota)_f\xi q(a^{-1})\eta =$$
$$= G_+(\mathbb{Q})q(\bar{b})_{\infty}^{-1}q(\iota)_f\xi q(a^{-1})\eta \quad .$$

Therefore

$$\psi(z'M_0) = \psi(q(\bar{b})_{\infty}^{-1}q(\iota)_f\xi q(a^{-1})\eta) = \psi(\xi\xi^{-1}q(\bar{b})_{\infty}^{-1}\xi q(\iota)_f q(a^{-1})\eta)$$
$$= \psi(\xi q(\iota)_f q(a^{-1})\eta^{-1}.\xi^{-1}q(\bar{b})_{\infty}\xi) = N_{\Sigma}^{\sim}(\bar{b}b^{-1})^w\psi(z')$$

according to the same calculations we have just completed for $\psi(zM_0)$. Hence,

$$\psi_{M_0}(z') = N\mathfrak{a}^{-2w}N_{\Sigma}^{\sim}\bar{b}^{2w} \quad .$$

By our earlier calculations, the polynomial function

$$v \longmapsto t - \psi_{M_0}(z) \quad , \quad z \in pr_K^{-1}(v) \quad ,$$

is of type $M(\mathbb{K})[t]$ on $\hat{B}$. What we now see from our calculations is that the function

$$N\mathfrak{a}^{-2w}N_{\Sigma}^{\sim}b^{2w} - \psi_{M_0} \quad ,$$

which belongs to the compositum $M(\mathbb{K}, K^*(\tilde{\Sigma}))$, is zero on B and $\neq 0$ on $\iota * B$ when $B \neq \iota * B$, provided

(d) $$N_{\Sigma}^{\sim}b^{2w} \neq N_{\Sigma}^{\sim}\bar{b}^{2w} \quad .$$

To achieve (d) we choose, for example, $b = \pi$, the generator of a principal unramified prime ideal of first degree in K , for which (d) must certainly be satisfied because the prime factorizations of $N_{\tilde{\Sigma}}b$ and of $N_{\tilde{\Sigma}}\bar{b}$ are disjoint. ([5], §6.3)

It follows that B is a zero cycle on $V_{\mathbb{K}}$ rational over the reflex field $K^*(\tilde{\Sigma})$.

4. ABELIAN EXTENSIONS GENERATED BY SPECIAL VALUES

Let K be an n.o.c. subgroup of $G(A_f)$ and B be the orbit under $I(K)_f$ of the image in $V_{\mathbb{K}}$ of $g = \xi\theta \in \Xi_\infty(\mathcal{O}, \tilde{\Sigma})$, $\theta \in \textcircled{H}$, $\xi \in \Xi_{\theta,\infty}(\mathcal{O}, \tilde{\Sigma})$. Thus $B \subset \Xi_{\mathbb{K}}(\mathcal{O}, \tilde{\Sigma})$. Let $\mathcal{L}_{\mathbb{K}}$ be the field obtained by adjoining to $K^*(\tilde{\Sigma})$ all the values

$$\{\Phi(z) \mid z \in B\ ,\ \Phi \in M(\mathbb{K})\ ,\ \Phi \text{ regular on } B\}.$$

Since B is a 0-cycle on $V_{\mathbb{K}}$ invariant under $\mathrm{Gal}(\bar{Q}/K^*(\tilde{\Sigma}))$, i.e., is a $K^*(\tilde{\Sigma})$-rational 0-cycle, $\mathcal{L}_{\mathbb{K}}$ is a finite normal extension of the reflex field $K^*(\tilde{\Sigma})$, and as in [11] is seen to be independent of B since $R(\omega).M(\mathbb{K}) = M(\mathbb{K})$ for every $\omega \in G(\hat{\mathbb{Z}})\mathbb{K}_\infty$, and every orbit of $I(K)_f$ in $\Xi_A(\mathcal{O}, \tilde{\Sigma})$ is of the form $B\omega$ for some $\omega \in G(\hat{\mathbb{Z}})K_\infty$ by Prop. 1 .

Let $\Psi_1, \ldots, \Psi_N$ be a set of generators of $M(\mathbb{K})$, as a direct sum of fields, which are all holomorphic at all points of $\Xi_K(\mathcal{O}, \tilde{\Sigma})$ and let $u_1, \ldots, u_N$ be indeterminates.

<u>Lemma 3.</u> <u>(Cf. [11] §5.2.1)</u> <u>If</u> $\Phi \in M(\mathbb{K})$ <u>and is regular on</u> B , <u>and if</u> $b \in I(K)_f$, <u>then there is a rational function</u>

$$R_{b,\Phi}(u^\vee) \in K^*(\tilde{\Sigma})(u_1, \ldots, u_N)$$

<u>such that for each</u> $z \in B$,

$$\Phi(b*z) = R_{b,\Phi}(\Psi_1(z), \ldots, \Psi_N(z)) ,$$

<u>with the understanding of course that the denominator of</u> $R_{b,\Phi}$ <u>is non-zero at the point</u> $(\Psi_1(z), \ldots, \Psi_N(z))$.

<u>Proof.</u> We may, without loss of generality, assume to begin with that the N-tuple $\Psi^\vee = (\Psi_1, \ldots, \Psi_N)$ is also holomorphic at every point of

$$\Xi_{\mathbb{K}}(\mathcal{O}, \tilde{\Sigma}) . \mathcal{C}_{\mathbb{K}}(q'(\hat{b})) ,$$

where $\mathcal{C}_{\mathbb{K}}(q'(\hat{b}))$ is the modular correspondence defined before. We may, and do, assume that for each $i = 1, \ldots, N$, Ψ_i is holomorphic on and separates the distinct points of the finite set S defined in §3.3 with $\mathbb{K}$ in place of $\mathbb{K}'$. Moreover, it suffices to prove the lemma for each of a set of generators $\Psi_1, \ldots, \Psi_N$ of the normal affine coordinate ring of an affine $K^*(\tilde{\Sigma})$-open subset of $V_{\mathbb{K}}$ containing S. Therefore, we shall assume Φ is one of these, say $\Phi = \Psi_i$. Finally, it suffices to treat the case when $\mathbb{K}$ is a principal congruence subgroup $\mathbb{K}(\mathfrak{n})$ of $G(\hat{\mathbb{Z}})$.

The ray class modulo $\mathfrak{n}$ of $\mathrm{id.}(b^{-1})$ contains an unramified prime ideal $\mathfrak{p}$ of first degree in K. Then $\mathfrak{p} = K \cap K_\infty \hat{\mathcal{O}} \pi$ for some

$$\pi \in \hat{\mathcal{O}} \cap K_\infty^\times K^\times U_{\mathfrak{n}}(\hat{\mathcal{O}})\, b^{-1}$$

(cf. [12], pp. 145-9). Since $\Phi(\pi^{-1} * z)$ is then equal to $\Phi(b*z)$, we may assume $b = \pi^{-1}$. We may also assume π belongs to the principal congruence subgroup modulo $\mathfrak{n}$ of $I(K)_f$, since $\mathfrak{p}$ and b are prime to $\mathfrak{n}$ ([12], Theorem 6, p. 317). This means that $\pi \equiv 1 \bmod^\times \mathfrak{n}$.

For an indeterminate t, we define the polynomial

$$T_{\pi,\Phi}(t\,;\,z) = \prod_{M \in C_{\mathbb{K}}(q'(\pi))/\mathbb{K}} (t - \Phi(zM)),$$

for $z \in G_+(A)$ such that Φ is holomorphic at all (images of the) points zM (in $V_{\mathbb{K}}$). We have $\Phi(zM) = (R(M).\Phi)(z)$ and $R(M)\beta(\sigma) = \beta(\sigma)R(M)$, $\sigma \in \mathrm{Gal}(\mathbb{Q}_{ab}/\mathbb{Q})$, by Prop. 1, §5.2 of [5]. Since $\beta(\sigma).\Phi = \Phi$ for all $\sigma \in \mathrm{Gal}(\mathbb{Q}_{ab}/\mathbb{Q})$, it follows that $\beta(\sigma)a_\nu = a_\nu$ for every coefficient a_ν of this polynomial. Moreover, the coefficients of $T_{\pi,\Phi}(t\,;\,z)$ as a polynomial in t are clearly right invariant under $\mathbb{K}$; therefore, those coefficients lie in $M(\mathbb{K})$ (as functions of z). Therefore, $T_{\pi,\Phi}(t\,;\,z)$ is a polynomial of type $M(\mathbb{K})[t]$ on all of $V_{\mathbb{K}}$ wherever its coefficients are holomorphic.

We have defined the ray-class polynomial

$$H_B^\Phi(t) = \prod_{z \in B} (t - \Phi(z)),$$

which has coefficients in $K^*(\tilde{\Sigma})$ as a consequence of Prop.9. We have assumed that B is the orbit $I(K)_f * \zeta_0$ of $\zeta_0 = \pi_{\mathbb{K}}(g)$, $g = \xi\theta$, where $(\tau) = \xi(i.e)$, $\xi \in \Xi_{\theta,\infty}(\mathcal{O}, \tilde{\Sigma})$, $\tau \in K$, and $\tilde{\Sigma}(\tau) = \tilde{\Sigma}$. It follows immediately from the definition of $H_B^\Phi(t)$ and from the choice of Φ, by which Φ takes distinct values at distinct points of B, that $H_B^\Phi(t)$ has no double roots. Therefore, for any z such that

$pr_{\mathbb{K}}(z) \in B$, the monic g.c.d. of $T_{\pi,\Phi}(t\,;\,z)$ and of $H_B^{\Phi}(t)$ is the monic whose roots are <u>simple</u> and are the numbers $\Phi(a*z) = \Phi(zM)$, with $a \in I(K)_f$ and $M = M_a \in C_{\mathbb{K}}(q'(\pi))$. Then we have necessarily (by the assumptions on Φ)

$$pr_{\mathbb{K}}(a*z) = pr_{\mathbb{K}}(zM) \ .$$

We may assume that $z = \xi q(b)\theta$, $b \in I(K)_f$. Then, in the notation of Prop. 3, $\omega = \gamma = 1$ and for some $\beta \in K^{\times}$, we have $M \in q'(a^{-1}\hat{\beta}).\mathbb{K}$. Since $M \in \mathbb{K}q'(\pi)\mathbb{K}$, this implies that $q'(a^{-1}\hat{\beta})$ has adelic integral entries and hence that $a^{-1}\hat{\beta} \in \hat{\mathcal{O}}$, while the divisor of $a^{-1}\hat{\beta}$ has the same norm $\mathcal{G}$ to k as the divisor $\mathcal{P}$ of π . Since $q'(a^{-1}\hat{\beta}) \in \mathbb{K}q'(\pi)\mathbb{K}$ and $\mathbb{K} = \mathbb{K}(\mathfrak{n})$, it follows that $a^{-1}\hat{\beta} \equiv \pi \bmod \mathfrak{n}\hat{\mathcal{O}}$. Let $c = a^{-1}\hat{\beta}$. Then $c \in \hat{\mathcal{O}}$, $(\mathrm{id.}(c),\mathfrak{n}) = (1)$, $c \equiv \pi \bmod \mathfrak{n}\hat{\mathcal{O}}$, and $\mathrm{id.}(N_{K/k}(c)) = \mathrm{id.}(N_{K/k}(\pi)) = \mathcal{G}$. So if $(\pi) = \mathrm{id.}(\pi) = \mathcal{P}$, we have $(c) = \mathcal{P}$ or $\bar{\mathcal{P}}$. Suppose $(c) = \mathcal{P} = (\pi)$, $c = \varepsilon\pi$, where $\varepsilon \in \hat{\mathcal{O}}^{\times}$, and $\pi(\varepsilon - 1) = c - \pi \in \mathfrak{n}\hat{\mathcal{O}}$; hence, since $(\mathfrak{n}, \pi) = (1)$, we have $\varepsilon \equiv 1 \bmod \mathfrak{n}$ and $q'(\varepsilon) \in K(\mathfrak{n})$. Another choice of, say, c_1 in place of c satisfying the same conditions as c with $c_1 = \varepsilon_1\pi$ also gives $q'(\varepsilon_1) \in K(\mathfrak{n})$; hence, if $M_1 \in q'(c_1)K(\mathfrak{n})$, we have $M^{-1}M_1 \in K(\mathfrak{n})$. If $(c) = \bar{\mathcal{P}} = (\bar{\pi})$, $c = \varepsilon'\bar{\pi}$ with $\varepsilon' \in \hat{\mathcal{O}}^{\times}$, then since $\bar{\pi} \equiv \pi \equiv c \equiv 1 \bmod \mathfrak{n}$, we have $\varepsilon' \equiv 1 \bmod \mathfrak{n}$ and $q'(\varepsilon') \in K(\mathfrak{n})$; and if in the same set-up we have c_1 in place of c with $(c_1) = \bar{\mathcal{P}} = (\bar{\pi})$, $c_1 = \bar{\pi}\varepsilon''$, then $\varepsilon'\varepsilon''^{-1} \in K(\mathfrak{n})$, so that if M_1 is defined in the same way with respect to c_1 , then $M^{-1}M_1 \in K(\mathfrak{n})$. Therefore, the only possible common roots of $H_B^{\Phi}(t)$ and of $T_{\pi,\Phi}(t\,;\,z)$ are $\rho_1(z) = \Phi((\varepsilon\pi)^{-1}*z)$ and $\rho_2(z) = \Phi((\varepsilon'\bar{\pi})^{-1}*z)$, and each of these is not only a simple root of H_B^{Φ} , but <u>also a simple root of</u> $T_{\pi,\Phi}(t\,;\,z)$ if $\rho_1(z) \neq \rho_2(z)$. We have

$$\rho_1(z) = \Phi(\pi^{-1}*z) \qquad \text{and} \qquad \rho_2(z) = \Phi(\bar{\pi}^{-1}*z)$$

because $q'(\varepsilon)$ and $q'(\varepsilon')$ both belong to $K(\mathfrak{n})$.

Suppose $\mathcal{P}$ and $\bar{\mathcal{P}}$ are in the same ray class modulo $\mathfrak{n}$. Then $\mathcal{P} = (\alpha)\bar{\mathcal{P}}$, $\alpha \equiv 1 \bmod^{\times}\mathfrak{n}$, $\alpha \in K^{\times}$, and hence $\pi = \varepsilon\hat{\alpha}\bar{\pi}$, where ε belongs to $\hat{\mathcal{O}}^{\times}$, which imply (since $\pi \equiv \bar{\pi} \equiv 1 \bmod \mathfrak{n}$) $\varepsilon \equiv 1 \bmod \mathfrak{n}$. Hence,

$$\pi^{-1}*z = \xi q(\pi b)\theta = q(\hat{\alpha})\xi q(\bar{\pi} b)\theta q'(\varepsilon) =$$
$$= q(\alpha)q(\alpha)_\infty^{-1}\xi q(\bar{\pi}b)\theta\kappa_{\mathfrak{n}} = q(\alpha)\xi q(\bar{\pi}b)\theta\kappa_\infty\kappa_{\mathfrak{n}} ,$$

where $\kappa_\infty \in \mathbb{K}_\infty$, $\kappa_{\mathfrak{n}} \in \mathbb{K}(\mathfrak{n})$ and, of course, $q(\alpha) \in G_+(\mathbb{Q})$. Since $\xi q(\bar{\pi}b)\theta = \bar{\pi}^{-1}*z$, this implies $\rho_1(z) = \Phi(\pi^{-1}*z) = \Phi(\bar{\pi}^{-1}*z) = \rho_2(z)$. Therefore, the monic g.c.d. of the two polynomials $H_B^\Phi(t)$ and $T_{\pi,\Phi}(t\,;z)$ is

$$t - \Phi((\varepsilon\pi)^{-1}*z) = t - \Phi(\pi^{-1}*z) ,$$

which is of type $K^*(\tilde{\Sigma})M(\mathbb{K})[t]$ on B since both of the first two polynomials are, on account of the Weber-Perron theorem. This proves Lemma 3 in case $\mathfrak{p}$ and $\bar{\mathfrak{p}}$ are in the same ray class modulo $\mathfrak{n}$.

Suppose on the other hand that $\mathfrak{p}$ and $\bar{\mathfrak{p}}$ lie in distinct ray classes modulo $\mathfrak{n}$; then for every $z \in B$ we have $\rho_1(z) \neq \rho_2(z)$ and the g.c.d.of $H_B^\Phi(t)$ and $T_{\pi,\Phi}(t\,;z)$, which is of type $K^*(\tilde{\Sigma})M(\mathbb{K})[t]$, is equal to

$$(t - \rho_1(z))(t - \rho_2(z)) = L_\Phi(t\,;\Psi^\vee(z)) ,$$

say, provided $q'(\bar{\pi})$ in fact belongs to $\mathbb{K}q'(\pi)\mathbb{K}$. In fact, this is automatically so, for $\det(q'(\pi))$ is an idele of k generating a prime ideal $\mathfrak{q}$ of first degree in k , which by elementary divisor theory over $k_{\mathfrak{q}}$ implies that

(*) $$q'(\bar{\pi}) = \kappa_1 q'(\pi)\kappa_2 ,\quad \kappa_1 , \kappa_2 \in \mathbb{K}(1) = G(\hat{\mathbb{Z}}) ;$$

and since (*) is a system of equations over the different finite places of k , and π and $\bar{\pi}$ are units $\equiv 1 \bmod \mathfrak{n}_{\mathfrak{q}}$ at every place $\mathfrak{q}|\mathfrak{n}$ (where $\mathfrak{n}_{\mathfrak{q}}$ is the power of $\mathfrak{q}$ dividing $\mathfrak{n}$), we may assume for each such $\mathfrak{q}$ that $\kappa_{1\mathfrak{q}}$ and $\kappa_{2\mathfrak{q}}$ both belong to $\mathbb{K}(\mathfrak{n})_{\mathfrak{q}}$, and for all other $\mathfrak{q}$, belong to $K(1)_{\mathfrak{q}}$ which is the same as $K(\mathfrak{n})_{\mathfrak{q}}$ if $\mathfrak{q}\nmid\mathfrak{n}$; therefore, we may take $\kappa_1 , \kappa_2 \in \mathbb{K}(\mathfrak{n}) = \mathbb{K}$ and $q'(\bar{\pi}) \in \mathbb{K}q'(\pi)\mathbb{K}$. Now each coefficient of $L_\Phi(t , \Psi^\vee(z))$ is in $K^*(\tilde{\Sigma})M(\mathbb{K})$, and in particular we have

(A) $$\Phi(\pi^{-1}*z) + \Phi(\bar{\pi}^{-1}*z) = R_0(\Psi_1(z),\ldots,\Psi_N(z))$$

for every $z \in B$, where R_0 is a fixed rational function in N variables of which the numerator and denominator D have coeffients in $K^*(\tilde{\Sigma})$ and such that $D(\Psi_1(z),\ldots,\Psi_N(z)) \neq 0$ for every $z \in B$.

We may now continue the argument of §5.2 of [11] to its end in essentially the same way as there, but now for the case of Hilbert modular functions, as follows:

Let e be a positive integer such that $\mathfrak{f}^e$ lies in the principal ray mod $\mathfrak{n}$, i.e., $\mathfrak{f}^e = \Pi.\mathcal{O}$, $\Pi \in K^\times$, $\Pi \equiv 1 \bmod \mathfrak{n}$. Then we have $\pi^e \in \Pi.\mathcal{U}_{\mathfrak{n}}(\hat{\mathcal{O}})$ and $\bar{\pi}^e \in \bar{\Pi}.\mathcal{U}_{\mathfrak{n}}(\hat{\mathcal{O}})$, where $\mathcal{U}_{\mathfrak{n}}(\hat{\mathcal{O}}) = \{\varepsilon \in \hat{\mathcal{O}}^\times \mid \varepsilon \equiv 1 \bmod \mathfrak{n}\}$. Let Φ be an arithmetic modular form in $A(\mathbb{K}, w)$ of weight $w > 0$, which is nowhere vanishing on the set S. For each $a \in I(K)_f$, we define the ideal

$$\mathfrak{U}_a = \mathfrak{U}(q'(a)^{-1}) \quad ,$$

where $\mathfrak{U}(M)$ and $\mathfrak{a} = \mathfrak{a}(M)$ are defined for $M \in G(A_f)$ as in §6.1 of [5], the ideal $\mathfrak{a}(M)$ being essentially the g.c.d. of the entries of M. Then for each $z \in \Xi_A(\mathcal{O}, \tilde{\Sigma})$ put

$$\Phi_a(z) = N\mathfrak{U}_a^{\,w}\,\Phi(a*z)/\Phi(z) \quad .$$

If $b \in I(K)_f$, then

$$\Phi_a(b*z) = N\mathfrak{U}_a^{\,w}\,\Phi((ab)*z)/\Phi(b*z)$$

Setting $b = a^i$, $i = 0,\ldots,e-1$ in turn, we get, analogously to the classical case

$$\Phi_a(z)\Phi_a(a*z)\cdot\ldots\cdot\Phi_a(a^{e-1}*z) = N\mathfrak{U}_a^{\,ew}\,\Phi(a^e*z)/\Phi(z) \quad .$$

If $a^e = \hat{\alpha}.\varepsilon$ with $\alpha \in K^\times$ and $\varepsilon \in \mathcal{U}_{\mathfrak{n}}(\hat{\mathcal{O}})$, then let $M_0 = q'(\hat{\alpha})^{-1}$ and we have $\mathfrak{a}(M_0) = \mathfrak{a}(q'(\pi^e)) = 1$ since $\mathfrak{f}^e$ has no proper integral $\mathfrak{o}$-divisor, and $\nu(M_0) = N_{K/k}\alpha^{-1} = N_{K/k}\Pi$, as ideals, if we take $a = \pi^{-1}$. Therefore,

$$\Phi_{a^e}(z) = N\mathfrak{U}_a^{\,ew}\,\Phi(a^e*z)/\Phi(z) = N_{K/Q}\Pi_a^{\,w}\,\Phi((\hat{\alpha}\varepsilon)*z)/\Phi(z) =$$

$$= N_{K/Q}\Pi_a^{\,w}\,\Phi(\alpha*z)/\Phi(z) \quad ,$$

if we take $a = \pi^{-1}$ resp. $\bar{\pi}^{-1}$ and $\Pi_a = \Pi$ resp. $\bar{\Pi}$,

since, for reasons explained earlier, $\Phi((\hat{\alpha}\varepsilon)*z) = \Phi(\varepsilon*(\hat{\alpha}*z)) = \Phi(\hat{\alpha}*z)$. But now if we put $k_\infty = \xi^{-1}q(\alpha)_\infty\xi$, then

$$\Phi(\hat{\alpha}*z) = \Phi(\xi q(\hat{\alpha}^{-1}b)\theta) = \Phi(q(\alpha)_\infty\xi q(b)\theta) =$$

$$= \Phi(\xi k_\infty q(b)\theta) = \chi_w(k_\infty)\Phi(z) \quad ,$$

while

$$\chi_w(k_\infty) = j(q(\alpha)_\infty,\tau)^w = N_{\tilde{\Sigma}}(\alpha^{-1}\bar{\alpha})^w = N_{\tilde{\Sigma}}(\Pi\bar{\Pi}^{-1})^w \quad ,$$

and therefore

$$\Phi_{a^e}(z) = N_{K/Q}\Pi^w N_{\Sigma}^{\sim}(\Pi\bar{\Pi}^{-1})^w = N_{\Sigma}^{\sim}\Pi^{2w} ,$$

with $a^{-1} = \pi$, so that we have

(B) $$\Phi_a(z)\Phi_a(a{*}z)\cdot\cdots\cdot\Phi_a(a^{e-1}{*}z) = N_{\Sigma}^{\sim}\Pi^{2w} .$$

In like manner we obtain a similar equation with a and Π replaced by their complex conjugates.

Now $\rho_1(z) = \Phi(\pi^{-1}{*}z)$ and $\rho_2(z) = \Phi(\bar{\pi}^{-1}{*}z)$ are distinct and therefore simple roots of $T_{\pi,\Phi}(t\,;\,z)$. We form the polynomial in t given by

$$P(t\,;\,\zeta) = \sum_{M\,\varepsilon\, C_{\mathbb{K}}(q'(\pi))/\mathbb{K}} \frac{T_{\pi,\Phi}(t\,;\,\zeta)}{t - \Phi(\zeta M)}\cdot\frac{\Phi(\zeta M)}{\Phi(\zeta)} ,$$

for $\zeta \in G_+(A)$.

It is easy to verify that as functions of $\zeta \in G_+(A)$, the coefficients of $P(t\,;\,\zeta)$ are in $M(\mathbb{K})$ and are regular on the set S . (Cf. Prop. 2.) Thus, as a polynomial in t , $P(t\,;\,\zeta)$ is of type $M(\mathbb{K})$. We know $T_{\pi,\Phi}(t\,;\,\zeta)$, hence also its derivative $T'_{\pi,\Phi}(t\,;\,\zeta)$ with respect to t , has coefficients in $M(\mathbb{K})$.

Again let a denote either π^{-1} or $\bar{\pi}^{-1}$, and let z be as before. If then we substitute $\Phi(a{*}z) = \Phi(zM_a)$ for t and use the fact that this is a simple root of $T_{\pi,\Phi}(t\,;\,z)$, then we have

$$P(\Phi(zM_a)\,;\,z) = T'_{\pi,\Phi}(\Phi(zM_a)\,;\,z)\frac{\Phi(a{*}z)}{\Phi(z)}$$

for every $z \in B$, and this is the same as

(C) $$\Phi_a(z) = N\mathfrak{U}_a\cdot\frac{P(\Phi(a{*}z)\,;\,z)}{T'_{\pi,\Phi}(\Phi(a{*}z)\,;\,z)} = R_1(\Phi(a{*}z);\Psi^{\vee}(z)) ,$$

where $\mathfrak{U}_a = \mathfrak{U}_{q'(a)}$ and $R_1(t\,;\,u^{\vee})$ is a rational function in t and $u^{\vee} = (u_1,\ldots,u_N)$ with coefficients in $\mathbb{Q}$, whose denominator does not vanish when we substitute

$$(\Phi(a{*}z)\,,\,\Psi_1(z),\ldots,\Psi_N(z)) \quad\text{for}\quad (t,u_1,\ldots,u_N) .$$

By our choice of Π , we see that $N_{\Sigma}^{\sim}\Pi^m \neq N_{\Sigma}^{\sim}\bar{\Pi}^m$ for any non-zero $m \in \mathbb{Z}$.

To deal with equation (A), which may be rewritten as

(D) $$\Phi(\pi^{-1}{*}z) = R_0(\Psi^{\vee}(z)) - \Phi(\bar{\pi}^{-1}{*}z) ,$$

let $\zeta \in Z(A)$, the center of $G_+(A)$. Then for any $\Phi \in M(\mathbb{K})$, $R(\zeta)\Phi$ also belongs to $M(\mathbb{K})$ and therefore is a rational function in $\Psi^\vee$ with rational coefficients. Since $\zeta = q(\pi\bar{\pi}) \in Z(A)$, this implies that we have

(D') $\quad \Phi(\pi^{-1}*z) = R_2(\Psi^\vee(z), \Psi^\vee(\pi*z))$,

or, for all $z \in \Xi_A(\mathcal{O}, \tilde{\Sigma})$,

(D") $\quad \Phi(\pi^{-2}*z) = R_2(\Psi^\vee(\pi^{-1}*z), \Psi^\vee(z))$.

Similarly, with the <u>same</u> rational function R_2,

(D"') $\quad \Phi(\bar{\pi}^{-2}*z) = R_2(\Psi^\vee(\bar{\pi}^{-1}*z), \Psi^\vee(z))$.

Here R_2 has coefficients in $K^*(\tilde{\Sigma})$. Replacing z by $a^{i-2}*z$ for $a^{-1} = \pi$ or $\bar{\pi}$, we obtain for $i \geq 2$,

$$\Phi(\pi^{-i}*z) = R_2(\Psi^\vee(\pi^{-i+1}*z), \Psi^\vee(\pi^{-i+2}*z))$$

and the same with $\bar{\pi}$ in place of π . Letting $\Phi = \Psi_i$, $i = 1,\ldots,N$, in turn, and iterating, we obtain rational mappings

(E) $\quad \Psi^\vee(\pi^{-i}*z) = R_i^\vee(\Psi^\vee(z) ; \Psi^\vee(\pi*z))$, $i \geq 1$,

where $R_i^\vee = (R_{i1},\ldots,R_{iN})$ and R_{ij} is a rational function in $2N$ variables with coefficients in $K^*(\tilde{\Sigma})$, $i,j = 1,\ldots, N$, such that the denominator of each R_{ij} is non-zero when the $2N$ variables are replaced by $(\Psi^\vee(z), \Psi^\vee(\pi^{-1}*z))$, $z \in \Xi_A(\mathcal{O}, \tilde{\Sigma})$, and one obtains the same result with $\bar{\pi}$ in place of π .

On the other hand, for each $i = 1,\ldots,N$, we may construct the polynomial in t :

$$P_i(t ; z) = \sum_{M \in C_{\mathbb{K}}(q'(\pi))/\mathbb{K}} \frac{T_{\pi,\Phi}(t ; z)}{t - \Phi(zM)} \cdot \Psi_i(zM) \quad ,$$

which is in manner similar to that in a previous case seen to be of type $M(\mathbb{K})[t]$. Then by the same kind of argument as that just now used, we obtain a rational expression for $\Psi_i(a*z)$ in terms of $\Phi(a*z)$ and $\Psi^\vee(z)$ with coefficients in $K^*(\tilde{\Sigma})$ (for some fixed choice of Φ as earlier), valid for $a^{-1} = \pi$ or $\bar{\pi}$:

(F) $\quad \Psi^\vee(a*z) = R_*(\Phi(a*z) ; \Psi^\vee(z))$.

Combining (B) , (C), (E), and (F), we obtain polynomials $P(t ; u^\vee)$ amd $Q(t ; u^\vee)$ with coefficients in $K^*(\tilde{\Sigma})$ such that $Q(\Phi(a*z) ; \Psi^\vee(z)) \neq 0$ for $a^{-1} = \pi$ or $\bar{\pi}$ and for every $z \in B$, such that

(G) $\quad N_{\tilde{\Sigma}}(\Pi_a)^{2w} = P(\Phi(a*z) ; \Psi^{\vee}(z)) / Q(\Phi(a*z) ; \Psi^{\vee}(z))$.

Note that P and Q are the <u>same</u> whether $a = \pi$ or $\bar{\pi}^{-1}$.

Put $\quad F_a(t ; u^{\vee}) = N_{\tilde{\Sigma}}(\Pi_a)^{2w} Q(t ; u^{\vee}) - P(t ; u^{\vee})$.

For every $z \in B$, $t = \Phi(b*z) = \rho_1(z)$ is a root of $F_{\pi^{-1}}(t ; \Psi^{\vee}(z)$, with $b = \pi^{-1}$, but since

$$N_{\tilde{\Sigma}}\Pi^{2w} \neq N_{\tilde{\Sigma}}\bar{\Pi}^{2w} ,$$

$\Phi(\bar{b}*z) = \rho_2(z)$ is not a root. Therefore, the monic g.c.d. of $F_{\pi^{-1}}(t ; \Psi^{\vee}(z))$ and of $L_{\Phi}(t ; \Psi^{\vee}(z))$ is $t - \Phi(b*z)$ for every $z \in B$. It follows from the Weber-Perron Theorem that for some rational function $R_{b,\Phi}(u^{\vee}) \in K^*(\tilde{\Sigma})(u^{\vee})$ we have

$$\Phi(b*z) = R_{b,\Phi}(\Psi^{\vee}(z)) \quad , \quad z \in B \quad ,$$

as required. This proves Lemma 3.

Now by exactly the same argument as that used in proving Prop. 5.2.2 of [11] one obtains at once:

<u>Proposition 11.</u> <u>For any</u> $z \in B$, $\mathcal{L}_{\mathbb{K}}$ <u>is the compositum of extensions</u> $K^*(\tilde{\Sigma})(\Phi(z))$, <u>with</u> $\Phi \in M(\mathbb{K})$ <u>regular on</u> B .

We consider again the action of $I(K)$ on $\Xi_{\mathbb{K}}(\mathcal{O}, \tilde{\Sigma})$. If the double coset

$$D(\xi\theta) = G_+(Q)\xi\theta\mathbb{K}\mathbb{K}_{\infty}$$

belongs to $\Xi_{\mathbb{K}}(\mathcal{O}, \tilde{\Sigma})$, with $(\tau) = \xi(i.e)$ and $q = q_{\tau}$ we have for $a \in I(K)_f$

$$a*D(\xi\theta) = D(\xi\theta)$$

if and only if for some $k \in \mathbb{K}$, $k_{\infty} \in \mathbb{K}_{\infty}$, and $\gamma \in G_+(Q)$,

$$q(a)\xi\theta = \gamma\xi\theta kk_{\infty} \quad .$$

By looking at the archimedean component of this, one obtains as usual $\gamma = q(\alpha)$ for some $\alpha \in K^{\times}$; hence, $q(a)\theta = q(\hat{\alpha})\theta k$ or $q'(\hat{\alpha}^{-1}a) = k \in \mathbb{K}$, which is to say that $\hat{\alpha}^{-1}a \in \tilde{H}$, or $a \in \hat{K}^{\times}\tilde{H}$, where $\tilde{H}$ is the open compact subgroup of $I(K)_f$ defined in §1.4. Therefore, in the action of $I(K)$ on the finite set $\Xi_{\mathbb{K}}(\mathcal{O}, \tilde{\Sigma})$, the stabilizer of $D(\theta\xi)$ is $K_{\infty}^{\times}K^{\times}\tilde{H}$. Since $I(K)$ is abelian, the stabilizer of any point z of the orbit B of $D(\theta\xi)$ is also $K_{\infty}^{\times}K^{\times}\tilde{H}$. Consequently the class group $C(\mathcal{O}, H) = I(K)/K^{\times}K_{\infty}^{\times}\tilde{H}$ acts faithfully and simply transitively on the orbit B .

Therefore, with the Gothic $\mathfrak{K}$ of [11] replaced by the reflex field $K^*(\tilde{\Sigma})$, one may repeat the formal proof of Lemma 5.3 of [11] to obtain

Lemma 4. Given $z_0 \in B$ and $\sigma \in Gal(\mathcal{L}_{\mathbb{K}}/K^*(\tilde{\Sigma}))$, there is exactly one class $A_{\tilde{H}}(\sigma) \in C(\mathcal{O}, H)$ such that for every $\Phi \in M(\mathbb{K})$ regular on B :

$$\Phi(z_0)^{\sigma} = \Phi(A_{\tilde{H}}(\sigma)*z_0) .$$

Similarly one obtains by the formal arguments of [11]:

Lemma 5. In the same notation as above, given $\sigma \in Gal(\mathcal{L}_{\mathbb{K}}/K^*(\tilde{\Sigma}))$, we have

$$\Phi(A_{\tilde{H}}(\sigma)*z) = \Phi(z)^{\sigma}$$

for every $z \in B$ and $\Phi \in M(\mathbb{K})$ regular on B .

In conclusion, one obtains as in [11], so in this case also:

Theorem 2. The map $\sigma \longmapsto A_{\tilde{H}}(\sigma)$ is a monomorphism from $Gal(\mathcal{L}_{\mathbb{K}}/K^*(\tilde{\Sigma}))$ into $C(\mathcal{O}, \tilde{H})$.

Corollary. $\mathcal{L}_{\mathbb{K}}$ is an abelian normal extension of the reflex field $K^*(\tilde{\Sigma})$.

The next problem is to recover by these and similar rather straightforward elementary considerations the known [13] reciprocity law for this extension. We intend to return to this question in a subsequent publication.

REFERENCES

1. Artin, E., Representatives of the Connected Component of the Idèle Class Group, Proc. International Symposium on Algebraic Number Theory, Tokyo(1955), pp. 51-54.

2. Baily, W.L.,Jr.,Galois action on Eisenstein series and certain abelian extensions of number fields, pre-print, 1980.

3. _______, On the Theory of Hilbert Modular Functions I. Arithmetic Groups and Eisenstein Series, J. Alg., 90(1984), 567-605.

4. _______, Arithmetic Hilbert Modular Forms, Automorphic Forms of Several Variables, Taniguchi Symposium, Katata, 1983, Birkhäuser, 1984.

5. _______, Arithmetic Hilbert Modular Functions II, Revista Matemática Iberoamericana, 1(1985), 85-119.

6. Chevalley, C., Deux Théorèmes d'Arithmétique, J. Math. Soc. Japan, 3(1951), 36-44.

7. Deligne, P., Travaux de Shimura, Séminaire BOURBAKI, 23(1970/71), No. 389, 123-165.

8. Hasse, H., Neue Begründung der komplexen Multiplikation, Jour. Reine Angew. Math. 157(1927), 115-139.

9.________, Neue Begründung der komplexen Multiplication II, Jour. Reine Angew. Math. 165(1931), 64-88.

10. Hecke, E., Höhere Modulfunktionen und ihre Anwendung auf die Zahlentheorie, Math. Ann. 71(1912), 1-37. (= No. 1 in Mathematische Werke).

11. Karel, M.L., Special Values of Elliptic Modular Functions, Proc. of International Symposium on Automorphic Forms of Several Variables, Katata 1983, Birkhäuser, 1984.

12. Lang, S., Algebraic Number Theory, Addison-Wesley, 1970.

13. Shimura, G., Construction of class fields and zeta functions of of algebraic curves, Ann. Math., 85(1967), 58-159.

14. _______, On canonical models of arithmetic quotients of bounded symmetric domains, 91(1970), 144-222; II, 92(1970), 528-549.

15. _______, and Taniyama, Y., Complex Multiplication of Abelian Varieties and its Applications to Number Theory, Publications of the Math. Soc. of Japan, No. 6, 1961/

16. Taniyama, Y., Jacobian Varieties and Number Fields, The Complete Works of Yutaka Taniyama, Yutaka Taniyama Complete Works Publication Society, 1961, 1-68 (originally mimeographed notes, University of Tokyo, September 1955).

17. Weber, H., Lehrbuch der Algebra (Kleine Ausgabe), Friedrich Vieweg & Sohn, Braunschweig, 1912.

18. Weil, A., Basic Number Theory, Springer-Verlag 1967.

The Heat Equation for the $\bar{\partial}$-Neumann Problem on Strictly Pseudoconvex Domains

Richard Beals*
Yale University
New Haven, Connecticut 06520

and

Nancy K. Stanton**
University of Notre Dame
Notre Dame, Indiana 46556

1. BACKGROUND

The heat equation for the $\bar{\partial}$-Neumann problem on strictly pseudoconvex domains is a complex analogue of a classical problem in Riemannian geometry. In this section, we will describe some of the classical Riemannian results. To keep things simple, we will only talk about domains.

Let Ω be a bounded domain in $\mathbb{R}^n$ with smooth boundary M. One powerful method of relating geometry and analysis in Ω is to study the heat equation. Let Δ denote the Laplacian on functions on Ω which vanish on M, i.e., which satisfy Dirichlet boundary conditions. A function $f(x,t) \in C^2(\bar{\Omega} \times \mathbb{R}^+)$ $(x \in \bar{\Omega}, t \in \mathbb{R}^+)$ solves the *heat equation with Dirichlet boundary conditions* if

$$\begin{aligned}(\frac{\partial}{\partial t} - \Delta) f &= 0 \\ f\big|_{M\times\mathbb{R}^+} &= 0.\end{aligned} \tag{1.1}$$

The *initial value problem* for the heat equation with Dirichlet boundary conditions is the following. Given $f_0 \in C^0(\bar{\Omega})$, find a solution $f(x,t)$ of the heat equation satisfying

$$\lim_{t\to 0} f(x,t) = f_0(x). \tag{1.2}$$

The solution of the initial value problem is given by applying the heat semigroup $e^{t\Delta}$, the semigroup generated by Δ, to the initial value,

---

* Research supported in part by NSF grant DMS8402637.

** Research supported in part by NSF grant DMS8200442-01 and the Alfred P. Sloan Foundation.

so $f = e^{t\Delta} f_0$. Let $0 > \lambda_1 \geq \lambda_2 \geq \cdots$ be the spectrum of Δ. Then

$$\operatorname{tr} e^{t\Delta} = \Sigma e^{t\lambda_i}. \tag{1.3}$$

McKean and Singer [9] proved the existence of an asymptotic expansion of this trace,

$$\operatorname{tr} e^{t\Delta} \sim (4\pi t)^{-n/2} \sum_{j\geq 0} c_j t^{j/2} \quad \text{as} \quad t \to 0, \tag{1.4}$$

where c_0 is the volume of Ω and for $j \geq 1$, c_j is the integral over M of a universal polynomial (depending only on j and n) in geometric invariants of M as a submanifold of $\mathbb{R}^n$. The calculation of c_0 is equivalent, via an Abelian theorem and Karamata's Tauberian Theorem, to an old result of Weyl's [12].

THEOREM 1.5 (Weyl). Let $N(\lambda)$ denote the number of eigenvalues (counted with multiplicities) of $-\Delta$ which are less than λ. Then

$$N(\lambda) \sim \frac{\text{Volume } (\Omega)}{(4\pi)^{n/2}\Gamma(n/2+1)} \lambda^{n/2} \quad \text{as} \quad \lambda \to \infty. \tag{1.6}$$

For the special case of a plane domain, $n = 2$, the first few terms of (1.4) are

$$\operatorname{tr} e^{t\Delta} \sim \frac{\text{Area } \Omega}{4\pi t} - \frac{\text{Length } M}{8\sqrt{\pi t}} + \frac{1}{12\pi}\int_M \varkappa + O(\sqrt{t}), \tag{1.7}$$

where $\varkappa$ is the curvature of M. By the Gauss-Bonnet Theorem

$$\frac{1}{12\pi}\int_M \varkappa = \frac{1}{6}(1-h) \tag{1.8}$$

where h is the number of holes.

The heat semigroup is given by integration against the heat kernel $p(x,y,t) \in C^\infty(\bar{\Omega} \times \bar{\Omega} \times \mathbb{R}^+)$ and

$$\operatorname{tr} e^{t\Delta} = \int_\Omega p(x,x,t)dx. \tag{1.9}$$

The proof of the asymptotic expansion (1.4) comes down to a detailed construction and analysis of a good enough approximation to p. To first approximation

$$p(x,y,t) \sim \left(\frac{1}{4\pi t}\right)^{n/2} e^{-|x-y|^2/4t},$$

the Euclidean heat kernel. This is Kac's "principle of not feeling the boundary" [8] and immediately gives c_0.

McKean and Singer worked in the more general context of Riemannian manifolds with boundary. Their work was generalized to elliptic boundary value problems by Greiner [6] and Seeley [11].

2. STRICTLY PSEUDOCONVEX DOMAINS

One analogous problem in several complex variables is the heat equation for the $\bar{\partial}$-Neumann problem. Let Ω be a bounded strictly pseudoconvex domain in $\mathbb{C}^{n+1}$, $n \geq 2$, with smooth boundary. For simplicity, we work on (0,1) forms. A form $u \in C^1(\Lambda^{0,1}(\bar{\Omega}))$ satisfies *$\bar{\partial}$-Neumann boundary conditions* if

$$u_{norm} = 0 = \bar{\partial}u_{norm} \quad \text{on} \quad M. \tag{2.1}$$

Let

$$\Lambda_b^{0,q} = \{u \in \Lambda^{0,q}|_M : u_{norm} = 0\} \tag{2.2}$$

and let

$$\nu : \Lambda^{0,q}|_M \to \Lambda_b^{0,q-1} \tag{2.3}$$

denote contraction with the vector field $N'' = \frac{1}{\sqrt{2}}(N + iJ_0 N)$ where N is the inward unit normal vector field on M and J_0 is the almost complex structure on $\mathbb{C}^{n+1}$. Then we can rewrite the $\bar{\partial}$-Neumann boundary conditions (2.1) as

$$\nu u = 0 = \nu\bar{\partial}u. \tag{2.4}$$

The $\bar{\partial}$-Laplacian $\square$ is defined by

$$\begin{aligned} &\text{Dom}\,\square = \{u \in C^2(\Lambda^{0,1}(\bar{\Omega})) : \nu u = \nu\bar{\partial}u = 0\} \\ &\square = \bar{\partial}\bar{\partial}^* + \bar{\partial}^*\bar{\partial} . \end{aligned} \tag{2.5}$$

If $u = \Sigma u_i d\bar{z}^i \in \text{Dom}\,\square$, then

$$\square u = -2 \sum_i \left(\sum_j \frac{\partial^2 u_i}{\partial z^j \partial \bar{z}^j} d\bar{z}^i\right). \tag{2.6}$$

Here $(z^1,\ldots,z^{n+1})$ are the complex coordinates on $\mathbb{C}^{n+1}$.

A (0,1) form $F(z,t) \in C^2(\Lambda^{0,1}(\bar{\Omega} \times R^+))$ solves the *heat equation for the $\bar{\partial}$-Neumann problem* if

$$\begin{aligned}&\text{for fixed } t,\ F(\cdot,t) \in \operatorname{Dom} \square \\ &(\frac{\partial}{\partial t} + \square)F = 0.\end{aligned} \tag{2.7}$$

The *initial value problem* for the heat equation for the $\bar{\partial}$-Neumann problem is the following. Given $f \in C^0(\Lambda^{0,1}(\bar{\Omega}))$, find a solution F of the heat equation (2.7) satisfying

$$\lim_{t\to 0} F(\cdot,t) = f. \tag{2.8}$$

The solution to the initial value problem is given by applying $e^{-t\square}$, the semigroup generated by $-\square$, to the initial value. This semigroup is given by integration against a smooth kernel $p(z,w,t)$ which belongs to $C^\infty(\Lambda^{0,1} \otimes \Lambda^{1,0}\ (\bar{\Omega} \times \bar{\Omega} \times \mathbb{R}^+))$,

$$F(z,t) = (e^{-t\square}f)(z) = \int_{\bar{\Omega}} p(z,w,t)\wedge * f(w). \tag{2.9}$$

Now,

$$\begin{aligned}\operatorname{tr} e^{-t\square} &= \Sigma e^{-t\lambda} \\ &= \int_\Omega \operatorname{tr} p(z,z,t)\,dV.\end{aligned} \tag{2.10}$$

The sum in the first line of (2.10) is over all eigenvalues of $\square$ (counted with multiplicity). In the second line, dV is Lebesgue measure and the local trace $\operatorname{tr}:\Lambda^{0,1} \otimes \Lambda^{1,0} \to \mathbb{C}$ is the linear map satisfying $\operatorname{tr} \mu \otimes \nu = \mu\wedge *\nu/dV$.

To study the asymptotic behavior of $\operatorname{tr} e^{-t\square}$ we need a good description of p. The difficulty is that $\bar{\partial}$-Neumann boundary conditions are non-elliptic so the methods used by McKean-Singer, Greiner and Seeley do not apply. We will describe our methods. The details are in [2]. First we use the classical method of reduction to the boundary. This was used by Greiner and Stein [7] in their work on the $\bar{\partial}$-Neumann problem. In our case, it reduces the construction of p to the inversion of a first order classical parabolic pseudodifferential operator on M.

To carry out this reduction, we write

$$e^{-t\square} = P = G + H \tag{2.11}$$

where G is the Green's operator for the heat equation, the fundamental solution of the initial value problem with boundary value 0, and H is a correction term. For $f \in C^0(\Lambda^{0,1}(\bar{\Omega}))$, Hf satisfies the heat equation with initial value 0. Thus it is given by the Poisson operator for the heat equation J applied to its boundary values $g = Hf|_M$,

$$Hf = Jg. \tag{2.12}$$

The first of the $\bar{\partial}$-Neumann boundary conditions is $\nu Pf = 0$. Since $Gf|_M = 0$, $\nu Gf = 0$ hence, by (2.11),

$$\nu g = \nu Hf = \nu Pf - \nu Gf = 0, \tag{2.13}$$

so $g \in \Lambda_b^{0,1}$. Also by (2.11) and the second $\bar{\partial}$-Neumann boundary condition,

$$\nu\bar{\partial}Jg = \nu\bar{\partial}Hf = -\nu\bar{\partial}Gf. \tag{2.14}$$

Let

$$\Box_h^+ = \nu\bar{\partial}J \tag{2.15}$$

acting on sections of $\Lambda_b^{0,1}$ over $M \times \mathbb{R}^+$. Then, we can rewrite (2.14) as

$$\Box_h^+ g = -\nu\bar{\partial}Gf. \tag{2.16}$$

The operator $\Box_h^+$ is a classical first order parabolic pseudodifferential operator on the space $\Lambda_b^{0,1}(M \times \mathbb{R}^+)$. If it is invertible, combining (2.16) with (2.11) and (2.12) gives

$$g = -\Box_h^{+-1}\nu\bar{\partial}Gf. \tag{2.17}$$

and

$$P = G - J\Box_h^{+-1}\nu\bar{\partial}G. \tag{2.18}$$

If we had an elliptic boundary condition, the operator analogous to $\Box_h^+$ would be invertible in the classical parabolic pseudodifferential calculus. Because of the non-ellipticity of the boundary condition, we cannot invert $\Box_h^+$ classically. However, following Greiner and Stein [7], we can find another first order classical parabolic operator $\Box_h^-$ such that

$$\Box_h^+ \circ \Box_h^- \approx \Box_b + \frac{\partial}{\partial t}. \tag{2.19}$$

Because $n \geq 2$, $\square_b + \frac{\partial}{\partial t}$ is invertible in the Heisenberg parabolic calculus of [1] and

$$\square_h^{+-1} \approx \square_h^{-} \circ (\frac{\partial}{\partial t} + \square_b)^{-1}. \qquad (2.20)$$

Using a mixture of classical and Heisenberg parabolic calculus, we can calculate a full local asymptotic expansion of the symbol of $\square_h^{+-1}$.

The asymptotic behavior of the trace of G is well known by [6], [9] and [11]. Thus, to know the asymptotic behavior of the trace of $e^{-t\square}$, by (2.18) it suffices to study the asymptotic behavior of the trace of

$$H = -J\square_h^{+-1}\nu\bar{\partial}G. \qquad (2.21)$$

Now, H vanishes to infinite order in Ω as $t \to 0$, uniformly on compact subsets of Ω, so the asymptotic behavior of its trace is concentrated near M.

Let r denote distance to M, and let $M_\varepsilon = \{z \in \bar{\Omega}: r(z) \leq \varepsilon\}$. For ε sufficiently small, $M_\varepsilon \simeq M \times [0,\varepsilon]$. We denote points of M_ε by (x,r), $x \in M$, $r \in [0,\varepsilon]$. Let $d\nu$ be the volume element on M. Then, in M_ε, $dV = g(x,r)dr\, d\nu(x)$. We write the kernel h of H on M_ε as $h(x,r,x',r',t)$, $x,x' \in M$, $r,r' \in [0,\varepsilon]$. Then as $t \to 0$

$$\operatorname{tr} H \sim \int_{M\times[0,\varepsilon]} \operatorname{tr} h(x,r,x,r,t)g(x,r)dr\, d\nu(x). \qquad (2.22)$$

To analyze this integral, we introduce an auxiliary operator K on $\Lambda^{0,1}(M \times \mathbb{R}^+)$ with kernel k given by

$$k(x,y,t) = \int_0^\varepsilon h(x,r,y,r,t)g(x,r)dr. \qquad (2.23)$$

Then as $t \to 0$

$$\operatorname{tr} H \sim \operatorname{tr} K = \int_M \operatorname{tr} k(x,x,t)d\nu(x). \qquad (2.24)$$

To analyze the right side of (2.24), we work locally, and now let x,y denote the local coordinates of points in an open set $U \subset M$, (ξ,ρ,τ) the variables dual to (x,r,t). Then

$$k(x,y,t) = \int_0^\varepsilon \check{q}(x,r,x-y,-r,t)g(x,r)dr \qquad (2.25)$$

where $q(x,r,\xi,\rho,\tau)$ is the symbol of H in $U \times [0,\varepsilon]$ and $\vee$ denotes the inverse Fourier transform in the variables (ξ,ρ,τ). Thus, the local symbol of K is

$$q_1(x,\xi,\tau) = \int_0^\varepsilon \tilde{q}(x,r,\xi,-r,\tau)dr \tag{2.26}$$

where $\sim$ denotes the inverse Fourier transform in the variable dual to r. Symbol calculus shows that q_1 has an asymptotic expansion as a sum of matrices where each matrix entry is the product of a symbol with classical parabolic homogeneity and a symbol with Heisenberg parabolic homogeneity. The kernel of the operator corresponding to such a product is then the convolution of two functions, each with a different kind of homogeneity. Of course, if we had an elliptic boundary condition, the kernel would be homogeneous and would contribute one term to the asymptotic expansion of the trace. Here we have the following theorem. (We have suppressed the first manifold variable in the kernel.)

THEOREM 1. Suppose

(i) $a(x,t)$, $b(x,t) \in C^\infty((\mathbb{R}^{2n+1} \times \mathbb{R})\setminus 0)$ and vanish for $t < 0$;

(ii) for some $m \in \mathbb{Z}$, $m \geq -2n-3$, and all $\lambda > 0$

$$a(\lambda x,\lambda^2 t) = \lambda^m a(x,t); \tag{2.27}$$

if $m = -2n-3$, a is a principal value distribution;

(iii) for some $m' \in \mathbb{Z}$, $m' \geq -2n-2$, and all $\lambda > 0$,

$$b(\lambda\cdot x,\lambda^2 t) = \lambda^{m'} b(x,t) \tag{2.28}$$

where $x = (x_0,x')$, $x_0 \in \mathbb{R}$, $x' \in \mathbb{R}^{2n}$, $\lambda\cdot x = (\lambda^2 x_0,\lambda x')$; then, as $t \to 0^+$

$$\int_{\mathbb{R}^{2n+2}} a(x,s)b(-x,t-s)ds\,dx \sim \tag{2.29}$$

$$t^{(2n+m+m'+4)/2} \sum_{j=0}^{\infty} [k_j + t^{(2n+m+2)/2}(k_j' + k_j'' \log t)]t^{j/2}.$$

Here, k_j, k_j' and k_j'' are constants determined by a and b, and if $m = -2n-3$, $m' = -2n-2$, $k_0'' = 0$. The integral in (2.29) is absolutely convergent for $m > -2n-3$ and is a principal value integral for $m = -2n-3$.

From this, we immediately obtain the following theorem.

THEOREM 2. As $t \to 0$,

$$\operatorname{tr} e^{-t\square} \sim t^{-(n+1)}(c_0 + \sum_{j\geq 1} (c_j + c_j' \log t)t^{j/2}). \tag{2.30}$$

We have not yet worked out examples, but the proof of Theorem 1 leads us to believe that the $\log t$ terms do occur. If this is correct, then the expansion (2.30) has two features not present for elliptic boundary value problems. First, $\log t$ enters the expansion. Logarithmic terms have been encountered recently in problems with singularities, for example, in the work of Cheeger [5], Callias and Uhlmann [4] and Brüning and Seeley [3]. Second, the coefficients in the expansion do not behave nicely with respect to scaling of the domain.

In the Riemannian case, it is easy to read off the leading term of $\operatorname{tr} e^{t\Delta}$ from the construction of the symbol of $e^{t\Delta}$. Here it is not immediate, but we have calculated c_0. It is the sum of two terms - the volume term coming from the trace of the Green's operator and the integral over the boundary of a positive function of the Levi form. Thus, Kac's "principle of not feeling the boundary" does not hold. To state the result about c_0, we need some notation. Let L denote the Levi form of M calculated using a defining function r of Ω, $\Omega = \{r < 0\}$, with $|dr| = 1$ on M. Let $\operatorname{tr}^- L$ be the sum of the negative parts of the eigenvalues of L. We denote the volume of a manifold N by $\operatorname{Vol} N$.

THEOREM 3.

$$c_0 = \frac{1}{(2\pi)^{n+1}}\{(n+1)\operatorname{Vol}\Omega + \int_M\int_0^\infty(\operatorname{tr} e^{-\sigma L})e^{-(\operatorname{tr}^- L)\sigma}\det(\sigma L(1-e^{-\sigma L})^{-1})\,d\sigma\,d\nu\}. \tag{2.31}$$

COROLLARY 4. Let $\Omega = B^{n+1}$, the unit ball. Then

$$c_0 = \frac{1}{(2\pi)^{n+1}}\{(n+1)\operatorname{Vol} B^{n+1} + n\Big(\int_0^\infty\Big(\frac{\sigma}{\sinh\sigma}\Big)^n e^{(n-2)\sigma}d\sigma\Big)\operatorname{Vol} S^{n+1}\}. \tag{2.32}$$

By Karamata's Tauberian Theorem, we have a new proof of a result of Métivier [10]. Métivier used completely different methods to prove his result.

THEOREM 5 (Métivier). Let $N(\lambda)$ denote the number of eigenvalues of $\square$ (counted with multiplicity) which are less than or equal to λ. Then

$$N(\lambda) \sim \frac{c_0}{(n+1)!}\lambda^{n+1} \quad \text{as } \lambda \to \infty$$

where c_0 is the constant of Theorem 3.

All of the results we have described generalize to (p,q) forms on compact complex Hermitian manifolds with boundary if the boundary satisfies Kohn's condition Y(q) [2].

BIBLIOGRAPHY

1. R. Beals, P.C. Greiner and N.K. Stanton, The heat equation on a CR manifold, J. Differential Geom., to appear.

2. R. Beals and N.K. Stanton, The heat equation for the $\bar{\partial}$-Neumann problem, in preparation.

3. J. Brüning and R. Seeley, Regular singular asymptotics, Advances in Math., to appear.

4. C.J. Callias and G.A. Uhlmann, Singular asymptotics approach to partial differential equations with isolated singularities in the coefficients, Bull. A.M.S. 11 (1984), 172-176.

5. J. Cheeger, Spectral geometry of singular Riemannian spaces, J. Differential Geom. 18 (1983), 575-657.

6. P. Greiner, An asymptotic expansion for the heat equation, Arch. Rational Mech. Anal. 41 (1971), 163-218.

7. P.C. Greiner and E.M. Stein, Estimates for the $\bar{\partial}$-Neumann problem, Math. Notes 19, Princeton Univ. Press, Princeton, N.J. 1977.

8. M. Kac, Can one hear the shape of a drum?, Amer. Math. Monthly 73 (1966), No. 4, Part II, 1-23.

9. H.P. McKean Jr. and I.M. Singer, Curvature and the eigenvalues of the Laplacian, J. Differential Geom. 1 (1967), 43-69.

10. G. Métivier, Spectral asymptotics of the $\bar{\partial}$-Neumann problem, Duke Math J. 48 (1981), 779-806.

11. R. Seeley, Analytic extension of the trace associated with elliptic boundary value problems, Amer. J. Math. 91 (1969), 963-983.

12. H. Weyl, Das asymptotische Verteilungsgesetz der Eigenwerte linearer partieller Differentialgleichungen (mit einer Anwendung auf die Theorie der Hohlraumstrahlung), Math. Ann. 71 (1912), 441-479.

Some Examples of the Twistor Construction

D. Burns
Department of Mathematics, University of Michigan
Ann Arbor, Michigan 48109 U.S.A.

To Professor Wilhelm Stoll

1. INTRODUCTION

A hyperkähler manifold X is a kähler manifold of even dimension with a holomorphic 2-form ω, everywhere non-singular, and covariant constant. In [2] Calabi constructed such metrics on the cotangent bundle of $\mathbb{P}^n$, with ω equal to the canonical symplectic 2-form ω_{can} on $T^*(\mathbb{P}^n)$, and gave an analytic construction of an associated complex manifold of dimension 2n+1. This construction, for n=1, is equivalent to the Atiyah-Hitchin-Singer version of Penrose's curved twistor space ([1], [4], [6]). Calabi asked for an algebraic or geometric description of this twistor space. On the other hand, it is widely known that the construction is reversible, i.e., there is an inverse reconstruction of a hyperkähler metric from geometric data. In examining the question of the geometric construction of the twistor space for Calabi's examples, it appeared that a simple enough procedure emerged to enable one to construct new hyperkähler metrics where Calabi's method no longer worked. The metrics constructed are on cotangent bundles again, and the constant, holomorphic, non-degenerate 2-form ω is the canonical symplectic form. They are local only, i.e., defined on a neighborhood of the zero section in $T^*(M)$. To make the necessary calculations feasible we assume M is a generalized flag manifold, i.e., a compact, simply-connected homogeneous kähler manifold.

Theorem: Let M be a generalized flag manifold, U its holomorphic isometry group. There exists a U-invariant hyperkähler metric on a neighborhood of the 0-section in $T^*(M)$ such that ω_{can} is covariant constant.

In case M in the theorem is symmetric, the general construction can be made more explicit and the metrics found above can be extended to be complete on all of T*(M).

There is much contact and overlap between the present paper and work of N. Hitchin and M. Roček, cf. [5]. I would like to thank B. Lawson for pointing this out, and especially Martin Roček for his explanation of some of their very interesting ideas.

The paper is arranged as follows. §1 sets down the twistor notation, much along the lines of [6], and gives a proof of the inversion theorem for the twistor construction in the hyperkähler case. In §2 we make our global geometric construction but verify the hypotheses of the inversion theorem only in a neighborhood of the 0-section. In §3 this is globalized for M a symmetric space, and a few questions/remarks are added in §4.

I would to thank E. Calabi for very useful discussions, and for drawing my interest to this problem.

§1. The Inverse Twistor Transform

We start by reviewing the characteristic properties of the twistor spaces associated to hyperkähler manifolds, cf. [2], [6]. For X a hyperkähler manifold of complex dimension 2n, Z(X) is a complex manifold of dimension 2n+1 with

i) a holomorphic map $\pi: Z(X) \to \mathbb{P}^1$ and smooth map $p: Z(X) \to X$ such that $Z(x) \xrightarrow{(\pi,p)} \mathbb{P}^1 \times X$ is a differmorphism.

ii) an anti-holomorphic involution $\sigma: Z(X) \to Z(X)$ covering the involution $\sigma_0: \mathbb{P}^1 \to \mathbb{P}^1$, given by $[z,w] \to [\bar{w},-\bar{z}]$ in homogeneous coordinates.

iii) The fiber $\mathbb{P}^1$'s of the map p in i) are holomorphic curves in Z(X), stable by σ.

iv) For a fiber $C_x = p^{-1}(x)$, the holomorphic normal bundle N_x of C_x in $Z(X)$ is isomorphic to $\bigoplus_{2n\text{-copies}} H$ where H is the line bundle of degree +1 on $C_x = \mathbb{P}^1$.

v) There exists a nowhere degenerate holomorphic section Φ of $\Omega^2_{Z(x)/\mathbb{P}^1} \otimes \pi^* H^{\otimes 2}$. Φ restricted to a fiber $\pi^{-1}(t)$ of π is identified, up to a constant, with a closed 2-form on $\pi^{-1}(t)$.

Finally, Φ must satisfy a reality condition with respect to σ. σ acts on relative differential forms by $\omega \to \overline{\sigma^*(\omega)}$, where the bar is complex conjugation on the coefficients of a differential form. σ also acts on $\pi^* H^{\otimes k}$, any k, compatibly via π with the action of σ_0 on $H^{\otimes k}$ over $\mathbb{P}^1$. We fix homogeneous coordinates $[z,\omega]$ on $\mathbb{P}^1$ so that, on $H^\circ(\mathbb{P}^1, \mathcal{O}(H))$, σ_0 acts conjugate linearly and $\sigma_0(z) = \omega$, $\sigma_0(\omega) = -z$. For each $x \in X$, Φ induces a non-degenerate skew-form on $H^\circ(C_x, \mathcal{O}(N_x \otimes \pi^* H^*))$, as follows from iv and v above. The reality conditions on Φ are:

vi) $\Phi(v_1, v_2) = \Phi(\sigma(v_1), \sigma(v_2))$, and

$\Phi(v_1, \sigma(v_1)) > 0$, if $v_1 \neq 0$,

where v_1, v_2 are sections in $H^\circ(C_x, \mathcal{O}(N_x \otimes \pi^* H^*))$.

We remark that these conditions are formally the same as those encountered in the case n=1 (X a kähler surface) except that in higher dimensions one has to check that Φ is closed on $\pi^{-1}(t)$ in (v), and that $\Phi(v_1, \sigma(v_1))$ is (positive) definite in (vi). These last facts are automatic for n=1.

Conversely, one has the following inversion theorem, due originally to Penrose, in a Minkowski form, for n=1, [4].

Theorem: Let Z be a complex 2n+1 manifold admitting maps π, p, σ and differential Φ as above, verifying (i) to (vi). Then there exists a unique hyperkähler metric on X such that Z=Z(X), the twistor space for that metric.

Proof: By a standard deformation theory argument (cf. [7], for example), one can identify X with a component of the σ-real points in the space of all sections of the map π. Under this identification, we can identify $T_x(X) \otimes \mathbb{C}$ with $H^\circ(C_x, \mathcal{O}(N_x))$, where $C_x = p^{-1}(x)$, and $T_x(X)$ with the σ-real vectors in $H^\circ(C_x, \mathcal{O}(N_x))$.

For $x \in X$, let $E_x = H^\circ(C_x, \mathcal{O}(N_x \otimes \pi^* H^*))$. By condition (iv), $\dim E_x = n$, and $E = \bigcup_{x \in X} E_x$ is a smooth vector bundle. Denote $H^\circ(\mathbb{P}^1, \mathcal{O}(H))$, by V and identify V with the sections in $H^\circ(Z, \mathcal{O}(\pi^* H))$ pulled-up from $\mathbb{P}^1$. By condition (iv), and the above, $T(X) \otimes \mathbb{C} \simeq E \otimes V$. Let ω_0 denote the skew-form on V such that $\omega_0(z,w)=1$. The form Φ in (v) defines a non-degenerate skew-form on E, which we continue to call just Φ, and $\Phi \otimes \omega_0$ is a non-degenerate symmetric form on $T(X) \otimes \mathbb{C}$. By condition (vi), $\Phi \otimes \omega_0$ is a real symmetric form on $T(X) \otimes \mathbb{C}$, i.e., is real on $T(X)$, and is positive definite on $T(X)$. Denote this metric by g.

Using this metric, we identify $T(X)$ with $T^*(X)$. Then $\Lambda^2 T^*(X) \otimes \mathbb{C} = \Lambda^2(E) \otimes S^2(V) + S^2(E) \otimes \Lambda^2(V)$. Under these identifications, we have a distinguished three dimensional space of 2-forms on X, of the form $\Phi \otimes S$, for $S \in S^2(V)$. Such a 2-form is real if and only if S is real with respect to σ_0.

Let S be real $\in S^2(V)$, with $|S|^2=2$: such an S defines an almost complex structure on X, via the 2-form $\Phi \otimes S$ and the metric constructed on X. More explicitly, for $t \in \mathbb{P}^1$, let s_t be a section in V such that $s_t(t)=0$, and $|s_t|^2 = \omega_0(s_t, \sigma_0(s_t)) = 1$. Then every such S can be written uniquely as

$$S = S_t \underset{\text{def.}}{=} -i(s_t \otimes \sigma_0(s_t) + \sigma_0(s_t) \otimes s_t)$$

for some $t \in \mathbb{P}^1$. Writing $\Phi \otimes S_t$ in terms of the metric g and a skew-symmetric transformation J_t on $T(X) \otimes \mathbb{C}$, one checks that J_t is $+i$ on the subspace $E \otimes \{\mathbb{C} \cdot \sigma_0(s_t)\}$ and $-i$ on $E \otimes \{\mathbb{C} \cdot s_t\}$.

Lemma: The map p, restricted to $Z_t=\pi^{-1}(t)$, is holomorphic to X with the almost complex structure J_t.

In particular, J_t is integrable, and $\Phi \otimes S_t$ is its kähler form for the metric g. We postpone briefly the proof of the lemma.

The complex 2-form represented by $\Phi\otimes(\sigma_0(s_t)\otimes\sigma_0(s_t))$ is of type (2,0) for the structure J_t, and its pull-back via p to Z_t agrees with the restriction of Φ on Z to Z_t. Since Φ is closed on Z_t by condition (v), Φ_t is closed on X. Since the 2-forms Φ_t, $t\in\mathbb{P}^1$, on X span the same three-dimensional space as the forms $\Phi \otimes S_t$, we conclude that $\Phi \otimes S_t$ is closed. Hence, our metric g is kähler for each J_t, and Φ_t is covariant constant, as a linear combination of kähler forms for various of the structures $J_{t'}$, $t' \in P^1$. Thus, g is hyperkähler.

Finally, the twistor space of g is the manifold $Z(X)=X \times \mathbb{P}^1$, where the complex structure on $X \times \{t\}$ is given by J_t. By the lemma, $Z \to Z(X)$ by $p \times \pi$ is biholomorphic, concluding the proof of the theorem.

Proof of the Lemma: We have only to make explicit the identification of $T(X)_x$ and the σ-invariants in $\Gamma(C_x,N_x)$ to evaluate the differential of p. Fix $q \in Z$ such that $p(q)=x$, $\pi(q)=t$. We identify geometrically $T(Z_t)_q$ and the real normal vectors $N_q^{\mathbb{R}}$ to C_x at q, and then $N_q^{\mathbb{R}}$ with the complex vector space N_x at q. The σ-real holomorphic sections of N_x are nowhere zero on C_x, explicitly, they are of the form

$$s = v \otimes \sigma_0(s_t) - \sigma(v) \otimes s_t$$

for $v \in E_x$. The differential of p sends s(t) to $s \in H^\circ(C_x,\mathcal{O}(N_x)) = T(X) \otimes \mathbb{C}$. So, $dp_*(v(t)\otimes\sigma_0(s_t)(t)) = v\otimes\sigma_0(s_t) - \sigma(v)\otimes s_t$, and $dp_*(i\, v(t)\otimes\sigma_0(s_t)(t)) = iv \otimes \sigma_0(s_t)+i\sigma(v)\otimes s_t$. On the other hand, since $\omega_0(s_t,\sigma_0(s_t))=|s_t|^2=1$, we calculate

$$J_t(v \otimes \sigma_0(s_t)) = i\, v \otimes \sigma_0(s_t),$$

and

$$J_t(v \otimes s_t) = -i\, v \otimes s_t.$$

Thus, p restricted to Z_t is holomorphic.

§2. The Geometric Construction

Let us fix some notation for describing the generalized flag manifolds. G is a complex semi-simple lie group, θ an anti-holomorphic Cartan involution fixing U, a maximal compact subgroup of G. Gothic letters denote corresponding Lie algebras. T is a maximal torus in U, Δ the roots of $\mathfrak{g}$ with respect to T, and $X_\alpha \in \mathfrak{g}$ a non-zero root vector for α. For $z_0 \in i\mathfrak{t} \subset \mathfrak{g}$, $\Delta_{\pm}(z_0)=\{\alpha\in\Delta \mid \pm\alpha(z_0)>0\}$, and $\mathfrak{n}_{\pm} = \bigoplus_{\alpha\in\Delta_+(z_0)} \mathbb{C}\cdot X_{\pm\alpha}$. Let $\mathfrak{c}(z_0)$ be the centralizer of z_0 in $\mathfrak{g}$, $\mathfrak{p}_{\pm}=\mathfrak{c}(z_0)\oplus\mathfrak{n}_{\pm}$, and $P_{\pm}$ the normalizer of $\mathfrak{p}_{\pm}$ in G under the adjoint representation. Our flag manifold M is the quotient space G/P_+. Since $\theta(z_0)=-z_0$, $\theta(P_+)=P_-$, and G/P_+ is conjugate biholomorphic to G/P_- via θ.

The holomorphic tangent bundle to M is the quotient of $G \times (\mathfrak{g}/\mathfrak{p}_+)$ by the action of P_+, $p\cdot(g,X \bmod \mathfrak{p}_+)=(gp^{-1},\mathrm{Ad}(p)X \bmod \mathfrak{p}_t)$. Let B be the Killing form of $\mathfrak{g}$. Under B, the dual of $\mathfrak{g}/\mathfrak{p}_+$ is $\mathfrak{n}_+$, and the cotangent bundle of M is $G \times \mathfrak{n}_+$ modulo the action of P_+, $p\cdot(g,Y)=(gp^{-1}, \mathrm{Ad}(p)Y)$. (Note that $\mathfrak{n}_+$ is normalized by P_+.) The tangent space to $T^*(M)$ at $(g,Y) \bmod P_+$ is $\mathfrak{g}\times\mathfrak{n}_+$ mod vectors of the form $(X,-[X,Y])$, $X \in \mathfrak{p}_+$ (these are the tangents to the P_+-orbit through (g,Y)). Thus, each tangent vector to $T^*(M)$ at (g,Y) has a unique representation as (X_-,X_+), with $X_+ \in \mathfrak{n}_+$. The canonical symplectic form ω on $T^*(M)$ is given at $(\bar{g},Y)$ by $\omega((X_-,X_+),(Y_-,Y_+))=B(X_-,Y_+)-B(Y_-,X_+)-B([X_-,Y_-],Y)$.

Finally, let $\mathfrak{a} \subset \mathfrak{t}\otimes\mathbb{C}$ be the center of $\mathfrak{c}(z_0)$. The U-invariant kähler metrics on M are given, at e mod $P_+ \in M$, by $(X,X)=-B(X,\mathrm{ad}(z)\theta(X))$, where $z \in \mathfrak{a}$ satisfies $\alpha(z)>0$, all $\alpha \in \Delta_+(z_0)$. Here, $x \in \mathfrak{g}/\mathfrak{p}_+$.

Note that the same constructions, mutatis mutandis, work for the homogeneous space G/P_- and its cotangent bundle.

We want to build a twistor space in two patches over $\mathbb{P}^1$. Let $U_0=\mathbb{P}^1-\{\infty\}$, $U_\alpha=\mathbb{P}^1-\{0\}$, t, the affine parameter on U_0, ζ the affine parameter on U_∞, $\zeta=1/t$ on $U_\alpha \cap U_\infty$. We want fiber spaces over U_0, U_∞ such that the fibers over 0 and ∞ will be $T^*(M)$ and its conjugate respectively. We view $\overline{T^*(M)}$ as $T^*(G/P_-)$. The problem will be to patch these two spaces over $U_0 \cap U_\infty$.

Let us fix z in the center of $\mathfrak{c}(z_0)$ such that $\alpha(z)>0$, all $\alpha \in \Delta_+(z_0)$. Consider the space $Z_0=G \times \{\mathbb{C}\cdot z + \mathfrak{n}_+\}/P_+$, where P_+ acts as above. Since z is central in $\mathfrak{c}(z_0)$, the map $(g,tz + X_+)\to t \in \mathbb{C}=U_0$ is well-defined on Z_0: call it π. The map π is a submersion. We also have another map

$$(g,\ tz + X_+) \longrightarrow (\mathrm{Ad}(g)\ (tz + X_+),\ t) \in \mathfrak{g}\times\mathbb{C}$$

which factors through Z_0. We have:

$$\begin{array}{ccc} Z_0 & \xrightarrow{q} & \mathfrak{g}\times\mathbb{C} \\ \downarrow\pi & & \downarrow p_2 \\ U_0 & = & \mathbb{C} \end{array}$$

On $\pi^{-1}(t)$, $t\neq 0$, $q|\pi^{-1}(t)$ is biholomorphic. The map q gives a resolution of singularities of the variety in $\mathfrak{g}\times \mathbb{C}$ given as the Zariski closure $\hat{Z}_0$ of $\{(X,t) \in \quad \times \mathbb{C}^*|\ X=\mathrm{Ad}(g)\ (tz)$, some $g \in G.\}$. (Note that $q|\pi^{-1}(0)$ is the moment map of the action of G on M.)

Over U_α, we construct $Z_\infty=G \times \{\mathbb{C}z + \mathfrak{n}_-\}/P_-$, and define π: $(g,\ \zeta z + X_-) \to \zeta \in U_\infty$, and

$$q\colon (g,\ \zeta z + X_-) \longrightarrow (\mathrm{Ad}(g)(\zeta z + X_-),\ \zeta) \in \mathfrak{g} \times \mathbb{C}.$$

The image $q(Z_\infty)$ is $\hat{Z}_\infty$. Consider the transformation f of $\mathfrak{g} \times \mathbb{C}^*$ sending (X,t) to $(t^{-2}\ X,\ t^{-1})$. This takes Z_0 to Z_∞ over $\mathbb{C}^*=U_0 \cap U_\infty$, if we set $\zeta=t^{-1}$. Denote by Z the manifold obtained by gluing Z_0, Z_∞ in this way. The map q is globally defined on Z to $\mathfrak{g} \otimes H^{\otimes 2}$ over $\mathbb{P}^1$. Note that Z has a G-action induced from $g\cdot(g',X) = (gg',X)$ on $G \times \mathfrak{g}$, and q is equivariant G acting on $\mathfrak{g} \otimes H^{\otimes 2}$ via the adjoint representation. Z will be our twistor space.

We must next construct a U-equivariant conjugation on Z, covering $t \to -1/\bar{t}$ on $\mathbb{P}^1$. On Z_0, σ is defined as

$$\sigma((g,\ tz + X_+) \bmod P_+)$$

$$= (\theta(g), -\bar{t}z + \theta(X_+)) \bmod P_-,$$

and similarly from Z_∞ to Z_0. For $t\neq 0$, one checks directly that $\pi(\sigma(\pi^{-1}(t))) = -1/\bar{t}$. (Note that $\theta(z)=-z$.) Since $\theta(ug) = u\,\theta(g)$, for $u \in U$, σ is U-equivariant.

The main question will hereafter be the existence of σ-real sections over $\mathbb{P}^1$ in Z. We compute this on $Z \subset \mathfrak{g}\otimes H^{\otimes 2}$, where σ acts $\theta \times \sigma_0$ (σ_0 as in §2 above). Thus, over U_0, a section of $\otimes H^{\otimes 2}$ is written as $X_0 + X_1 t + X_2 t^2$, $X_i \in \mathfrak{g}$, and the reality conditions are

$$\theta(X_0)=X_2$$
$$\theta(X_1)=-X_1.$$

On the other hand, for any $u \in U$, $X_0=X_2=0$, $X_1=\mathrm{Ad}(u)z$ defines such a real curve. Its lift to Z_0 is represented as $(u,tz) \bmod P_+$. At $t=0$, these sections pass through the zero section of $T^*(M)=\pi^{-1}(0)$, and likewise at $t=\infty$, or $\zeta=0$ in U_∞. Before calculating the normal bundle to these sections, we describe the twisted two form Φ on Z.

In $\mathfrak{g} \times \mathbb{C}^*$, note that $q(t^{-1}(t))$, $t\neq 0$, is G-orbit $(\mathrm{Ad}(g)(tz),t), g \in G$, of (tz,t). We've identified $\mathfrak{g}$ with $\mathfrak{g}^*$ by B, and on $\hat{Z}_t=q(\pi^{-1}(t))$ we are going to set Φ_0 equal to the negative of the Kirillov 2-form. More specifically, at $\xi \in Z_t$, a holomorphic tangent vector is a vector $[X,\xi] \in \mathfrak{g}$, for $X \in \mathfrak{g}$ (since $\hat{Z}_t$ is G-homogeneous), and for $X,Y \in \mathfrak{g}$, set

$$\Phi_{0,\xi}([X,\xi],[Y,\xi]) = -B(\xi,[X,Y]).$$
$$= B([X,\xi],Y) = -B(X,[Y,\xi]).$$

It is clear that $\Phi_{0,\xi}$ is well-defined, i.e., the right hand side is 0 if $[X,\xi]=0$, that $\Phi_{0,\xi}$ varies holomorphically with ξ in $\mathfrak{g}$, is G invariant of type (2,0) on Z_t, and is readily checked to be closed on each Z_t. To represent this on Z_0, take $(g,tz + Y_0) \in Z_0$, when $Y_0 \in \mathfrak{n}_+$, $g \in G$, $t\neq 0$, and consider Lie algebra elements X_-, X'_- in $\mathfrak{n}_-$, Y_+, Y'_+ in $\mathfrak{n}_+$. Let $\tilde{Y}_+ \in \mathfrak{n}_+$ verify $[\tilde{Y}_+, tz + Y_0] = Y_+$: $\tilde{Y}_+$ exists, since $t\neq 0$ and ad(z) is invertible on $\mathfrak{n}_+$. Under q, the tangent vectors $(X_-,0)$ and $(0,Y_+)$ to Z_0 are represented by $[Ad(g)(X_-),Ad(g)(tz + Y_0)]$ and $Ad(g)\, Y_+ = [Ad(g)\, \hat{Y}_+, Ad(g)\; (tz + Y_0)]$ at $\xi = Ad(g)\; (tz + Y_0)$ in similarly for X'_-, Y'_+. Hence, on Z_0, we have:

1) $\Phi_{0,\xi}((0,Y_+), (0,Y'_+)) =$

$$-B(Ad(g)(\tilde{Y}_+), Ad(g)(Y'_+)) = -B(\tilde{Y}_+,Y'_+) = 0.$$

2) $\Phi_{0,\xi}((X_-,0), (0,Y_+)$

$$= -B(Ad(g)(X_-), Ad(g)(Y_+)) = -B(X_-,Y_+)$$

3) $\Phi_{0,\xi}((X_-,0), (X'_-,0)) =$

$$= -B(Ad(g)(tz + Y_0), [Ad(g)(X_-), Ad(g)(X'_-)])$$

$$= -B(tz + Y_0, [X_-, X'_-])$$

From these formulas it is clear that Φ_0 extends holomorphically across t=0 in Z_0, and agrees with the canonical 2-form on $\pi^{-1}(0) = T^*(M)$. One constructs Φ_∞ on $Z_\infty - \pi^{-1}(\infty)$, and extends across Z_∞, in a completely similar fashion. Under the patching map f, we have

$$f^*\Phi_\infty = t^2\Phi_0 ,$$

so that Φ_0, Φ_∞ patch to give a section of $\Omega^2_{Z/P^1} \otimes \pi^*(H^{\otimes 2})$

globally. The reality condition (vi) of §1 follows directly from the fact $B(\theta(X),\theta(Y)) = \overline{B(X,Y)}$, $X,Y \in$.

Next we calculate the normal bundles of the real curves (u, tz), $u \in U$ (represented in Z_0), and to verify the positivity condition in (vi), §1. Without loss of generality, assume $u=e$. By a theorem of Grothendieck, the normal bundle to our section, call it C, splits as a sum of line bundles:

$$\bigoplus_{i=1}^{2n} \pi^* H^{\otimes d_i} = N.$$

By the non-degeneracy of Φ constructed above,

$$N_C \otimes \pi^*(H^*) = N_C^* \otimes \pi^*(H).$$

Since the integers d_i are uniquely determined, we see that each d_i is 0, 1 or 2, and the number of 0's is equal to the number of 2's. To show that all $d_i=1$, it suffices to show that for every vector v in the fiber of N at $t=0$, there exists a global section V of N over C such that $V(0)=v$, and $V(\infty)=0$. (This will guarantee that no $d_i=0$, and hence no $d_i=2$.) For $X \in \mathfrak{g}$, let V_X denote the vector field on Z, the derivative of the action of $\exp(sX) \in G$ on Z. V_X along C gives a section of N_C. If $X=X_- \in \mathfrak{n}_-$, $V_{X-}(0) = (X_-,0) \in \mathfrak{n}_- \times \mathfrak{n}_+ = T(T^*(M))$ at $(e,0) = N_C$ at $(e,0)$. On the other hand, if $X=X_+ \in \mathfrak{n}_+$, $V_{X_+}(0) = (X_+,0) \equiv 0$ in N_C at $(e,0)$. Thus, $\frac{1}{t} V_{X_+}$ defines a holomorphic section of N_C, which vanishes at $t=\infty$, and whose value at $t=0$ can be computed at

$$\lim_{t\to 0} \frac{1}{t} \frac{d}{ds}(\exp(s X_+), t z)\Big|_{s=0}$$

$$\equiv \lim_{t\to 0} \frac{1}{t} \frac{d}{ds}(e, \mathrm{Ad}(s X_+)(t z))\Big|_{s=0}$$

$$= (0,[X_+, z])$$

Since ad z is invertible on $\mathfrak{n}_+$, this is an arbitrary vector in $0 \times \mathfrak{n}_- \subset N_C$ at $(e,0)$. Finally, for $X_- \in \mathfrak{n}_- \subset \mathfrak{p}_-$, V_{X-} is 0 in N_c at $t=\infty$, since in Z_∞, $G \times \mathfrak{n}_-$ is divided by P_-.

The preceeding argument shows that the (nowhere vanishing) holomorphic sections of $N_c \otimes \pi^*(H^*)$ are spanned by

1) V_{X_-} over U_0, equivalent to $\frac{1}{\xi} V_{X_-}$ over U_∞,

2) $\frac{1}{t} V_{X_+}$ over U_0, equivalent to V_{X_+} over U_∞,

for $X_\pm$ in $\mathfrak{n}_\pm$. The action of σ on global sections is given by

$$\sigma(V_{X_-}) = \frac{1}{t} V_{\theta(X_-)},$$

$$\sigma(\frac{1}{t} V_{X_+}) = -V_{\theta(X_+)},$$

over U_0. Thus, $\Phi_0(V_{X_-}, \sigma(V_{X_-}))$ is a constant, which we evaluate pointwise over U_0 as

$$- \frac{1}{t} B(t\,z, [X_-, \theta(X_-)])$$

$$= B(X_-, \theta([z,x_-])) > 0, \text{ if } X_- \neq 0.$$

Similarly, $\Phi_0(\frac{1}{t} V_{X_+}, \sigma(\frac{1}{t} V_{X_+}))$

$$= - \frac{1}{t} B(tz, [X_+, -\theta(X_+)])$$

$$= -B(X_+, \theta([z,X_+])) > 0, \text{ if } X_+ \neq 0.$$

Finally, $\Phi_0(\frac{1}{t} V_{X_+}, \sigma(V_{X_-}))=0$. Since Φ is everywhere non-degenerate, this suffices to verify the positivity of $\Phi(v,\sigma(v))$ everywhere, concluding the proof of the main theorem.

§3. Global Calculations

The metrics constructed in the preceeding § are rather algebraic in nature. Thus, although the existence proof is local, one expects these metrics to have global extensions to $T^*(M)$, to be complete there, perhaps asymptotically locally flat at infinity, and unique subject to these conditions. In this § we work out the global extension in the case where M is a compact hermitian symmetric space. Without loss of generality, we assume M is irreducible, and G is $Sl(n,\mathbb{C})$, $Sp(n,\mathbb{C})$ or $SO(n,\mathbb{C})$. We fix the notation briefly.

1) $G = Sl(n,\mathbb{C})$. P_+ is given by block upper triangular matrices

$$\begin{matrix} p\,\{ \\ q\,\{ \end{matrix} \begin{pmatrix} * & * \\ 0 & * \end{pmatrix}, \quad p + q = n$$

z_0 is of the form $\begin{pmatrix} a\, I_p & 0 \\ 0 & b\, I_q \end{pmatrix}$, a, b real, where

$pa + qb = 0$ and $a-b>0$.

2) $G = Sp(n,\mathbb{C})$ preserves the skew-form represented by

$$J = \begin{pmatrix} 0 & I_n \\ -I_n & 0 \end{pmatrix}.$$

P_+ are as in 1), if $p=q$, restricted to the subgroup $Sp(n,\mathbb{C})$.

3) $G = SO(n,\mathbb{C})$ preserves the symmetric form represented by

$$Q = \begin{pmatrix} 0 & 0 & 1 \\ 0 & I_{n-2} & 0 \\ 1 & 0 & 0 \end{pmatrix} \begin{matrix} \}1 \\ \}n-2 \\ \}1 \end{matrix}$$

P_+ is of the form

$$\begin{pmatrix} 1 & a & 0 \\ 0 & I_{n-2} & -a' \\ 0 & 0 & 1 \end{pmatrix}, \quad a \in \mathbb{C}^n,$$

where a' denotes transpose. z_0 is of the form

$$\begin{pmatrix} \lambda & & \\ & 0 & \\ & & -\lambda \end{pmatrix}, \quad \lambda>0.$$

For all three cases $\theta(A)=-\bar{A}'$. For symmetric spaces $\mathfrak{n}_+$ is abelian.

We will describe the real curves as in §2 above, but take advantage of the following ansatz, which is immediately valid only in the case of symmetric M: given $Y_0 \in \mathfrak{n}_+$, we seek $C=C(Y_0) \in \mathfrak{n}_-$ so that $(e^{tC}, tz_0 + Y_0)$ mod P_+, or equivalently

$$\mathrm{Ad}\,(e^{tC})\ (tz_0 + Y_0) \in \mathfrak{g},$$

describes a real section of Z passing through (e,Y_0) at $t=0$. We, of course, want C to vary real analytically with Y_0, etc., all of which will be clear by the construction.

Note that $(\mathrm{ad}C)^3=0$ in our cases, so that

$$\begin{aligned} &\mathrm{Ad}(e^{tC})\ (tz_0 + Y_0) \\ &\quad = e^{t\,\mathrm{ad}C}\ (tz_0 + Y_0) \\ &\quad = tz_0 + Y_0 + t[C,Y_0] + t^2[C,z_0] + \frac{t^2}{2}(\mathrm{ad}\ C)^2\ (Y_0) \end{aligned}$$

which is automatically holomorphic at ∞. For such a section to be real, one needs

$$\theta(z_0 + [C,Y_0]) = -z_0 - [C,Y_0],$$

and

$$\mathrm{ad}\ C(z_0 + \tfrac{1}{2}[C,Y_0]) = \theta(Y_0)$$

In our specific cases, this means $[C,Y_0]$ should be hermitian and

$$(3.1)\qquad [C,z_0] + \frac{1}{2}\,[C,[C,Y_0]] = -\bar{Y}_0'.$$

For Y_0 close to zero, this last equation clearly has a unique solution, $C(Y_0)$, such that $C(0)=0$. This verifies the hermitian condition as well, and is the solution described in the previous §. It thus suffices to continue the solution to 3.1 to all $Y_0 \in n_+$. We'll treat the cases of $Sl(n,\mathbb{C})$, $Sp(n,\mathbb{C})$, which are formally identical, and describe the corresponding result for $SO(n,\mathbb{C})$.

We'll seek a solution to 3.1 of the form $C=D\overline{Y}_0'$, where C and $\overline{Y}_0'$ are viewed as $q \times p$ matrices, and where D is a $q \times q$ hermitian matrix, a spectral function $f(Y_0'Y_0)$ of the non-negative hermitian matrix $\overline{Y}_0'Y_0$. (3.1) becomes

$$(a-b)D\overline{Y}_0' - D\overline{Y}_0'\, Y_0\, D\, \overline{Y}_0' = -\overline{Y}_0'$$

$$= ((a-b)D - D^2\overline{Y}_0'Y_0)\overline{Y}_0'$$

The solution is given by $D=f(\overline{Y}_0'Y_0)$, where f of the real variable μ is

$$f(\mu) = \frac{(a-b) - \sqrt{(a-b)^2 + 4\mu}}{2\mu}$$

which is real analytic for $\mu > -\frac{1}{4}(a-b)^2$, and for which $f(0)=-1/(a-b)$ and $\frac{-1}{a-b}=f(0) < f(\mu) < 0$, for $\mu>0$, and so $D=f(\overline{Y}_0'Y_0)\leq 0$.

We use the notation of the proof of the local existence theorem in §2. We first note that along any of the real sections found above, the vector fields V_{X_+}, resp. V_{X_-}, as normal vectors along the section, have zeroes at $t=0$, $t=\infty$, respectively. Hence, $\frac{1}{t} V_{X_+}$ is a holomorphic section of the normal bundle. So, the bundle $E|_C$ is trivial with sections as in §2,

1) V_{X_-} over U_0, $\frac{1}{\xi} V_{X_-}$ over U_∞

2) $\frac{1}{t} V_{X_+}$ over U_0, V_{X_+} over U_∞

Over U_0, $\sigma(V_{X_-}) = \frac{1}{t} V_{\theta(X_-)}$, $\sigma(\frac{1}{t} V_{X_+}) = -V_{\theta(X_+)}$. We calculate the value of $\frac{1}{t} V_{X_+}$ at $t=0$, taking the limit as $t\to 0$ along our curve, represented by $(e^{tC_0}, tz_0 + Y_0) \bmod P_+$. Thus,

$$\lim_{t\to 0} \frac{1}{t} V_{X_+} = \lim_{t\to 0} \frac{1}{t} (\mathrm{Ad}(e^{-tC_0})X_+, 0) \bmod \mathfrak{p}_+$$

$$= \lim_{t\to 0} \frac{1}{t} (X_+ - t\,\mathrm{ad}(C_0)(X_+) + \frac{t^2}{2}\mathrm{ad}(C_0)^2(X_+), 0) \bmod \mathfrak{p}_+$$

$$= \lim_{t\to 0} \frac{1}{t} (\frac{t^2}{2}\,\mathrm{ad}(C_0)^2(X_+), [X_+ - t\,\mathrm{ad}(C_0)(X_+),$$

$$tz_0 + Y_0]) \bmod \mathfrak{p}_+$$

$$= \lim_{t\to 0} \frac{1}{t} (\frac{t^2}{2}\,\mathrm{ad}(C_0)^2(X_+),\ t[X_+, z_0] - t[\mathrm{ad}(C_0)(X_+), Y_0]$$

$$-t^2[\mathrm{ad}(C_0)(X_+), z_0]) \bmod \mathfrak{p}_+$$

$$= (0, [X_+, z_0] + \mathrm{ad}(Y_0)\mathrm{ad}(C_0)(X_+)) \bmod \mathfrak{p}_+.$$

$$= (0, (b-a)X_+ + \mathrm{ad}([Y_0, C_0])(X_+)) \bmod \mathfrak{p}_+.$$

Setting $W_{X_+} = (b-a)X_+ + \mathrm{ad}([Y_0, C_0])(X_+)$, we conclude that our metric, at $(e, Y_0) \bmod P_+$ in $T^*(M)$, is given by

$$(W_{X_+}, W_{X_+}) = \Phi(\frac{1}{t} V_{X_+}, \sigma(\frac{1}{t} V_{X_+}))$$

$$= -\Phi(\frac{1}{t} V_{X_+}, V_{\theta(X_+)})$$

$$= B(\theta(X_+), W_{X_+})$$

$$= -(a-b)B(\theta(X_+), X_+)$$

$$+ B(\theta(X_+), \mathrm{ad}([Y_0, C_0])(X_+)).$$

$$(W_{X_+}, V_{X_-}) = \Phi(\frac{1}{t} V_{X_+}, \sigma(V_{X_-}))$$

$$= \Phi(\frac{1}{t} V_{X_t}, \frac{1}{t} V_{\theta(X_-)})$$

$$= B(W_{X_+}, W_{\theta(X_-)}) = 0.$$

$$\begin{aligned}(V_{X_-},V_{X_-}) &= \Phi(V_{X_-},\sigma(V_{X_-}))\\ &= B(X_-,W_{\theta(X_-)}).\\ &= -(a-b)B(X_-,\theta(X_-))\\ &\qquad + B(X_-,\mathrm{ad}([Y_0,C_0])(\theta(X_-))).\end{aligned}$$

From the properties above, it is easy to see that $T^*(M)$ is complete in this metric. Presumably, the sectional curvative decays to zero at infinity, but I have not verified this.

The same calculations are valid for the quadrics, i.e., $G=SO(n,\mathbb{C})$. Here, for the real curve $(e^{tC_0}, tz_0 + Y_0)$ through (e,Y_0) with

$$Y_0 = \begin{pmatrix} 0 & a & 0 \\ 0 & 0 & -a' \\ 0 & 0 & 0 \end{pmatrix}$$

one solves for

$$C_0 = \begin{pmatrix} 0 & 0 & 0 \\ -\mu\bar{a}' & 0 & 0 \\ 0 & \mu a & 0 \end{pmatrix}$$

where $\mu = \dfrac{2\lambda - \sqrt{4\lambda^2+4|a|^2}}{2|a|^2}$. (Recall that the eigen-values of ad z_0 are 0, $2\lambda>0$ and -2λ).

The reader can check that the same arguments as above treat, less explicitly, the two exceptional compact hermitian symmetric spaces. One uses a matrix representation of G in which z_0 is real and diagonal, the root ordering in $\mathfrak{g}$ is compatible with the usual one in $\mathfrak{sl}(n,\mathbb{C})$, and $\theta(A)=-\bar{A}'$.

§4. Final Remarks

The general result of §2 should be globalized as in §3. Pursuing the present method, this might mean a more sophisticated ansatz than that used in §3. A twistor approach would almost necessarily involve higher order algebraic equations to solve for real sections of π.

If these metrics are globalized in the case P_+ is a Borel subgroup, one should try to find embedded twistor subspaces representing the Ricci flat Einstein metrics on the resolutions of 2-dimensional rational double points.

REFERENCES

[1] Atiyah, M., N. Hitchin, I. Singer, Self-duality in four-dimensional Riemannian geometry, Proc. R. Soc. Lond. A 362, 425-461 (1978).

[2] Calabi, E., Métriques kähleriennes et fibrés holomorphes, Ann. scient. Ec. Norm, Sup., 4e série, t. 12, 269-294 (1979).

[3] Hitchin, N., Polygons and gravitons, Math. Proc. Comb. Phil. Soc. 85, 465-476 (1979).

[4] Penrose, R., Non-linear gravitons and curved twistor theory, Gen. Rel. Grav. 7, 31-52 (1976).

[5] Roček, M., Supersymmetry and non-linear σ-models, to appear, proceedings of the conference "Supersymmetry in Physics," Los Alamos, NM, Dec. 1983.

[6] Salamon, S., Quaternionic kähler manifolds, Inv. Math. 67, 143-171 (1982).

[7] Burns, D., Some background and examples in deformation theory, in D. Lerner, P. Sommers (eds.), Complex Manifold Techniques in Theoretical Physics, London 1979.

Complete Kähler Domains. A Survey of Some Recent Results

Klas Diederich

1. Introduction

One of the major aspects of complex analysis consists in the investigation of the implications between geometric properties of complex analytic manifolds (or complex spaces) and the nature of certain complex analytic objects on them. Again, an important part of this program is the famous Levi problem, a general somewhat imprecise version of which can be formulated in the following way:

Generalized Levi problem: Characterize Stein manifolds or, more generally, manifolds satisfying certain cohomology vanishing or finiteness theorems by geometric conditions. (Here the word "geometric" is meant in a very broad sense.)
The following, nowadays classical solution of this problem in the Stein case is well-known:

Theorem 1 (Docquier-Grauert 1960 [11]) A complex manifold M is Stein if and only if there is a strictly plurisubharmonic C^∞ function φ on M such that

$$M_c := \{ x \in M : \varphi(x) < c \} \subset\subset M$$

for all $c \in \mathbb{R}$ (i.e. φ is exhaustive from above).

The questions dealt with in this survey article immediately arise from the following observation:

Let φ on M be as in the theorem. We may suppose that $\varphi \geqq 0$ on M. Then the form

$$\omega := i\, \partial\bar\partial\, \varphi^2$$

induces a complete Kähler metric on M . Consequently, we have

Proposition. Any Stein manifold M carries a complete Kähler metric.

This fact was observed for the first time by H. Grauert in his thesis, published in 1956 [14]. At the same time, H. Grauert asked whether the converse also holds:

Question. Let M be a complete non-compact Kähler manifold. Is M necessarily Stein?

There are several reasons why the answer to this question is, in general, negative. In order to eliminate some of them from the beginning we will specialize our situation somewhat more by considering at first only the

Question. Let M be a Stein manifold and $D \subset M$ an open subdomain. Suppose D is complete Kähler. Is D necessarily Stein?

The answer to this question is, in general, still "no". The following class of counterexamples is, essentially, already contained in [14]:

Theorem 2. Let M be a Stein manifold and $A \subset M$ a closed complex analytic subvariety. Then $D := M \setminus A$ is complete Kähler. (If $\operatorname{codim}_{\mathbb{C}} A \geqq 2$ D is, of course, not Stein.)

A small variation of the proof of H. Grauert [2] goes as follows. Put for $t > 0$

$$\rho(t) := \int_0^t \frac{1}{\tau} \left(\frac{1}{\ell n\, \tau}\right)^2 \left(\frac{1}{\ell n\, |\ell n(\tau/e)|}\right)^2 d\tau$$

and

$$\sigma(t) := \int_0^t \frac{1}{\tau}\, \rho(\tau)\, d\tau \quad \text{for} \quad 1 > t > 0 \tag{1}$$

Then the function

$$\tilde{\psi}(z) := \sigma(z\bar{z}) \qquad (2)$$

is subharmonic and continuous on the unit disc $\Delta \subset \mathbb{C}$, strictly subharmonic and C^∞ on $\Delta^* := \Delta \setminus \{0\}$ and a simple calculation shows that the Kähler metric $ds^2 = dd^c\tilde{\psi}$ on Δ^* is complete at 0 (For more details see [5]). After changing $\tilde{\psi}$ slightly near $\partial\Delta$ it extends to a subharmonic function ψ on $\mathbb{C}$ which is C^∞ and strictly subharmonic on $\mathbb{C}^*$. Finally, in the situation of thm. 2 we choose functions $f_1, \ldots, f_m \in \mathcal{O}(M)$ with $A = \{x \in M : f_1(x) = \cdots = f_m(x) = 0\}$, a complete Kähler metric $d\sigma^2$ on M and put on D

$$ds^2 = d\sigma^2 + dd^c\psi\left(\left(\sum_{j=1}^{m} |f_j|^2\right)^{1/2}\right)$$

This metric has the desired properties.

Remark. Notice that ψ is continuous across A.

Theorem 2 shows that the property of being complete Kähler does not imply holomorphic convexity. Additional assumptions have to be made. They can go into two different directions:

A) Curvature assumptions on the complete Kähler metric.

B) Regularity assumptions on the boundary of D.

In section 2 we will discuss some recent results concerning A). Section 3 then deals with B). In section 4 we will study the weak 1-completeness of certain complete Kähler domains $D \subset\subset M$ where M is not Stein.

2. Curvature conditions

Let M be a complex manifold and $D \subset\subset M$ an (open) domain. We can ask the following questions:

a) Which curvature conditions on a complete Kähler metric ds^2 on D guarantee or even characterize that D is locally Stein in M (and, therefore, globally Stein if M is Stein)?

b) Let M be non-compact. Which curvature conditions on a complete Kähler metric ds^2 on M guarantee or even characterize that M is Stein?

Let us first consider the question a). It has been clarified very much by the work of Cheng and Yau [3] and Mok and Yau [18].

In [18] it is shown:

Theorem 3. Let $D \subset\subset M$ be a (non-compact) domain on a complex manifold M. Suppose there is a complete hermitian metric on D such that

$$- C \leqq \text{Ricci curvature} \leqq 0 ,$$

then D is locally Stein (and, hence, Stein, if M is Stein or $M = \mathbb{P}^n$). The same statement also holds, if $\pi : D \to M$ is a Riemann domain over a complex manifold M, such that $\pi(D) \subset\subset M$ and there is a metric on D as above.

Remark. Since a Stein Domain D always can be imbedded into some $\mathbb{C}^N$, it always carries a complete Kähler metric with non-positive Ricci curvature (see f.i. Kobayashi, Nomizu [17], prop. 9.4).

In generalizing previous work of Griffiths [15] and Shiffman [25] it also was proved in [18] that local Steinness follows from a condition on the holomorphic sectional curvature:

Theorem 4. Let $\pi : D \to M$ be a Riemann domain over a complex manifold M with $\pi(D) \subset\subset M$. Suppose, there exists a complete Kähler metric ds^2 on D which satisfies the following condition : There is a function $\lambda : \mathbb{R} \to \mathbb{R}$ with $\lim_{t\to\infty} \lambda(t) = 0$ such that one has for the holomorphic sectional curvature K of ds^2

$$K(T) \leqq \lambda\,(d(q,q_o))$$

for all $q \in D$, $T \in T_q^{10} M$, where $d(\cdot, q_o)$ denotes the geodesic distance with respect to ds^2 from a fixed point $q_o \in D$. Then D is locally Stein.

Remark. Again, since a Stein domain D can be imbedded, it always carries a complete Kähler metric with non-positive holomorphic sectional curvature.

The proofs of both theorems 3 and 4 are given by showing that the Kontinuitätssatz of Hartogs holds on D as a consequence of a Schwarz lemma for volume forms (in the case of the Ricci condition) due to Yau.

The existence statement in the remark after theorem 3 can considerably by sharpened for Riemann domains over Stein manifolds. Cheng, Mok and Yau proved in [3] and [18] together:

Theorem 5. Let $\pi : D \to M$ be a Riemann domain over a Stein manifold M such that $\pi(D) \subset\subset M$. Then D carries a complete Kähler-Einstein metric. (It automatically has negative Ricci curvature.)

For the highly complicated proof of this theorem, which starts with the solution of the Monge-Ampère equation

$$\det\left(\frac{\partial^2 u}{\partial z_i \cdot \partial \bar{z}_j}\right) = e^{(n+1)u}$$

with boundary values ∞ on strictly pseudoconvex domains in $\mathbb{C}^n$, the reader is refered to the original articles.

We now come to the discussion of question b) from above. We only can mention a few results in this direction. Elencwaig considered in [12] the situation of a pseudoconvex domain D in a Kähler manifold M and showed:

Theorem 6. If $D \subset\subset M$ is a pseudoconvex domain in a Kähler manifold M with positive holomorphic bisectional curvature, then D is holomorphically convex.

Remark. Of course, M need not be holomorphically convex.

R.E. Greene and H.H. Wu [15a] gave already in 1974 the following list of different properties which imply the property of being Stein:

Theorem 7. Let M be a complete Kähler manifold. Then M is Stein, if it satisfies one of the following properties:

1) M is simply connected and its sectional curvature is $\leqq 0$;

2) M is noncompact and its sectional curvature is $\geqq 0$ and strictly positive outside a compact set;

3) M is noncompact, its sectional curvature is $\geqq 0$ and its holomorphic bisectional curvature is strictly positive.

4) M is noncompact, its Ricci curvature is strictly positive, its sectional curvature is $\geqq 0$ and its canonical bundle is trivial.

In all these cases the hypotheses are essentially made such that it is possible to construct strictly plurisubharmonic exhaustion functions on M using the geodesic distance with respect to the given Kähler metric.

3. The boundary of complete Kähler domains

We now come back to the approach of the thesis of H. Grauert [14] by considering case B) of the introduction. More precisely:
Let M be an open complex manifold and $D \subset\subset M$ a complete Kähler domain. We want to investigate the nature of the obstructions in ∂D that might prevent D from being locally Stein. H. Grauert proved in [14] in this direction:

Theorem 8. If D as above has real-analytic smooth boundary, then D is locally Stein.

This shows that the obstruction can, indeed, be ruled out by regularity assumptions on ∂D . In fact, the nature of the counterexamples of theorem 2 might be considered as a motivation for the question whether all obstructions are located on the so-called "thin complement" of D , namely the set

$$A := \overset{\circ}{\bar{D}} \setminus D \qquad (1)$$

This is indeed the case. One has

Theorem 9. [4] Let $D \subset\subset M$ be complete Kähler and suppose that D is locally Stein near all points $x \in A$ defined as in (1). Then D is locally Stein everywhere.

Remark. As a consequence one obtains that any complete Kähler domain $D \subset\subset M$ which is topologically fat, i.e. $\overset{\circ}{\bar{D}} = D$, is everywhere locally Stein. This result was first proved by T. Ohsawa [21] under the additional assumption that ∂D is C^1-smooth. The proof of [4] follows in so-far the same line as T. Ohsawa as the L^2-theory of $\bar{\partial}$ on complete Kähler manifolds is the main tool. In [4] this is combined with the techniques of H. Skoda from [27] ([28] might also be used) in order to prove:

Proposition. Let $D \subset\subset \mathbb{C}^n$ be a complete Kähler domain and $z^o \notin \bar{D}$. Then there are holomorphic functions $h_1, \ldots, h_n$ on D such that

$$\sum_{j=1}^{n} h_j(z)(z_j - z_j^o) \equiv 1 \quad \text{on } D .$$

Theorem 9 is then a simple consequence of this proposition.

The next question which now, obviously, has to be asked is:

Question: What is the nature of thin complements $A = \overset{\circ}{\bar{D}} \setminus D$ of complete Kähler domains $D \subset\subset M$?

In particular, because of theorem 2 one might ask: Are all thin complements A as above complex analytic? It turns out that this is indeed the case under certain additional regularity assumptions on A . At first, T. Ohsawa proved 1980 in [22]:

Theorem 10. Let A be a thin complement of a complete Kähler domain D and suppose that A is a C^1-smooth real submanifold of $\overset{\circ}{\bar{D}}$ of real codimension 2. Then A is a complex-analytic hypersurface in $\overset{\circ}{\bar{D}}$.

Ohsawa's method of proof, which, again, makes essential use of the L^2-theory of $\bar{\partial}$ for solving $\bar{\partial}$-closed $(n,1)$-forms on complete Kähler domains, does not seem to be generalisible to the cases of higher (real) codimension of A . Therefore, new methods had to be used in order to show:

Theorem 11. [5] Let A be a thin complement of a complete Kähler domain D and suppose that A is a real-analytic subvariety of $\overset{\circ}{\bar{D}}$ of (real) codimension ≥ 3 . Then A is complex-analytic.

Remarks. 1) Notice, that smoothness of A is not assumed. This does, however, not cause essential new difficulties, since it is not difficult to prove (see [5]) that real-analytic subvarieties which are complex-analytic at all regular points, are necessarily complex-analytic everywhere.

2) Theorem 11 says, in particular, that real-analytic subvarieties of odd codimension > 1 never occur as thin complements of complete Kähler domains.

Because of remark 1 one may assume in the proof of theorem 11 that A is smooth. Then the idea of the proof is as follows: If A is not complex-analytic at a point $z^o \in A$, a small neighborhood U can be chosen and a closed 2-dimensional complex submanifold $X \subset U$ such that $\gamma := X \cap A$ is a real-analytic curve. Then the restriction of the given Kähler metric ds^2 to $X \setminus \gamma$ is changed into a new Kähler metric $d\sigma^2$ in such a way that there is a plurisubharmonic function Φ on X which is C^∞ on $X \setminus \gamma$ and induces $d\sigma^2$. Furthermore, $d\sigma^2$ can be kept complete along γ . The complexification of γ now contains a complex regular disc $\Delta_r \subset X$ with $\Delta_r \cap X = \Delta_r \cap \gamma$. Let $\psi := \Phi \mid \Delta_r$. Then the fact that $\Delta_r \cap \gamma$ is at infinite distance with respect to $dd^c\psi$ on $\Delta_r \setminus \gamma$ gives an immediate contradiction to the other fact, that the Laplacian of ψ on Δ_r is locally integrable.

This proof is very simple. But compared to Ohsawa's theorem 11 the result also seems to be much weaker because of the assumption that A is C^ω . One has to ask whether Ohsawa's result also holds in higher codimension for real C^1 submanifolds. Surprisingly enough this is not the case. Namely, one has:

Theorem 12. [5] For each integer $k \geq 3$ there is a closed real C^∞-submanifold A of pure codimension k in a ball B which is nowhere complex-analytic and such that $B \setminus A$ is, nevertheless, complete Kähler.

The construction of such manifolds $A \subset B$ is naturally rather technical, since they may not be C^ω anywhere because of theorem 11. Let us indicate some basic ideas of it for the case of a 2-dimensional non-complex C^∞-submanifold $A \subset B \subset \mathbb{C}^3$ (i.e. the case of real codimension 4. Here A is constructed as a graph over the (x_1,x_2)-plane of a $\mathbb{C}$-valued C^∞-function f :

$$A = \{ (x_1,x_2,f(x_1,x_2)) \; : \; (x_1,x_2,0) \in (\mathbb{R}^2 \times \{0\}) \cap B\} .$$

The function f is, of course, constructed by an approximation by complex polynomials F_k on $\mathbb{C}^2$. But in order to avoid that A has a kind of "complexification"

$$\hat{A} = \{ (z_1,z_2,F(z_1,z_2)) : (z_1,z_2,0) \in B\}$$

which would, indeed, prevent $B \setminus A$ from being complete Kähler, one has to make sure that the sequence $F_k / \mathbb{R}^2$ converges in the C^∞-sense near 0, but that $F_k(z_1,z_2)$ diverges for all $(z_1,z_2) \in \mathbb{C}^2 \setminus \mathbb{R}^2$ near 0. The complete Kähler metric ds^2 on $B \setminus A$ is obtained by constructing a continuous plurisubharmonic function Φ on B, C^∞ on $B \setminus A$ and such that $ds^2 := dd^c \Phi$ is complete on $B \setminus A$. For this the continuous subharmonic function $\tilde{\Psi}$ of (2) and the complete Kähler metric ds_k^2 on $B \setminus \hat{A}_k$ constructed from it in the proof of theorem 2 is used $\hat{A}_k = \{(z_1,z_2,F_k(z_1,z_2)) \; : \; (z_1,z_2) \in \mathbb{C}^2\}$. In fact, suitable smoothing of the corresponding potential functions across $\hat{A}_k$ multiplied with small constants can essentially be added up.

The difference between the situation of then complements A of real codimension 2 (Ohsawa's theorem 10) and the case of real codimensions $\gneq 2$ (theorem 11 and 12), which just became clear, is nevertheless quite unsatisfying. It might be considered as a contradiction to "the beauty of the mathematical univers", unless it can be better understood. But this can be achieved if one uses the following notion:

<u>Definition</u>. Let A be a real C^1-submanifold of an open subset $U \subset \mathbb{C}^n$. Then A is called (linearly) generating at a point $p \in A$ if $\mathbb{C} \otimes T_pA = \mathbb{C}^n$

(here T_pA is the real tangent space to A at p).

Remark. Notice that the property of being linearly generating is an open condition on A .

Since any real submanifold $A \subset U$, for which T_pA is complex linear for all $p \in A$, is complex-analytic, one has the following simple observation:

Lemma. A real C^1-submanifold $A \subset U$ of (real) codimension 2 is complex-analytic if it is nowhere linearly generating.

With it the theorem 10 of Ohsawa can now be reformulated as saying, that any thin complement $A \subset \overset{\circ}{\bar{D}}$ which is a real C^1-submanifold of codimension 2 is nowhere linearly generating. Although this formulation is not as nice as the original one, it is in a certain sense the right one, because we can now say that Ohsawa's theorem holds in all real codimensions $\geqq 2$ despite of the "counterexamples" of theorem 12. Namely, one has:

Theorem 13. [9] Let $A \subset \overset{\circ}{\bar{D}}$ be a thin complement of a complete Kähler domain D . If A is a C^∞ real submanifold of codimension $\geqq 2$, it is nowhere linearly generating.

Remark. The "counterexamples" constructed for the proof of theorem 12 are also C^∞ .

Again, we will not go into the technical details of the proof; but we want to indicate at least its main ideas. Namely, we assume that A is somewhere (and we may, therefore, assume everywhere) linearly generating and that the codimension of A is $\geqq 3$. It then follows from results of Harvey ([16]) and El-Mir ([13]) (see also Sibony [26]) that the Kähler form ω of a complete Kähler metric ds^2 on D continues as a closed positive $(1,1)$ - current to $\overset{\circ}{\bar{D}}$. Therefore, any point $p \in A$ has a neighborhood $\Omega \subset \overset{\circ}{\bar{D}}$ with a plurisubharmonic function φ on Ω such that $\varphi / \Omega \setminus A$ is C^∞ and $i\partial\bar{\partial}\varphi = \omega$ on $\Omega \setminus A$. - In a next step one shows that this function φ cannot be $-\infty$ on a large part of $A' = A \cap \Omega$. More precisely, one has: There is a $c \in \mathbb{R}$ such that for $m := \dim_{\mathbb{R}} A$ and the m-dimensional Hausdorff measure Λ_m one has

$$\Lambda_m (\{z \in A' : \varphi(z) > c\} > 0 .$$

This follows by foliating A' locally by arcs of fixed length on the boundaries of holomorphic Sadullaev discs [24] contained in $\Omega \setminus A'$ (C^∞ up to the boundary) and using the mean value property for φ on these discs. - The real tangent space T_pA to A at p has the following structure after a suitable affine coordinate change:

$$T_pA = \mathbb{R}^k \times (i0)^k \times \mathbb{C}^{n-k}$$

(because A is linearly generating). Therefore, one can construct a nice smooth real $(2n-2)$-dimensional family of holomorphic discs which intersect A' transversally at their centers and are pairwise disjoint on sectors of fixed size. We then can apply the following uniform estimate for subharmonic functions on the unit disc Δ to the restriction of φ to those discs of the family for which the value of φ at the center is $> c$ with the c from the above lemma:

<u>Lemma</u>. For any $\varepsilon > 0$ there exists an r_o, $0 < r_o < 1$ such that any subharmonic function $\psi < 0$ on the unit disc Δ with $\psi(0) > -1$ satisfies: The set

$$P := \{\theta \in [0,2\pi) : \psi(r e^{i\theta}) < -1000 \text{ for some } 0 < r \leqq r_o\}$$

has linear measure $\leqq \varepsilon$.

By this method we have ensured that there are many real disjoint curves γ running in $\Omega \setminus A'$ towards A' on which $\varphi > -1000\,|c|$ everywhere. In fact, the union W of these curves has positive measure. The plurisubharmonic function $\hat{\varphi} := \sup\{-2000\,|c|, \varphi\}$ agrees with φ on an open neighborhood of W in $\Omega \setminus A'$. Since it is bounded one has according to Bedford and Taylor [1]

$$\int_\Omega (dd^c \hat{\varphi})^n < \infty \tag{2}$$

On the other hand, by using the fact, that all the above curves γ have infinite length with respect to the Kähler metric induced by $\hat{\varphi}$ on W one obtains

$$\int_W \frac{\sqrt{m(z)}}{\operatorname{dist}^{k-1}(z,A)}\, d v_z = \infty \tag{3}$$

where $m(z)$ is the maximal eigenvalue of the Levi form $L_{\hat{\varphi}}(z)$ of $\hat{\varphi}$ at $z \in W$. In a final step, one essentially replaces $\hat{\varphi}$ by

$$\sigma(z) := \hat{\varphi}(z) + C\,|z|^2 + \operatorname{dist}^{1+\varepsilon}(z,A)$$

with $C > 0$ very large. Then σ remains bounded and plurisubharmonic, the curves γ still have infinite length, but one can show in addition that on W

$$MA(\sigma) := \det\left(\frac{\partial^2\sigma}{\partial z_i \partial \bar{z}_j}\right)_1^n \geqq \tilde{c}\, \frac{m(z)}{\operatorname{dist}^{(k-1)(1-\varepsilon)}(z,A)} \tag{4}$$

By putting together (2), (3) and (4), one, finally, arrives at the following contradiction $(d(z) := \operatorname{dist}(z,A))$:

$$\infty = \int \frac{\sqrt{m(z)}}{d^{k-1}(z)}\, d v_z = \int_W \frac{\sqrt{m(z)}}{d^{(k-1)(1-\varepsilon)/2}(z)} \frac{d v_z}{d^{(k-1)(1+\varepsilon)/2}(z)}$$

$$\leqq \left(\int_W \frac{m(z)}{d^{(k-1)(1-\varepsilon)}(z)}\, d v_z \int_W \frac{d v_z}{d^{(k-1)(1+\varepsilon)}(z)}\right)^{1/2}$$

$$\leqq \left(\tilde{c}^{-1} \int_W MA(\sigma)(z)\, d v_z \int_W \frac{d v_z}{d^{(k-1)(1+\varepsilon)}(z)}\right)^{1/2}$$

$$\leqq \left(\tilde{c}^{-1} \int_W (dd^c\sigma)^n \int_{\Omega \setminus A} \frac{d v_z}{d^{(k-1)(1+\varepsilon)}(z)}\right)^{1/2} < \infty$$

where the last integral is finite if $\varepsilon < \frac{1}{k-1}$, because codim A = k . □

The last theorem obviously can be reformulated in the following stronger form:

Theorem 14. Let $A \subset \overset{\circ}{\bar{D}}$ be a thin complement of a complete Kähler domain D . If A is a C^∞ real submanifold of codimension ≥ 2 and $p \in A$, then there is no neighborhood $U = U(p)$ with a complex submanifold $M \subset U$, $p \in M$, such that $A \cap M \subset U$ is a real submanifold, $A \cap M \subsetneq M$, but

$$\mathbb{C} \otimes T_p(M \cap A) = T_p M \ .$$

Remark. This reformulation now also contains the statement of theorem 11 for real-analytic thin complements. Namely, in this case we may take near any regular point $p \in A$ as M the complexification of A . If A would not be complex-analytic near p , we would, indeed, have on a neighborhood $U = U(p)$

$$A \cap U \subsetneq M \quad \text{and} \quad \mathbb{C} \otimes T_p A = T_p M \ .$$

The examples constructed for the proof of theorem 12 show that the structure of thin complements can be very complicated even if they are C^∞ real submanifolds. Therefore, it is important to know more examples and to study their properties. A whole class of examples arises in the following way:

Proposition. ([6]) Let $A \subset \Omega$ be a closed complete pluripolar subset of the pseudoconvex domain $\Omega \subset \mathbb{C}^n$, such that there is a continuous plurisubharmonic function $\varphi : \Omega \to \mathbb{R} \cup \{-\infty\}$ with

$$A = \{ z \in \Omega : \varphi(z) = -\infty \} \ .$$

Then the domain $D := \Omega \setminus A$ is complete Kähler.

Proof. We may assume that φ is in fact strictly plurisubharmonic and φ / D is C^∞ ([23]) . Furthermore, we choose a complete Kähler-metric

$d\sigma^2$ on Ω . Then

$$ds^2 := d\sigma^2 + i\,\partial\bar{\partial}\,(|z|^2 - \log(1+\varphi^2))$$

is complete Kähler on D . □

Furthermore, one has the following

<u>Theorem 15.</u> ([9]) Any real closed C^∞ CR-manifold $A \subset \Omega \subset \mathbb{C}^n$, which is complete pluripolar, is maximally foliated by complex submanifolds.

<u>Proof.</u> This follows again from the existence of certain holomorphic discs. Namely, if A is not maximally foliated, then the vector-valued Leviform of A is not $\equiv 0$. Therefore, there exists according to Boggess and Polking [2] a C^∞-map $h : \bar{\Delta} \to \Omega$, $h|\Delta$ holomorphic, with $h(\partial\Delta) \subset A$ and $h(\Delta) \subset \Omega \setminus A$. This gives a contradiction if one considers for a plurisubharmonic function φ on Ω with $A = \{\varphi = -\infty\}$ the submeanvalue property for $\varphi \circ h$ on $\bar{\Delta}$. □

Because of the above proposition the last theorem might be considered as an indication of an additional property of thin complements A which are C^∞ real CR-submanifolds. At the same time one might ask whether all thin complements are automatically complete pluripolar.

Unfortunately, the answer to both questions is "no", because one has:

<u>Theorem 16.</u> [9] There exists a real closed C^∞ CR-submanifold $A \subset B$ of the unit ball in $\mathbb{C}^n$, $n \geq 5$, such that A is not maximally foliated by complex submanifolds, but, nevertheless, $B \setminus A$ admits a complete Kähler metric. In particular, A is not complete pluripolar in B .

The construction of these examples is a variation of the construction in the proof of theorem 12. In order to avoid the existence of the foliation, A is constructed as a CR-graph over a part of the unit sphere $S^{2n-7} \subset \mathbb{C}^{n-3} \subset \mathbb{C}^n$. The manifold A then has real dimension $2n-7$ and the rank of its holomorphic tangent bundle is $n-4$. For details see [9].

Remark. Whether it is possible, to give a complete characterization of arbitrary thin complements (without additional regularity assumptions) by simple geometric properties, is an open question.

3. Pseudoconvexity and weak 1-convexity

Let M be a complex manifold and $D \subset\subset M$ a domain with smooth C^2-boundary. If ∂D is strictly pseudoconvex, then D is (strictly) 1-convex (as is well-known), i.e., there is a C^∞ function $\varphi : D \to \mathbb{R}$ which is exhaustive from above and strictly plurisubharmonic outside a compact subset $K \subset D$. Furthermore, as is also well-known ([1]), strictly 1-convex manifolds are holomorphically convex and, therefore, proper modifications of Stein spaces.

It is, therefore, a natural question to ask, what can be said about domains $D \subset\subset M$ with smooth C^2-boundaries which are only known to be weakly pseudoconvex. Of course, they are always locally Stein and, therefore, even Stein, if M is Stein. But it is well-known from examples of H. Grauert [20] that without this hypothesis D need not be strictly 1-convex. One, therefore, has to ask, whether such D are at least weakly 1-convex in the sense of the following

Definition. A non-compact complex manifold M is called weakly 1-convex, if there is a C^∞ plurisubharmonic function $\varphi : M \to \mathbb{R}$ which is exhaustive from above.

Such weakly 1-complete complex manifolds M still have interesting analytic properties as can, for instance, be seen from the following theorem.

Theorem 17. (Nakano [19]) Let M be a weakly 1-convex manifold, $\dim M = n$, and (E,h) a hermitian vector bundle over M which is Nakano-positive. Then one has

$$H^q(M, \Omega^n(E)) = 0 \quad \forall\, q \geq 1 .$$

With respect to the last question it must, unfortunately, be said that pseudoconvexity of a domain $D \subset\subset M$ is, unfortunately, in general not a sufficient condition for its weak 1-convexity, even not, if ∂D is C^{ω}-smooth. Namely, one has

<u>Theorem 18</u>. [7] There exists a locally trivial holomorphic disc bundle D over the Hopf surface H given by

$$H = (\mathbb{C}^2 \setminus \{0\}) / (z \sim 2z) ,$$

such that D cannot be exhausted by an increasing sequence of relatively compact pseudoconvex domains. In particular, D is not weakly 1-convex.

<u>Remarks</u>. 1) $D \to H$ obviously extends to a locally trivial holomorphic $\mathbb{P}^1$-bundle $M \to H$ and $D \subset\subset M$ has a C^{ω}-smooth boundary and is pseudoconvex.

2) Such counterexamples are impossible of $\dim M = 2$, $D \subset\subset M$ is pseudoconvex with ∂D C^{ω}-smooth and if on each connected component of ∂D there is at least one strictly pseudoconvex point [8]. In this case D is even holomorphically convex.

In the direction of the questions dealt with in this article it has to be asked whether a domain $D \subset\subset M$ with ∂D C^{∞}-smooth (or only $D = \mathring{\bar{D}}$) is weakly 1-complete if it is complete Kähler (notice that it is then automatically pseudoconvex according to theorem 9). The answer to this question is, in general, not known.

It might, however, be considered as a hint in the positive direction that one has the following result:

<u>Theorem 19</u>. [10] Any locally trivial holomorphic disc bundle D over a compact Kähler manifold M is weakly 1-convex.

References

[1] BEDFORD, E., TAYLOR, B.A.: The Dirichlet problem for the complex Monge-Ampère equation. Invent. Math. 37 (1976), 1-44

[2] BOGGESS, A., POLKING, J.: Holomorphic extension of CR-functions. Duke Math. J. 49 (1982), 757-784

[3] CHENG, S-Y, YAU, S.-T.: On the existence of a complete Kähler metric on non-compact complex manifolds and the regularity of Fefferman's equation. Comm.Pure Appl.Math. 33 (1980), 507-544

[4] DIEDERICH, K., PFLUG, P.: Über Gebiete mit vollständiger Kählermetrik. Math.Ann. 257 (1981), 191-198

[5] DIEDERICH, K., FORNAESS, J.E.: Thin complements of complete Kähler domains. Math.Ann. 259 (1982), 331-341

[6] DIEDERICH, K., FORNAESS, J.E.: Smooth, but not complex-analytic pluripolar sets. Manuscripta math. 37 (1982), 121-125

[7] DIEDERICH, K., FORNAESS, J.E.: A smooth pseudoconvex domain without pseudoconvex exhaustion. Manuscripta math. 39 (1982), 119-123

[8] DIEDERICH, K., OHSAWA, T.: A Levi problem on two-dimensional complex manifolds. Math.Ann. 261 (1982), 255-261

[9] DIEDERICH, K., FORNAESS, J.E.: On the nature of thin complements of complete Kähler metrics. Math. Ann. 268 (1984), 475-495

[10] DIEDERICH, K., OHSAWA, T.: Harmonic mappings and disc bundles over compact Kähler manifolds. Publ.RIMS, Kyoto Univ., 21 (1985), 819-833

[11] DOCQUIER, F., GRAUERT, H.: Levisches Problem und Rungescher Satz für Teilgebiete Steinscher Mannigfaltigkeiten. Math.Ann. 140 (1960), 94-123

[12] ELENCWAJG, G.: Pseudoconvexité locale dans les variétés kählériennes. Ann.Inst.Fourier 25 (1975), 295-314

[13] EL-MIR, M.H.: Sur le prolongement des courants positifs fermés. Acta math. 153 (1984), 1-45

[14] GRAUERT, H.: Charakterisierung der Holomorphiegebiete durch die vollständige Kählersche Metrik. Math.Ann. 131 (1956), 38-75

[15] GRIFFITHS, P.A.: Two theorems on extension of holomorphic mappings. Invent. Math. 14 (1971), 27-62

[15a] GREENE, R.E., WU, H.: Analysis on noncompact Kähler manifolds. Proc. Symp. Pure Math. 30 (1977), 69-100

[16] HARVEY, R.: Removable singularities for positive currents. Am.J. Math. 96 (1974), 67-78

[17] KOBAYASHI, S., NOMIZU, K.: Foundations of differential geometry II. New York 1969. Interscience Publishers

[18] MOK, N., YAU, S.-T.: Completeness of the Kähler-Einstein metric on bounded domains and the characterization of domains of holomorphy by curvature conditions. Proc.Symp,Pure Math. 39 (1983), 41-59

[19] NAKANO, S.: Vanishing theorems for weakly 1-complete manifolds. Number theory, algebraic geometry and commutative algebra, Kinokuniya, Tokyo (1973)

[20] NARASIMHAN, R.: The Levi problem in the theory of functions of several complex variables. Proc.Int.Congr.Math. (Stockholm 1962), 385-388

[21] OHSAWA, T.: On complete Kähler domains with C^1-boundary. Publ. RIMS, Kyoto 16, 929-940 (1980)

[22] OHSAWA, T.: Analyticity of complements of complete Kähler domains. Proc. Japan Acad. 56, Ser. A, 484-487 (1980)

[23] RICHBERG, R.: Stetige streng pseudokonvexe Funktionen. Math.Ann. 175 (1968), 257-286

[24] SADULLAEV, A.: A boundary uniqueness theorem in $\mathbb{C}^n$. Math. USSR Sb. 30 (1976), 501-514

[25] SHIFFMAN, B.: Extension of holomorphic maps into hermitian manifolds. Math.Ann. 194 (1971), 249-258

[26] SIBONY, N.: Quelques problemes de prolongement de courants en analyse complexe. Duke Math.J. 52 (1985), 157-197

[27] SKODA, H.: Application des techniques L^2 à la théorie des idéaux d'une algébre de fonctions holomorphes avec poids. Ann.Sci.Ecole Norm. Sup. 5 (1972), 545-580

[28] SKODA, H.: Morphismes surjectifs de fibrés vectoriels semi-positifs. Ann. Sci. École Norm. Sup. 11 (1978), 577-611

Klas Diederich
Bergische Universität
GHS Wuppertal
Mathematik
Gaußstr. 20
D-5600 Wuppertal 1
FRG

On the Minimality of Hyperplane Sections of Gorenstein Threefolds

Maria Lucia Fania

and

Andrew John Sommese

To Wilhelm Stoll on his Sixtieth Birthday

Let X be a normal irreducible three dimensional projective variety whose local rings are Cohen Macaulay and whose dualizing sheaf, K_X is invertible (see §0 for more details). We will call such a variety a Gorenstein threefold throughout this article.

We say that a pair (X,L) with L an ample line bundle on a Gorenstein threefold has non negative logarithmic Kodaira dimension [I] if there is some integer $n > 0$ such that $h^0(X,(K_X \otimes L)^n) > 0$.

Assume that (X,L) is such a pair and that there is at least one smooth element of the linear system $|L|$. This condition of course implies that X has at most isolated singularities. The logarithmic Kodaira dimension condition on (X,L) implies that all smooth elements of $|L|$ have non negative Kodaira dimension (see §2).

The main theorem of this paper (which generalizes [So4], [So5], [So6]) is the following.

__Main Theorem__. __Let__ (X,L) __be as above__. __There exists a pair__ (X',L') __with__ L' __an ample line bundle on a Gorenstein threefold__ X __such that there is a holomorphic surjection__ $\pi: X \longrightarrow X'$, __expressing__ X __as__ X' __with a finite set__ $F \subseteq X'_{reg}$ __blown up, which satisfies the following conditions:__

a) __given a smooth__ $S \in |L|$, $\pi_S: S \longrightarrow S' = \pi(S)$ __is the map of__ S __onto its minimal model,__

b) $L' = [\pi(S)]$ __for smooth__ $S \in |L|$ __and there is a one to one correspondence between smooth__ $S' \in |L'-F|$ __and smooth__ $S \in |L|$ __gotten by sending such__ S' __to their proper transforms in__ X,

c) $K_{X'} \otimes L'$ __is numerically effective__.

All the corollaries of the smooth version of this result from [So4], [So5], [So6], and [So7] carry over with little work; we will discuss these results in another place.

As remarked earlier, the smooth version of this result was proved by the second author [So4], [So5]. In [So6], he further classified all the pairs (X ,L) with L an ample line bundle on a smooth threefold X and the logarithmic Kodaira dimension of (X ,L) negative. These last results were never published because N. Shepherd-Barron showed that they were easy consequences of Mori's Theory of extremal rays (cf. [Mo], [Ka_2]); for explicit details on the use of Mori's Theory see [Be+Pa]. We needed (e.g. [Fa]) the results of [So4], [So5], and [So6] for local complete intersections with isolated singularities. In this case Mori's theory does not apply but the methods of [So6] combined with the results of [L+So] (see (0.9)) work.

We use these methods to prove the theorem stated above. A substantial part of this paper is identical with [So6]. In a sequel we will deal with the pairs (X ,L) where the logarithmic Kodaira dimension is negative.

Let us give a detailed description of this paper.

In §0 we give background material and results for which we don't know a good reference. In §1 we recall the basic results on the Fano-Morin 3 dimensional adjunction process. We work in more generality than needed since we will use the results in a sequel to classify pairs with log (X ,L) < 0.

In §2 we prove the main theorem.

The authors would like to thank the Max Planck Institut für Mathematik for its support. The second author would like to thank the University of Notre Dame, the National Science Foundation [NSF Grants #MCS80-03257 and #MCS82-00629], and the Sloan Foundation for their support.

§0 NOTATION AND BACKGROUND MATERIAL

For the convenience of the reader, we will review our notation, which agrees with that of [So_3] and [So_4]. We suggest that the reader look over (0.1) to (0.4) and (0.7) and then go to §1 returning to this section when needed.

(0.1) Given a sheaf $\mathcal{S}$ of abelian groups on a topological space X , we denote the global sections of $\mathcal{S}$ over X by $\Gamma(\mathcal{S})$, or, when some confusion can result, by $\Gamma(X , \mathcal{S})$.

(0.2) All spaces and manifolds are complex analytic unless otherwise specified; all dimensions are over $\mathbb{C}$. We often ab-abreviate complex analytic to analytic. Given an analytic space X, we denote its structure sheaf by $\mathcal{O}_X$ and its smooth points by X_{reg}. We do not distinguish between a holomorphic vector bundle E on a complex space, X, and its sheaf of germs of holomorphic sections. Thus when a coherent analytic sheaf and a holomorphic vector bundle are being tensored together, the meaning is clear; the appropriate sheaves are being tensored over $\mathcal{O}_X$.

Given a coherent analytic sheaf, $\mathcal{S}$, on an analytic space, X, we let $\chi(X,\mathcal{S})$, or $\chi(\mathcal{S})$ for short denote its Euler characteristic. We let $h^i(X,\mathcal{S})$, or $h^i(\mathcal{S})$ for short, denote $\dim H^i(X,\mathcal{S})$. We let $h^{0,i}(X) = h^i(\mathcal{O}_X)$. If X is in addition a complex manifold, then $h^{p,q}(X) = h^q(\Lambda^p T_X^*)$ where T_X^* is the holomorphic cotangent bundle of X.

Let X be a normal complex analytic space. By K_X we denote the dualizing sheaf on X. By definition $K_X = \det(T_X^*)$ if X is smooth and $i_*(K_{X_{reg}})$ where $i: X_{reg} \longrightarrow X$ is the inclusion, in general. A normal complex analytic space X is called Gorenstein if its local rings of holomorphic functions are Cohen Macaulay and K_X is invertible. The main example for us of Gorenstein varieties are local complete intersections. Serre duality in its usual form holds for Gorenstein varieties. The following corollary (see [Sh+So]) of the Kawamata-Viehweg-Kodaira-Ramanujam vanishing theorem is very useful.

(0.2.1) Kawamata-Viehweg-Kodaira-Ramanujam Vanishing <u>Theorem</u>. <u>Let</u> X <u>be an n dimensional normal irreducible Gorenstein projective variety. Let</u> L <u>be a numerically effective line bundle on</u> X, <u>i.e.,</u> $L \cdot C \geq 0$ <u>for all effective curves on</u> X. <u>If</u> X <u>has isolated singularities and</u> $\underbrace{L \cdot L \cdots L}_{n \text{ copies}} > 0$ <u>then</u>

$$H^i(X,L^{-1}) = H^{n-i}(X,K_X \otimes L) = 0$$

<u>for</u> $i < \dim X$.

(0.3) Let X be an irreducible projective variety. Let D be an effective Cartier divisor on X. Denote by $[D]$ the holomorphic line bundle associated to D. If L is a holomorphic

line bundle on X, let $|L|$ denote the linear system of all Cartier divisors associated to L. Given a finite set $F \subseteq X$, $|L-F|$ denotes the set $\{D \in |L| \,\big|\, F \subseteq D\}$. We often let L denote either the homology class associated to a $D \in |L|$ or its Poincare dual $c_1(L)$, e.g. given a curve $C \in X$, $L \cdot C = D \cdot C = c_1(L)[C]$. In particular if $\dim X = 2$ and $D \in |L|$ is a smooth connected curve, then the genus of D is given by
$\frac{1}{2}(D \cdot K_X + D \cdot D + 2) = \frac{1}{2}(L \cdot K_X + L \cdot L + 2)$.

We denote the restriction of a vector bundle E on an analytic space X to an analytic subspace S, by E_S. One slight exception is when E is the normal bundle N_T of a submanifold of X and $S \subseteq T$, then E_S is denoted $N_{T,S}$. We denote the restriction of a holomorphic map $\pi: X \longrightarrow Y$ where Y is a complex analytic space, by $\pi_S: S \longrightarrow Y$. We often denote isomorphism of two vector bundles E and F on an analytic space X by $E \approx F$.

(0.4) We often denote complex projective space by $\mathbb{P}_{\mathbb{C}}$ when its exact dimension is irrelevant. A line bundle L on a projective variety, X, is <u>very ample</u> if $\Gamma(L)$ spans L and the map $\phi: X \longrightarrow \mathbb{P}_{\mathbb{C}}$ associated to $\Gamma(L)$ is an embedding. A line bundle L on a projective variety X is <u>ample</u> if L^n is very ample for some $n > 0$. Good references for the standard facts about ample line bundles are [Ha_1, Ha_2].

(0.5) A smooth connected projective surface S' is said to be a minimal model or a minimal surface if S' has no smooth rational curves, E, with $E \cdot E = -1$. It is well known [Z_1, cf. B+H also] that given any smooth connected projective surface S, there is a holomorphic birational map $\pi: S \longrightarrow S'$ from S onto a minimal model S'. If S is of non-negative Kodaira dimension then S' is uniquely determined by the function field of S, and the map π is uniquely determined by S. For us there are two important consequences of this for a surface of non-negative Kodaira dimension, S. The first is that two distinct smooth rational curves E and F on S with $E \cdot E = -1 = F \cdot F$ are disjoint. The second is that there are only a finite number of distinct smooth rational curves on S with self-intersection equal to -1.

A minimal surface, S, of non-negative Kodaira dimension has the important property [cf. Mu, B+H] that K_S^n is spanned by global sections for some $n > 0$. This implies that $K_S \cdot C \geq 0$ for all effective divisors C on S. This implies the following important lemma.

(0.5.1) Lemma. Let S be a connected smooth projective surface of non-negative Kodaira dimension. Then either $K_S^t = \mathcal{O}_S$ for some $t > 0$ or

$$(K_S + L) \cdot (K_S + L) \geq L \cdot L + 1 .$$

Proof. If $K_S \cdot L = 0$ then since some power K_S^t with $t > 0$ of K_S has a non-trivial section, $K_S^t = \mathcal{O}_S$. Therefore it can be assumed without loss of generality that $K_S \cdot L > 0$.
Let $\pi\colon S \longrightarrow S'$ be the map of S onto its minimal model. Then $K_S = \pi^* K_{S'} + P$ where P is an effective divisor satisfying $-k = P \cdot P = K_S \cdot K_S - K_{S'} \cdot K_{S'}$. Since S is obtained from S' by a sequence of blowups, we see that $k' = c_2(S) - c_2(S')$ equals the number of reduced and irreducible components of P. By the invariance of $K_{S'} \cdot K_{S'} + c_2(S')$ under birational transformations $k' = k$. By the ampleness of L, $L \cdot P \geq k'$. Therefore: $(K_S+L) \cdot (K_S+L) = (\pi^* K_{S'} + P+L) \cdot (\pi^* K_{S'} + P+L)$

$= K_{S'} \cdot K_{S'} + P \cdot P + L \cdot L + 2 (\pi^* K_{S'}) \cdot L + 2 L \cdot P$

$\geq -k + L \cdot L + 2(\pi^* K_{S'}) \cdot L + L \cdot P + k' = L \cdot L + 2(\pi^* K_{S'} \cdot L) \cdot L + L \cdot P \geq L \cdot L$ with equality in the last inequality only if $L \cdot P = 0$ and $(\pi^* K_{S'} \cdot L) = 0$. The former equality implies that $S = S'$ and the latter inequality combined with this implies that $K_S \cdot L = 0$. This contradiction proves the lemma. □

(0.6) The Hirzebruch Surfaces [Ha_2, pg. 369ff; Nag]:
By F_r with $r \geq 0$ we denote the rth Hirzebruch surface. F_r is the unique holomorphic $\mathbb{P}^1_{\mathbb{C}}$ bundle over $\mathbb{P}^1_{\mathbb{C}}$ with a section E satisfying $E \cdot E = -r$. Let $\pi\colon F_r \longrightarrow \mathbb{P}^1_{\mathbb{C}}$ denote the bundle projection. In the case $r = 0$, F_0 is simply $\mathbb{P}^1_{\mathbb{C}} \times \mathbb{P}^1_{\mathbb{C}}$. In the cases $r \geq 1$, E is the unique irreducible curve on F_r with negative self intersection. By $\tilde{F}_r$ for $r \geq 1$, we denote the normal surface obtained from F_r by blowing down E. In case $r = 1$, $\tilde{F}_1$ is $\mathbb{P}^2_{\mathbb{C}}$. A basis for the second integral homology of F_r is given by E and f, a fibre of π; of course $f \cdot f = 0$ and $f \cdot E = 1$. The line bundles on F_r are given by $[E]^a \otimes [f]^b$ and the latter is ample if and only if it is very ample, and it

is very ample if and only if $a > 0$ and $b \geq ar + 1$. $[E]^a \otimes [f]^b$ is spanned by global sections if and only if $b \geq ar$ and $a \geq 0$. Given a line bundle L on $\tilde{F}_r$, the pullback to F_r is of the form $([E] \otimes [f]^r)^A$ for some integer A.

We will use the following generalization of a result of Kobayashi-Ochiai [K+O] many times.

(0.6.1) Theorem. *Let X be an n dimensional connected normal irreducible Gorenstein projective variety with isolated singularities. Let L be an ample line bundle on X. Assume that:*

$$K_X^a \otimes L^b \approx \mathcal{O}_X \quad \text{where} \quad b < a < 0 .$$

Then there is an ample line bundle M on X such that $M^t = K_X^{-1}$, *and* $M^q = L$ *where* $bq = ta$. *In particular* $t > q \geq 1$ *and* $t \leq n + 1$. *If* $t = n + 1$, $(X,L) \simeq (\mathbb{P}^n_{\mathbb{C}}, \mathcal{O}_{\mathbb{P}^n_{\mathbb{C}}}(1))$. *If* $t = n$, *then X is biholomorphic to an irreducible quadratic in* $\mathbb{P}^{n+1}_{\mathbb{C}}$ *and L is isomorphic to the restriction of* $\mathcal{O}_{\mathbb{P}^{n+1}_{\mathbb{C}}}(1)$.

Proof. First note that $H_1(X,\mathbb{Z}) = 0$. To see this, we use a pretty argument that the second author has heard attributed to Kobayashi. Assume that $H_1(X,\mathbb{Z}) \neq 0$ and let $\tilde{X}$ be a connected finite k sheeted cover of X. Since $K_X^{-a} \simeq L^b$, it follows that K_X^{-1} and hence $K_{\tilde{X}}^{-1}$ are ample. By the Kodaira vanishing theorem:

$$*) \quad H^{0,i}(X) = h^{0,i}(\tilde{X}) = 0 \quad \text{for} \quad i > 0 .$$

Therefore $\chi(\mathcal{O}_X) = 1 = \chi(\mathcal{O}_{\tilde{X}})$. But by functoriality of Chern classes and the Hirzebruch-Riemann Roch it follows that:

$$1 = \chi(\mathcal{O}_{\tilde{X}}) = k\chi(\mathcal{O}_X) = k$$

implying that $H_1(X,\mathbb{Z}) = 0$.

By the above, *), and the exponential sequence, it follows that:

$$\mathrm{Pic}(X) \approx H^2(X,\mathbb{Z}) \text{ is torsion free.}$$

Therefore $M \in \mathrm{Pic}(x)$ exists such that:

$$M^t = K_X^{-1} \text{ and } M^q = L \text{ where } bq = ta.$$

The rest of the theorem follows along the lines of [K+O]. □

(0.7) We need a few facts from deformation theory.

The next lemma is a special case of lemma (0.7.2) of $[SO_4]$.

(0.7.1) Lemma. Let E be a smooth rational curve contained in the smooth points of an irreducible projective threefold, X. If $\Gamma(N_E)$ spans N_E, the normal bundle of E in X, then the union of the deformations of E in X is dense in X.

(0.7.2) Lemma. Let $\mathfrak{D}$ be a smooth surface contained in the smooth points of an irreducible projective threefold X. Let E be a smooth rational curve on $\mathfrak{D}$. Let $N_{\mathfrak{D}}$ be the normal bundle of $\mathfrak{D}$ in X. If $E \cdot E \geq 0$ in and the degree of $N_{\mathfrak{D},E}$ is ≥ 0, then the union of the deformations of E in X is dense in X.

Proof. Let $N_{\mathfrak{D}/E}$ be the normal bundle of E in $\mathfrak{D}$, and let N_E denote the normal bundle of E in X. Note that by hypothesis $N_{\mathfrak{D}/E}$ and $N_{\mathfrak{D},E}$ are spanned by global sections. It, therefore, follows from:

$$0 \longrightarrow N_{\mathfrak{D}/E} \longrightarrow N_E \longrightarrow N_{\mathfrak{D},E} \longrightarrow 0$$

that $\Gamma(N_E)$ spans N_E. Use (0.7.1). □

(0.7.3) Lemma. Let S be a smooth ample divisor on a connected three dimensional normal projective variety, X. Let $\mathfrak{E}$ be a smooth rational curve on S satisfying $\mathfrak{E} \cdot \mathfrak{E} = -1$ on S. If the union of all smooth deformations of $\mathfrak{E}$ is not dense in X, then there exists a divisor $\mathfrak{D} \subseteq X$ with sing $(\mathfrak{D}) \subseteq \text{sing}(X)$, meeting S transversely in $\mathfrak{E}$. Let $\bar{\mathfrak{D}}$ denote the normalization of $\mathfrak{D}$, $p\colon \bar{\mathfrak{D}} \longrightarrow \mathfrak{D}$.

a) $\bar{\mathfrak{D}}$ is biholomorphic to F_r, $p^*[S]_{\mathfrak{D}} \approx [E] \otimes [f]^k$ for $k \geq r + 1$ in the notation of (0.6),

or,

b) $\bar{\mathfrak{D}}$ is biholomorphic to $\mathbb{P}^2_{\mathbb{C}}$, $p^*[S]_{\mathfrak{D}} \approx \mathcal{O}_{\mathbb{P}^2_{\mathbb{C}}}(1)$;

$\deg(N_{\mathfrak{D}_{reg}}|_E) = -1$.

Proof. The proof of theorem (1.2) of $[So_4]$ minus the last sentence of that proof with lemma (0.7.1) above replacing (0.8) of $[So_4]$, shows that there exists a divisor $\mathfrak{D}$ with the normalization $\mathfrak{D}$ biholomorphic to F_r, $\mathbb{P}^2_{\mathbb{C}}$ or $\tilde{F}_r$ in the notation of (0.6). The non-normal points of $\mathfrak{D}$ must belong to sing(X) since sing($\mathfrak{D}$) in an analytic set in X-S.

Since $K_S \cdot \mathfrak{E} = -1$ it follows that:

$$*)\quad -1 = (K_X+[S]) \cdot \mathcal{E} = (K_{X,\mathcal{D}}+[S]_{\mathcal{D}}) \cdot [S]_{\mathcal{D}} .$$

A simple argument based on (0.6) shows that given two Cartier divisors A and B on $\tilde{F}_r$, it follows that $A \cdot B$ is a multiple of r. Therefore *) precludes $\bar{\mathcal{D}}$ being biholomorphic to $\tilde{F}_r$ with $r \geq 2$.

Since $\mathcal{E} = \mathcal{D} \cap S$ is smooth, rational, and an ample divisor on $\mathcal{D}$, it follows from (0.6) that a) is true if $\bar{\mathcal{D}}$ is biholomorphic to F_r.

If $\bar{\mathcal{D}}$ is biholomorphic to $\mathbb{P}^2_{\mathbb{C}}$, then $\deg(N_{\mathcal{D}_{reg}}|_E) = -1$, where $N_{\mathcal{D}_{reg}}$ is the normal bundle of $\mathcal{D}_{reg}$ in X, follows from $E \cdot E = -1$ on S. This implies that $p^*[S]_{\mathcal{D}} = \mathcal{O}_{\mathbb{P}^2}(1)$. □

The following is a simple but important notion.

(0.7.4) Definition. Let L be a holomorphic line bundle on an irreducible Gorenstein projective threefold, X. A reduction (X',L') of (X,L) is a pair consisting of a line bundle, L', on irreducible Gorenstein projective threefold, X', such that:

a) X is the blowup, $\pi: X \longrightarrow X'$ of X' at a finite set of smooth points, F,

b) $\pi^*(K_{X'} \otimes L'^2) \simeq K_X \otimes L^2$.

Let us make some observations.

(0.7.4.1) b) is equivalent to:

$$L \simeq \pi^*L' \otimes [\pi^{-1}(F)]^{-1} .$$

(0.7.4.2) If S is a smooth element of $|L|$, then $\pi(S)$ is smooth and $[\pi(S)] \simeq L'$. Further sending smooth $S' \in |L'-F|$ to their proper transforms on X, sets up a one to one correspondence between smooth $S' \in |L'-F|$ and smooth $S \in |L|$.

(0.7.4.3) Let X and L be as in the definition of (0.7.4). Let $\{\mathcal{D}_1,\ldots,\mathcal{D}_n\}$ be a set of disjoint divisors $\subseteq X_{reg}$ such that:

a) each $\mathcal{D}_i$ is biholomorphic to $\mathbb{P}^2_{\mathbb{C}}$,

b) $L_{\mathcal{D}_i} \simeq \mathcal{O}_{\mathbb{P}^2_{\mathbb{C}}}(1)$ and $N_{\mathcal{D}_i} \simeq \mathcal{O}_{\mathbb{P}^2_{\mathbb{C}}}(-1)$ where $N_{\mathcal{D}_i}$ is the normal bundle of $\mathcal{D}_i$ in X.

Let X' be the threefold gotten by blowing down the $\mathfrak{D}_i$. Let L' be the unique line bundle on X' such that b) of (0.7.4) holds. Then (X',L') is a reduction of (X,L).

(0.7.4.4) Let (X',L') be a reduction of (X,L). If L is ample, then so is L'. This is an almost immediate consequence of [Z_2, theorem (6.2)]; cf. [F_1, lemma (5.7)] for a proof of a more general fact.

(0.7.5) Lemma. Let X be a normal irreducible Gorenstein projective threefold. Let L be an ample line bundle on X such that there is a smooth $S \in |L|$ and the pair (X,L) has non-negative logarithmic Kodaira dimension. Let $\mathfrak{E}$ be a smooth rational curve on S satisfying $\mathfrak{E} \cdot \mathfrak{E} = -1$ on S. Then the union of all the smooth deformations of $\mathfrak{E}$ is not dense in X and the linear $\mathfrak{D}$ of lemma (0.7.3) satisfies (0.7.3b).

Proof. If the union of all the deformations of $\mathfrak{E}$ were dense in X it would follow that $(K_X+S) \cdot \mathfrak{E} \geq 0$. This contradicts the fact that $-1 = \text{degree}(K_{S,\mathfrak{E}}) = (K_X+S) \cdot \mathfrak{E}$.

To see that a) is not possible, choose a smooth curve $C \subseteq \mathfrak{D}$ with homology class E + (k-1)f in the notation of (0.7.3). Since in homology

$$\mathfrak{E} = C + f$$

and $\deg(N_{\mathfrak{D}_{reg},\mathfrak{E}}) = -1$ it follows that either:

1) $\text{degree}(N_{\mathfrak{D}_{reg},C}) \geq 0$

or

2) $\text{degree}(N_{\mathfrak{D}_{reg},f}) \geq 0$.

If 1) happened then by (0.7.2) the union of the deformation of C are dense in X and thus $(K_X+S) \cdot C \geq 0$. But:

$$-2 = K_X \cdot C + \text{degree}(N_{\mathfrak{D}_{reg},C}) + C \cdot C \text{ on } \mathfrak{D}$$

$$= (K_X+S) \cdot C + \text{degree}(N_{\mathfrak{D}_{reg},C}) - C \cdot f$$

which is ≥ -1 by 1) and $(K_X+S) \cdot C \geq 0$. A similar contradiction happens in case 2) . □

(0.8) Theorem. Let L be an ample line bundle on a smooth connected projective surface, S. Assume either that L is very ample or that $\Gamma(L)$ spans L, $h^{\circ}(L) \geq 4$, and $L \cdot L \geq 5$. Then $K_S \otimes L$ is spanned by global sections unless:

a) $(S,L) = (\mathbb{P}^2_{\mathbb{C}}, \mathcal{O}_{\mathbb{P}^2_{\mathbb{C}}}(e))$ where e = 1 or 2,

b) S is a $\mathbb{P}^1_{\mathbb{C}}$ bundle and $L_F \simeq \mathcal{O}_{\mathbb{P}^1_{\mathbb{C}}}(1)$ where F is a fibre of the bundle.

This second condition of b) is equivalent to smooth $C \in |L|$ being sections of the bundle.

Proof. The above is proved for L very ample by Sommese [So_3], and Van de Ven [VdV]. The argument in [VdV] has two parts. The first part shows that $L \cdot L \geq 5$, $h^{\circ}(L) \geq 4$, and the two connectedness of $C \in |L|$ imply $\Gamma(K_S \otimes L)$ spans $K_S \otimes L$. This part of the argument, which is based on a method of Bombieri [B] applies with no change to the situation of the theorem. The second part of Van de Ven's argument shows that, with the exceptions listed in b), $C \in |L|$ are two connected for all very ample L satisfying $L \cdot L \geq 5$. It is a straightforward check that the second part of Van de Ven's argument works for L ample, spanned by global sections, and satisfying $L \cdot L \geq 5$. □

(0.8.1) Lemma. Let L be an ample line bundle on a smooth connected projective surface S. Assume that $\Gamma(K_S^A \otimes L^N)$ spans $K_S^A \otimes L^N$ where A and N are positive integers.

Let $\phi: S \longrightarrow \mathbb{P}_{\mathbb{C}}$ be the map associated to $\Gamma(K_S^A \otimes L^N)$.

a) If $\dim \phi(S) = 2$, then any connected component, $\mathfrak{E}$, of a positive dimensional fibre of ϕ is a smooth rational curve satisfying $\mathfrak{E} \cdot \mathfrak{E} = -1$ on S and:

*) $$A = N(L \cdot \mathfrak{E}) .$$

In particular for large enough n, the map ϕ': $S \longrightarrow \phi'(S)$ associated to $\Gamma(K_S^{An} \otimes L^{Nn})$ expresses S as a smooth surface, $\phi'(S)$, with a finite set blown up.

b) If $\dim \phi(S) = 1$, then a connected component F of a general fibre of ϕ is a smooth rational curve and:

**) $$2A = N(L \cdot F) .$$

For large enough n, the maps $\phi'\colon S \longrightarrow \phi'(S)$ associated to $(K_S^{An} \otimes L^{Nn})$ has connected fibres and maps S onto a smooth curve, $\phi'(S)$.

c) If $\dim \phi(S) = 0$, i.e. $K_S^A \otimes L^N \simeq \mathcal{O}_S$, then either $A \geq N$, or (S,L) is as in a) of (0.8), or S is biholomorphic to a smooth quadratic in $\mathbb{P}^3_{\mathbb{C}}$ and L is isomorphic to the restriction of $\mathcal{O}_{\mathbb{P}^3_{\mathbb{C}}}(1)$.

Proof. In case a), the assertions about C are immediate consequences of lemma (2.3.3) of [So_3]; the reader can check that the proof of (2.3.3) of [So_3] still holds under the hypotheses of (0.8.2) above. The rest of part a) is standard.

In case b), let F be a connected component of general fibre of ϕ. Since $K_{S,F} \simeq K_F$ it follows that:

$$A \deg(K_F) + N \deg(L_F) = 0 .$$

Since L_F is ample and {A,N} are positive, this implies that K_F^{-1} is ample. Therefore $F \simeq \mathbb{P}^1_{\mathbb{C}}$ and $\deg K_F = -2$ giving **) of b). The rest of part b) is standard.

Assume that $K_S^A \otimes L^N \simeq \mathcal{O}_S$. If $N > A$, then by (0.6.1), we are done.

The following is a very slight modification of [L + So, (2.3)].

(0.9) Theorem (Lipman and Sommese). Let V be a three dimensional irreducible normal Gorenstein variety. Let L and H be ample line bundles in V. Assume that there is a subvariety $P \subseteq V$ which is biholomorphic to $\mathbb{P}^2_{\mathbb{C}}$ and assume that $K_{V,P} \simeq \mathcal{O}_{\mathbb{P}^2}(-2)$. Assume that there exists a map $p\colon V^* \longrightarrow A$ where:

a) A is affine and V* is a Zariski neighborhood of P,

b) $p\colon V^* - P \longrightarrow A - p(P)$ is a biholomorphism.

Assume that there is a smooth $S \in |L|$ and H is spanned by global sections in a neighborhood of $S \cup P$ and $H_P \simeq \mathcal{O}_{\mathbb{P}^2}(1)$. Assume that $S + \mathfrak{D} \in |H|$ where $\mathfrak{D} \not\supseteq P$. Then P does not meet the singular set of V.

Proof. This will follow from the proof of [L + So, (2.3)] if for any $x \in P \cap \mathrm{Sing}(V)$ we can find an irreducible $D \in |H|$ with isolated singularities such that $x \in D$.

Let $|H-x|$ denote the set of $D \in |H|$ that contains $x \in P \cap \mathrm{Sing}(V)$. Note that the base locus of $|H-x|$ does not meet S. Indeed if it did then since H is spanned in a neighborhood of S, it would follow that $|H-x| = |H-y|$ where $y \in S$. Since $S + \mathfrak{D}$ y but not x by $\mathfrak{D} \not\supseteq P$ this is absurd. Thus by Bertini's theorem there is $D \in |H-x|$ with D meeting S transversely in a smooth ample curve. Thus D is irreducible and $\mathrm{Sing}(D) \subseteq V-S$ which implies that $\mathrm{Sing}(D)$ is finite. □

(0.10) Lemma. Let A be an effective ample divisor on a connected projective manifold X. Let $\phi: X \longrightarrow \mathbb{P}_{\mathbb{C}}$ be a holomorphic map with $\dim \phi(A) < \dim A$. Then $\phi(A) = \phi(X)$.

Proof. Let $a = (\dim \phi(A))+1$ and assume that $\phi(A) \neq \phi(X)$. Then $\dim \phi(A) < a$ and:

$$*) \qquad \underbrace{L \cdot \cdot \cdot L}_{a \text{ times}} \cdot A = 0 \text{ in homology}$$

where $L = \phi^*\mathcal{O}_{\mathbb{P}_{\mathbb{C}}}(1)$. L is spanned by global sections and thus $L \cdot \cdot \cdot L$ can be represented by an effective cycle $\mathfrak{D}$ that meets A in a cycle representing $\mathfrak{D} \cdot A$. Further $\mathfrak{D}$ is a non-trivial union of $\dim X - a \geq 1$ dimensional analytic sets since $\dim \phi(X) \geq \dim \phi(A) + 1 = a$ and $\dim X = \dim A + 1 \geq \dim \phi(A) + 2$. Therefore *) contradicts the ampleness of A. □

(0.1) Lemma. Let A be an ample divisor on an irreducible projective local complete intersection X. Assume that there is a continuous map $r: X \longrightarrow A$ such that $r_A: A \longrightarrow A$ is a homotopy equivalence. Then $\dim X \leq 2$.

Proof. Assume that $\dim X \geq 3$. Then by the first Lefschetz theorem:

$$*) \qquad 0 \longrightarrow H^2(X,\mathbb{C}) \longrightarrow H^2(A,\mathbb{C}) .$$

Since r_A is a homotopy equivalence:

$$**) \qquad r_A^*: H^2(A,\mathbb{C}) \longrightarrow H^2(A,\mathbb{C}) \text{ is an isomorphism.}$$

Combining *) and **), it follows that:

$$r^*: H^2(A,\mathbb{C}) \longrightarrow H^2(X,\mathbb{C}) \text{ is an isomorphism.}$$

Therefore a Kaehler class ω on X can be written $r^*\eta$ where $\eta \in H^2(A,\mathbb{C})$. This implies that $\omega^{a+1} = r^*(\eta^{a+1}) = 0$ where $a = \dim A$. This is absurd since ω raised to the dimension of X must be non-trivial. □

§1 The Adjunction Process

(1.0) Throughout this section L is an ample line bundle on an irreducible three dimensional normal Gorenstein projective variety, X. It is further assumed that there is at least one smooth $S \in |L|$.

The adjunction process that we use in this paper is a modification of the adjunction process Morin [Ro, pg. 66] used to reprove Fano's classification of threefolds with rational hyperplane sections. This process used by Morin was based on the Castelnuovo-Enriques adjunction process for surfaces [C+E]. The following lemma is at the heart of the process.

(1.0.1) Lemma. Let $\mathcal{L}$ be a holomorphic line bundle on a smooth, connected, projective threefold, X. Let S be a smooth ample divisor on X. Let $\mathcal{L}(d)$ denote $K_X^d \otimes [S]^d \otimes \mathcal{L}$. Assume that:

(1.0.1.1) $\mathcal{L}$ is spanned by global sections,

(1.0.1.2) $\mathcal{L}(d)_S$ is spanned by global sections for $0 \leq d \leq N$.

Then there is an integer $N' > 0$ such that either:

a) $(\mathcal{L}(N))^{N'}$ is spanned by global sections,

or,

b) $(\mathcal{L}(d'))^{N'}$ is spanned by global sections for some non-negative $d' < N$, and the map associated to $\Gamma((\mathcal{L}(d'))^{N'})$ has an image of less than 3 dimensions.

Proof. Let d' be the largest integer less than or equal to N such that there is an $N' > 0$ such that:

$$\Gamma((\mathcal{L}(d'))^{N'}) \text{ spans } (\mathcal{L}(d'))^{N'}.$$

Since $\mathcal{L}$ is spanned $d' \geq 0$. If $d' = N$ there is nothing to prove. Therefore it can be assumed that $d' < N$. It can be assumed also that the map associated to $\Gamma((\mathcal{L}(d'))^{N'})$ has a three dimensional image, or else the statement b) would be true. Note that this implies by the Kodaira vanishing theorem, e.g. [(0.9.1) or Mu_1] that:

$$*) \quad H^1(X, K_X \otimes \mathcal{L}(d')) = 0 .$$

Consider the residue sequence for S tensored with $\mathcal{L}(d')$:

$$o \longrightarrow K_X \otimes \mathcal{L}(d') \longrightarrow \mathcal{L}(d'+1) \longrightarrow \mathcal{L}(d'+1)_S \longrightarrow o .$$

By assumption $\mathcal{L}(d'+1)_S$ is spanned by global sections. By *) and the above exact sequence, it follows that:

$$**) \quad \Gamma(\mathcal{L}(d'+1)) \text{ spans } \mathcal{L}(d'+1)_S.$$

Since S is ample the statement **) implies that the set where $\mathcal{L}(d'+1)$ is not spanned is finite. Therefore by a theorem of Zariski [Z_2, theorem (6.2)], there is an N" > 0 such that:

$$(\mathcal{L}(d'+1))^{N''} \text{ is spanned by global sections.}$$

This last statement contradicts my choice of d'. This absurdity proves the lemma. □

(1.0.2) Remark. The previous proof shows a little more:

$$\Gamma(\mathcal{L}(d')) \text{ spans } \mathcal{L}(d')_S.$$

This observation has one very important consequence. Assume that a smooth $S' \in |[S]|$ can be chosen to pass through a preassigned point $x \in X$; e.g. assume that $(X,[S])$ is a reduction (cf. (0.7.4)) of a pair $(\tilde{X},\tilde{L})$ where $\tilde{L}$ is a very ample line bundle on a smooth, projective threefold, $\tilde{X}$. Then, if (1.0.1.2) is true for all smooth $S' \in |[S]|$, it follows that N' can be chosen to be 1 in the above lemma.

To utilize the above lemma, we need criteria for (1.0.1.2). The next theorem does this. We are more general than necessary because we will use this result in a sequel.

(1.1) Theorem. Let L be an ample line bundle on a smooth, connected, projective surface, S. There exist arbitrarily large N with the property that $\Gamma(K_S^n \otimes L^N)$ spans $K_S^n \otimes L^N$ for all non-negative integers $n \leq A$ for some $A \geq 0$ where:

a) $A \geq N$ and $K_S^A \otimes L^N$ is not ample,

or,

b) S is a $\mathbb{P}^1_{\mathbb{C}}$ bundle $r: S \longrightarrow C$ over a smooth curve C, and $L_F \approx \mathcal{O}_{\mathbb{P}^1_{\mathbb{C}}}(1)$ for any fibre, F, of r; in this case

$2A = N$

or,

c) $K_S^a \otimes L^N$ is trivial for some $a > 0$.

Proof. We adopt the notation $\mathcal{L}(n) \approx K_S^n \otimes L^N$ where N is a positive integer to be specified. Note that if S has negative Kodaira dimension, then $h^{\circ}(\mathcal{L}(n)) = 0$ for all large n, [B+H; Proposition (3.3), pg. 371]. If S is not a minimal model, then there is a smooth rational curve $\mathcal{E}$ such that $K_S \cdot \mathcal{E} = -1$ and hence regardless of the Kodaira dimension of S, $\Gamma(\mathcal{L}(n))$ can't span $\mathcal{L}(n)$ for $n > N(L \cdot \mathcal{E})$.

Next assume that S has non-negative Kodaira dimension. We have the following lemma.

(1.1.1) Lemma. Let $\mathcal{L}$ be an ample line bundle on S, a smooth connected projective surface on non-negative Kodaira dimension. Assume that $\mathcal{L}$ is spanned by global sections, $h^{\circ}(\mathcal{L}) \geq 4$, and $\mathcal{L} \cdot \mathcal{L} \geq 7$. Then $h(K_S \otimes \mathcal{L}) \geq 4$ and $(K_S+\mathcal{L}) \cdot (K_S+\mathcal{L}) \geq 7$.

Proof. By (0.5.1), $(K_S+\mathcal{L}) \cdot (K_S+\mathcal{L}) \geq 7$. By the Kodaira vanishing theorem and the Riemann-Roch theorem for $K_S \otimes \mathcal{L}$:

$$h(K_S \otimes \mathcal{L}) = \chi(\mathcal{O}_S) + \frac{1}{2}(K_S+\mathcal{L}) \cdot \mathcal{L}.$$

Since S is of non-negative Kodaira dimension, $\chi(\mathcal{O}_S) \geq 0$ and by (0.5) $K_S \cdot \mathcal{L} \geq 0$. Thus $h^{\circ}(K_S \otimes \mathcal{L}) \geq 4$. □

Choose N large enough so that L^N is very ample, $N^2 L \cdot L \geq 7$, and $h^{\circ}(L^N) \geq 4$. Theorem (0.8) and lemma (1.1.1) immediately yield that either $\Gamma(\mathcal{L}(n))$ spans $\mathcal{L}(n)$ for all $n \geq 0$ or there is a finite smallest non-negative integer, A, such that $\Gamma(\mathcal{L}(A+1))$ doesn't span $\mathcal{L}(A+1)$, and $\mathcal{L}(A)$ is not ample. By *) of a) of lemma (0.8.1), it follos that there is a smooth rational curve $\mathcal{E}$ on S such that:

$$\mathcal{E} \cdot \mathcal{E} = -1 \text{ on S and } A = N(L \cdot \mathcal{E}).$$

Therefore by the first equality S is not a minimal model and by the second $A \geq N$. Therefore the theorem is proved if S has non-negative Kodaira dimension.

Next assume that S is rational. Choose for N an even

number that is large enough so that L^N is very ample. We have the following lemma.

(1.1.2) Lemma. Let $\mathcal{L}$ be a very ample line bundle on a smooth connected surface S surface satisfying $h^{1,0}(S) = 0$. Assume that $K_S \otimes \mathcal{L}$ is ample and spanned by global sections. Then $K_S \otimes \mathcal{L}$ is very ample unless:

a) $\mathcal{L} = K_S^{-2}$, i.e. $K_S^2 \otimes \mathcal{L} \simeq \mathcal{O}_S$

or

b) $\mathcal{L} \simeq K_S^{-3}$, i.e. $K_S^3 \otimes \mathcal{L} \simeq \mathcal{O}_S$.

Proof. In [So_3, §3], the second author studied the mapping associated to $\Gamma(K_S \otimes \mathcal{L})$. He showed that under the hypotheses of the lemma, $K_S \otimes \mathcal{L}$ is very ample except in the two cases given by (2.5.1) and (2.5.2) of [So_3].

In the first case S is a two sheeted branched cover of $\mathbb{P}^2_{\mathbb{C}}$. A direct computation using the description in (2.5.1) of [So_3] shows that $K_S^{-2} \simeq \mathcal{L}$.

In the second case S is a two sheeted branched cover of a singular quadratic. A direct computation using the description in (2.5.2) of [So_3] shows that $K_S^{-3} \simeq \mathcal{L}$. □

Choose the smallest non-negative integer A such that $\mathcal{L}(A)$ is not very ample; by the first paragraph of this proof such a finite A exists. We claim that either the theorem is true or:

(1.1.3) $\mathcal{L}(A)$ is spanned by global sections.

To see this note that since $\mathcal{L}(A-1)$ is very ample, it follows from theorem (0.8) that (1.1.3) can fail only if:

α) $S \simeq \mathbb{P}^2_{\mathbb{C}}$ and $\mathcal{L}(A-1) \simeq \mathcal{O}_{\mathbb{P}^2_{\mathbb{C}}}(e)$ for e = 1 or 2,

or,

β) S is a $\mathbb{P}^1_C$ bundle $r: S \longrightarrow \mathbb{P}^1_{\mathbb{C}}$ and

$\mathcal{L}(A-1)_F \simeq \mathcal{O}_{\mathbb{P}^1_{\mathbb{C}}}(1)$ where F is a fibre of r.

If case α) occurs then (S,L) is as in case c) of the conclusions of the theorem. If case β) occurs then since

$K_{S,F} \simeq K_F \simeq \mathcal{O}_{\mathbb{P}^1_C}(-2)$:

$$-2(A-1) + N(L \cdot F) = 1$$

which contradicts the fact that N is even.

We claim that either the theorem is true or:

(1.1.4) $\mathcal{L}(A)$ is ample.

To see this assume that $\mathcal{L}(A)$ is not ample. Let $\phi: S \to \mathbb{P}_{\mathbb{C}}$ be the map associated to $\Gamma(\mathcal{L}(A))$. Since according to (1.1.3), $\Gamma(\mathcal{L}(A))$ spans $\mathcal{L}(A)$, we can use (0.8.1).

If $\dim \phi(S) = 2$, then exactly as in the case when S had non-negative Kodaira dimension, we can use (0.8.1) to conclude that $A \geq N$ and therefore (S,L) is as in part a) of the conclusions of the theorem that we are proving. If $\dim \phi(S) = 1$, we conclude from part b) of (0.8.1) that (S,L) is as in part b) of the conclusions of the theorem that we are proving. If $\dim \phi(S) = 0$, the we conclude from part c) of (0.8.1), that (S,L) is as in part c) of the conclusion of the theorem we are proving. Therefore without loss of generality, we can assume that (1.1.4) is true. But by lemma (1.1.2), we conclude that either (S,L) is as in part c) of the conclusions of the theorem we are proving or $\mathcal{L}(A)$ is very ample. Since by the choice of A, $\mathcal{L}(A)$ is not very ample, it follows that the theorem is proven if S is rational.

Finally assume that S is birationally ruled and $h^{1,0}(S) > 0$.

Choose $N = (12!)^2 \cdot 4 \cdot d \cdot N'$ where $d = \max\{1, |K_S \cdot K_S|\}$ and N' is chosen large enough so that:

(1.1.5) L^N is very ample, $N^2(L \cdot L) \geq 5$, and $h^\circ(L^N) \geq 4$

By the first paragraph $h^\circ(\mathcal{L}(n)) = 0$ for all large enough n. Choose the largest non-negative integer A such that:

(1.1.6) $\Gamma(\mathcal{L}(A-1))$ spans $(A-1)$, $(A-1) \cdot (A-1) \geq 5$, $h^\circ(\mathcal{L}(A-1) \geq 4$, and $\mathcal{L}(A-1)$ is ample.

By the same argument as in the case of rational S, it can

be assumed that:

(1.1.7) $\Gamma(\mathcal{L}(A))$ spans $\mathcal{L}(A)$ and $\mathcal{L}(A)$ is ample.

Since S is birationally ruled:

(1.1.8) $$h^{2,0}(S) = 0.$$

By (1.1.6) and the Kodaira vanishing theorem for $\mathcal{L}(A-1)$:

(1.1.9) $$h^i(\mathcal{L}(A)) = 0 \text{ for } i > 0.$$

Using (1.1.8), (1.1.9) and the residue sequence for smooth $C \in |\mathcal{L}(A-1)|$:

$$0 \longrightarrow K_S \longrightarrow \mathcal{L}(A) \longrightarrow K_C \longrightarrow 0 .$$

We conclude that:

(1.1.10) $$g(C) = h^\circ(\mathcal{L}(A)) + h^{1,0}(S)$$

where g(C) is the genus of C.

Since S is birationally ruled and satisfies $h^{1,0}(S) > 0$ it follows that:

(1.1.11) $$K_S \cdot K_S \leq 8 - 8h^{1,0}(S) \leq 0 .$$

Combining (1.1.11) with (1.1.10) we get:

$$\mathcal{L}(A) \cdot \mathcal{L}(A) = (K_S+\mathcal{L}(A-1))(K_S+\mathcal{L}(A-1)) = K_S \cdot K_S + 4g(C) - 4$$
$$- \mathcal{L}(A-1) \cdot \mathcal{L}(A-1) \leq 8 - 8h^{1,0}(S) + 4g(C) - 4 - \mathcal{L}(A-1) \cdot \mathcal{L}(A-1)$$
$$= 4 + 4h^\circ(\mathcal{L}(A)) - 4h^{1,0}(S) - \mathcal{L}(A-1) \cdot \mathcal{L}(A-1) .$$

By (1.1.5) and the fact that $h^{1,0}(S) \geq 1$, the last inequality becomes:

$$\tfrac{1}{4}(\mathcal{L}(A) \cdot \mathcal{L}(A)) + 1 \leq h^\circ(\mathcal{L}(A))$$

Therefore if I show that:

(1.1.12) $$\mathcal{L}(A) \cdot \mathcal{L}(A) \geq 13.$$

We will conclude that $h^\circ(\mathcal{L}(A)) \geq 4$. This combined with (1.1.7) would show the contradiction that (1.1.6) is true with A+1 in the place of A. Therefore to prove the theorem, we must only show that (1.1.12) is true.

Therefore assume that $\mathcal{L}(A) \cdot \mathcal{L}(A) \leq 12$. Let:
$x = \mathcal{L}(A) \cdot \mathcal{L}(A) = (AK_S+NL) \cdot (AK_S+NL) = A^2K_S \cdot K_S + NT$
for some integer T. If $A = 0$ or $K_S \cdot K_S = 0$, then N divides x

and therefore $x \geq N \geq 13$. Therefore we can assume that:

$$A \cdot K_S \cdot K_S \neq 0 .$$

Recalling that $K_S \cdot K_S \leq 0$, and using the definition of N, we conclude that:

$$*) \qquad y = -A^2 + \tilde{N} \cdot T$$

where $x = y|K_S \cdot K_S|$, and $N = \tilde{N} \; |K_S \cdot K_S|$.

If $\underset{\sim}{y} \leq 12$, then from the form of *) and the fact that $(12!)^2$ divides $\tilde{N}$, we conclude that y is of the form z^2. We can re-write *):

$$**) \qquad 1 + \tilde{A}^2 = \tilde{N}'T$$

where $z\tilde{A} = A$, and $z^2\tilde{N}' = \tilde{N}$.
Since 4 divides $\tilde{N}'$, we get a contradiction from **).

Therefore (1.1.12) is true. □

§2 Proof of the Main Theorem

Assume that L is an ample line bundle on a normal irreducible Gorenstein projective threefold, X. Assume that there is a smooth $S \in |L|$ and that $\log (X,L) \geq 0$.

It is easy to see that such S have non-negative Kodaira dimension by means of the argument used in [So4]. Let us go through it. Since $\log (X,L) \geq 0$, there is an effective $D \in |N(K_X+L)|$ for sure $N > 0$. If S is not a component of D we are done by the adjunction formula. Therefore we can assume without loss of generality that $D = r S + \mathfrak{X}$ where $r > 0$ and S is not a component of $\mathfrak{X}$. Thus $K_S^N = L_S^r \otimes [\mathfrak{X} \cap S]$. Since some power of K_S is a product of an ample divisor and an effective divisor, it follows that S is of non-negative Kodaira dimension.

By theorem (1.1) and the fact that S has non-negative Kodaira dimension, it is true for all sufficiently large $N > 0$ that $\Gamma(K_S^n \otimes L_S^N)$ spans $K_S^n \otimes L_S^N$ for:

a) all n if S is minimal

or

b) for $0 \leq n \leq A$ where $A \geq N$ and $K_S^A \otimes L_S^N$ is not ample.

Since log $(X,L) \geq 0$ it follows that if $\Gamma(K_X^a \otimes L^b)$ spans $K_X^a \otimes L^b$ with $b > a$ then the map associated to $K_X^a \otimes L^b$ has a 3 dimensional image. Therefore by lemma (1.0.1) there is for each $n > 0$ an integer N' such that $\Gamma(K_X^{nN'} \otimes L^{(N+n)N'})$ span $K_X^{nN'} \otimes L^{(N+n)N'}$.

In case a) we see that $K_X \otimes L$ is numerically effective. Indeed let C be an effective curve. We have

$$(K_X + L) \cdot C + \frac{N}{n} L \cdot C \geq 0$$

Letting n go to ∞ gives the result.

Therefore the main theorem is proven if all smooth $S \in |L|$ are minimal and therefore we can assume without loss of generality that we are in case b).

By (0.8.1), (0.7.5) and the fact that $A \geq N$ we conclude:

*) For each $n > 0$ and $\leq N$ there exists an N' such that $\Gamma(K_X^{nN'} \otimes L^{(n+N)N'})$ spans $K_X^{nN'} \otimes L^{(n+N)N'}$ and the assorted map $\phi_{n,N'}$ has a 3 dimensional image. When $n = N$, the map has some positive dimensional fibres.

We can choose the N' large enough so that the maps $\phi_{n,N'}$ have normal images and connected fibres. Denote this map from X to $\phi_{n,N'}(X)$ by ϕ_n.

We claim that $\phi_n : X \longrightarrow X' = \phi_N(X)$ is the reduction with the properties of the main theorem. We proceed by analyzing the positive dimensional fibres of ϕ_N.

(2.1) Lemma. $H^1(K_X^a \otimes L^b) = 0$ if $b \geq 2a - 2 \geq 0$.

Proof. By *) there is an $N'' > 0$ such that $K_X^{(a-1)N''} \otimes L^{bN''}$ spanned by global sections and has a 3 dimensional image. Therefore by (0.2.1) the lemma is proven. □

From this we see that

$$\Gamma(K_X^a \otimes L^b) \longrightarrow \Gamma(K_S^a \otimes L_S^{b-a}) \longrightarrow 0$$

for $b \geq 2a > 0$. Therefore by choosing the N' large enough it can be assumed that

$$\phi_{n,S} : S \longrightarrow \phi_n(S)$$

has connected fibres with $\phi_n(S)$ normal. Further $\phi_{n,S}$ is an embedding for $0 < n < N$.

By (0.7.5) there are Weil divisors $\mathfrak{D}_1,\ldots,\mathfrak{D}_k$ on X such that each $\mathfrak{D}_i$ meets S transversely in a smooth rational curve $\mathfrak{E}_i$ such that $\mathfrak{E}_i \cdot \mathfrak{E}_i = -1$ on S and these $\mathfrak{E}_i$ are precisely the fibre of $\phi_{N,S}$. By the last paragraph $\phi_N(\mathfrak{E}_i) \neq \phi_N(\mathfrak{E}_j)$ for $i \neq j$ implies that the $\mathfrak{D}_i$ are disjoint.

Note $K_X^{N-1} \otimes L^{N-1} \otimes L^N$ is spanned in a neighborhood of S. This implies that $\Gamma(K_X^{N-1} \otimes L^{2N-1})$ is spanned in a neighborhood of each $\mathfrak{D}_i$. Indeed let $p_i\colon \mathbb{P}^2 \longrightarrow \mathfrak{D}_i$ be the map from the normalization of $\mathfrak{D}_i$ to $\mathfrak{D}_i$. Since $p_i^*(K_X^{N-1} \otimes L^{2N-1}) \simeq \mathcal{O}_{\mathbb{P}^2}(1)$ and since $K_X^{N-1} \otimes L^{2N-1}$ is spanned by global sections in a neighborhood of S, it suffices to find a global section of $K_X^{N-1} \otimes L^{2N-1}$ whose restriction to $\mathfrak{D}_i$ vanishes only on $S \cap \mathfrak{D}_i = \mathfrak{E}_i$. Choose any section t of $K_X^{N-1} \otimes L^{2N-2}$ which does not vanish identically on $S \cap \mathfrak{D}_i$ and let s be the tautological section of S vaishing on S. Then $t \otimes s$ is the desired section.

Since $K_X^{N-1} \otimes L^{2N-1}$ is spanned by global sections in a neighborhood of $\mathfrak{D}_i$ and $p_i^*(K_X^{N-1} \otimes L^{2N-1})$ is $\mathcal{O}_{\mathbb{P}^2}(1)$ we easily see that the p_i are biholomorphisms.

(2.2) <u>Lemma</u>. <u>There exist large</u> t <u>such that</u> $K_S^t \otimes L_S^{t-1} \otimes \bigotimes_i [\mathfrak{E}_i]^{-1}$ <u>is spanned by global sections</u>.

<u>Proof</u>. Consider $\phi_{N,S}\colon S \longrightarrow S' = \phi_{N,S}(S)$. Note that there is an ample line bundle $\mathfrak{L}$ on S' such that $\phi_{N,S'}^*\mathfrak{L} = K_S \otimes L_S$. Thus by (0.8.1) and since:

$$K_S^t \otimes L_S^{t-1} \otimes \bigotimes_i [\mathfrak{E}_i]^{-1} \simeq \phi_{N,S'}^*(K_{S'} \otimes \mathfrak{L}^{t-1})$$

the lemma is clear. □

We claim that ϕ_N has only the $\mathfrak{D}_i$ as positive dimensional fibres. If we show this then by theorem (0.9) $\phi_N\colon X \longrightarrow X' = \phi_N(X)$ is a reduction.

If there was any other irreducible positive dimensional variety V such that $\phi_N(V)$ is a point, then $V \cap S$ is non-trivial. We claim $V \cap S$ belongs to $\bigcup_i \mathfrak{E}_i$. Indeed this will follow from

$d\phi_N$ being of rank 3 on $T_{X,x}$ for $x \in S-\bigcup_i \mathfrak{E}_i$. By construction $d\phi_N$ is of rank 2 in $T_{S,x}$ for $x \in S-\bigcup \mathfrak{E}_i$. Therefore $d\phi_N$ will be of rank 3 on $T_{X,x}$ if we can produce a section of $K_X^{NN'} \otimes L^{2NN'}$ for some N' of the form $t \otimes s$ where t is a section of $K_X^{NN'} \otimes L^{2NN'-1}$ with $t(x) \neq 0$ and s is the tautological section of $[S] = L$. This is clear by lemma (2.2) and lemma (2.1).

Further by lemma (2.2) and (2.1) given $x \in \mathfrak{E}_i$, t can be chosen so that t_S vanishes only to the first order on $\mathfrak{E}_i$ in a neighborhood of x on S. Therefore by the implicit function theorem t vanishes only on a manifold in a neighborhood of x on X. Assume x was chosen so that $V \cap S \not\ni x$. Since $s \otimes t$ vanishes on V it follows that $V \subseteq S \cup \mathfrak{D}_i$ which implies that the only positive dimensional fibres of ϕ_N are the $\mathfrak{D}_i$.

Let $L' = [\phi_N(S)]$ and $F = \phi_N(\bigcup_i \mathfrak{D}_i)$.

We claim that S' is minimal for all smooth $S' \in |L'-F|$ Indeed if not then there is by repeating the whole argument we find a smooth rational curve $\mathfrak{E}$ in $S' = \phi_N(S)$ such that $\mathfrak{E} \cdot \mathfrak{E} = -1$ on S', $L'_{\mathfrak{E}} \approx \mathcal{O}_{\mathbb{P}^1}(1)$ and $\phi_N(\mathfrak{E}_i) \in \mathfrak{E}$ for some $\mathfrak{E}$.

This is absurd since $L_{\tilde{\mathfrak{E}}} \approx \mathcal{O}_{\tilde{\mathfrak{E}}}$ where $\tilde{\mathfrak{E}}$ is the proper transform of S'. This proves that the smooth $S' \in |L'-F|$ are minimal. The argument at the beginning of the proof shows $K_{X'} \otimes L'$ is numerically effective. □

(2.3) Corollary. Let L be an ample line bundle on an irreducible projective Gorenstein threefold, X. Assume there is a smooth $S \in |L|$. There is a polynomial p(n) such that:

$$h^\circ(K_X^n \otimes L^n) = p(n) \text{ for } n > 0 .$$

Proof. If log (X,L) < 0 the corollary is trivial with p(n) = 0.

If log (X,L) $\geq$ 0 let $\pi: X \longrightarrow X'$ be the reduction of the main theorem. Since $K_{X'} \otimes L'$ is numerically effective it follows that $K_X^{n-1} \otimes L^n$ is ample for $n > 0$. By (0.2.1) $H^i(K_X^n \otimes L^n) = 0$ for $i > 0$ and therefore by the Riemann-Roch theorem the above is true. □

In a sequel we will give a detailed structure theorem for (X',L') in terms of the degree of p(n). The reader can consult [So5], [So6] for a description of these results in the case of X smooth.

References

[Bo] E. Bombieri, Canonical models of surfaces of general type, Publ. Math. I.H.E.S. 42 (1973), 171-219.

[Be+Pa] M. Beltrametti, M. Palleshi, On threefolds with low sectional genus, preprint.

[B+H] E. Bombieri and D. Husemoller, Classification and embeddings of surfaces, Proc. Symp. Pure Math. 29, ed. by R. Hartshorne, (1975), 329-420.

[C+E] G. Castelnuovo and F. Enriques, Sur quelques resultats nouveaux dans la theorie des surfaces algebrique, Note V in P+S below.

[F_1] T. Fujita, On the hyperplane section principle of Lefschetz, J. Math. Soc. Japan 32 (1980), 153-169.

[F_2] T. Fujita, On the structure on polarized manifolds of total deficiency, I, J. Math. Soc. Japan 32 (1980), 709-725.

[Ha_1] R. Hartshorne, Ample Subvarieties of Algebraic Varieties, Lecture Notes in Math. 156 (1970), Springer-Verlag, New York (1977).

[Ha_2] R. Hartshorne, Algebraic Geometry, Springer-Verlag, New York (1977).

[Hi] F. Hirzebruch, New Topological Methods in Algebraic Geometry, 3rd Edition, Springer-Verlag, Berlin (1966).

[I_1] S. Iitaka, On D dimensions of algebraic varieties, J. Math. Soc. Japan 23 (1971), 356-373.

[I_2] S. Iitaka, On logarithmic Kodaira dimension of algebraic varieties, <u>Complex Analysis and Algebraic Geometry</u>, ed. by W.L. Baily, Jr. and T. Shioda, (1977), 175-189, Iwanami Shoten.

[I+S] V.A. Iskovskih and V.V. Sokurov, Biregular theory of Fano 3-folds, <u>Proceedings Algebraic Geometry Conference</u>, Copenhagen, 1978, edited by K. Lonsted, Lect. Notes in Math. 732 (1979), 171-182.

[Ka] Y. Kawamata, A generalization of Kodaira-Ramanujam's vanishing theorem, Math. Ann. 261 (1982), 43-46.

[Ka_2] Y. Kawamata, Elementary contractions of algebraic 3-folds, Ann. of Math 119 (1984) 95-110.

[Kl] S. Kleiman, Towards a numerical theory of ampleness, Ann. of Math. 84 (1966), 293-344.

[K+O] S. Kobayashi and T. Ochiai, Characterizations of complex projective spaces and hyperquadrics, J. Math Kyoto Univ. 13 (1972), 31-47.

[Ko] K. Kodaira, Pluricanonical systems on algebraic surfaces of general type, J. Math. Soc. Japan, 20 (1968), 170-192.

[L+So] J. Lipman and A.J. Sommese, On the contraction of projective spaces on singular varieties, preprint.

[Mo] S. Mori, Threefolds whose canonical bundles are not numerically effective, Ann. Math 116 (1982), 133-176.

[Mu] D. Mumford, The canonical rings of an algebraic surface, Appendix in Z_2 below.

[Nag] M. Nagata, On rational srufaces I, Mem. Coll. Sci. Kyoto (A) 32 (1960), 351-370.

[Nak] S. Nakano, On the inverse of monoidal transformation, Publ. R.I.M.S. Kyoto Univ. 6 (1971), 483-502.

[P+S] E. Picard and G. Simart, <u>Theories des Fonctions Algebriques de Deux Variables Independantes</u>, Chelsea Publ. Co., Bronx, New York (1971).

[Ra] C.P. Ramanujam, Remarks on the Kodaira Vanishing theorem, Jour. of the Indian Math. Soc. 36 (1972), 41-51; Supplement to the article "Remarks on the Kodaira vanishing theorem," J. Indian Math. Soc., 38 (1974), 121-124.

[Ro] L. Roth, <u>Algebraic Threefolds</u>, Springer-Verlag, Heidelberg, (1953).

[Sa_1] F. Sakai, Semi-stable curves on algebraic surfaces and logarithmic pluricanonical maps, Math. Ann. 254, 89-120 (1980).

[Sa_2] F. Sakai, D-dimensions of algebraic surfaces and numerically effective divisors, preprint.

[Sh+So] B. Shiffman and A.J. Sommese, Vanishing Theorems on complex manifolds, to appear.

[So_1] A.J. Sommese, On manifolds that cannot be ample divisors, Math. Ann. 221 (1976), 55-72.

[So_2] A.J. Sommese, Non-smoothable varieties, Com. Math. Helv. 54 (1979), 140-146.

[So_3] A.J. Sommese, Hyperplane sections of projective surfaces: I-the adjunction mapping, Duke Math. J. 46 (1979), 377-401.

[So_4] A.J. Sommese, On the minimality of hyperplane sections of projective threefolds, Journal für die reine und angewandte Mathematik, 329 (1981), 16-41.

[So_5] A.J. Sommese, Ample divisors on 3-folds, Algebraic Threefolds, Springer Lecture Notes in Math 947 (1982), 229-240.

[So_6] A.J. Sommese, On the birational theory of hyperplane sections of projective threefolds, unpublished 1981 manuscript.

[So_7] A.J. Sommese, Configurations of -2 rational curves on hyperplane sections of projective threefolds, Classification of Algebraic and Analytic Manifolds, ed. by K. Ueno, Progress in Mathematics 39 (1983) Birkhäuser Boston.

[VdV] A. Van De Ven, On the 2 connectedness of very ample divisors on a surface, Duke Math. J. 46 (1979), 403-407.

[Vi] E. Viehweg, Vanishing theorems J. reine angew. Math. 335 (1982), 1-8.

[Z_1] O. Zariski, Introduction to the Problem of Minimal Models in the Theory of Algebraic Surfaces, Pub. Math. Soc. of Japan 4 (1958).

[Z_2] O. Zariski, The theorem of Riemann-Roch for high multiples of an effective divisor of an algebraic surface, Ann. of Math. 76 (1962), 560-615.

Department of Mathematics
University of L'Aquila
L'Aquila, Italy

Department of Mathematics
University of Notre Dame
Notre Dame, Indiana 46556

On Meromorphic Equivalence Relations

Hans Grauert

Mathematisches Institut, Universität Göttingen,
Bunsenstr. 3-5, D-34oo Göttingen, West Germany

Introduction.

1. We denote by X a weakly normal (see § 2.3.) complex space with countable topology and by $R \subset X \times X$ an analytic set with the following two properties:

1) R contains the diagonal $D \subset X \times X$,

2) R is mapped by the reflexion $(x_1, x_2) \to (x_2, x_1)$: $X \times X \tilde{\to} X \times X$ onto itself.

Such an analytic set defines a fibration in X. The fibre X_x through $x \in X$ is defined as $p_1(R \cap (X \times x))$ where p_1 denotes the projection $X \times X \to X$ onto the first component (as p_2 will denote the projection onto the second). Here, X_x always is considered as a set, not as a complex subspace with a nilpotent structure.

Definition: R *is a normal complex equivalence relation if:*

1) R *is an equivalence relation in* X,

2) *the codimension of the fibres is constant everywhere, equal to* $c \geq 0$.

3) *the projections* $p_i : R \to X$ *are open.*

We shall prove in § 5 that under this assumption the quotient space X/R is a weakly normal complex space of pure dimension c .

2. But the main purpose of this paper is to prove something for meromorphic equivalence relations in normal complex spaces:

Definition: R *is a meromorphic equivalence relation in* X *of codimension* c *if:*

1) *there is a nowhere dense analytic set* $P \subset X$ *(polar set), such that* $R \cap (X \times P)$ *is nowhere dense in* R,

2) $R|X \setminus P = R \cap ((X \setminus P) \times (X \setminus P))$ *is a normal complex equivalence relation of codimension* c *in* $X \setminus P$.

We denote by $\pi : \tilde{R} \to R$ the normalization of R. We have the holomorphic map $\varphi : \tilde{R} \overset{\pi}{\to} R \overset{p_2}{\to} X$. The analytic set R has pure codimension c. The normal complex space X decomposes into connected components X_i of dimension n_i. For $x \in X_i$ the generic fibre $\varphi^{-1}(x)$ has codimension c+n. We look at the degeneration set $E := \{(x_1,x_2) \in \tilde{R} : \operatorname{codim}_{(x_1,x_2)} \varphi^{-1}(x_2) < c + n_i,\ x_i \in X_i\}$, which is a nowhere dense analytic set of $\tilde{R}$ (see [Re]). We put $\tilde{R}' = \tilde{R} \setminus E - \varphi^{-1}(P)$ and $\varphi' = \varphi|\tilde{R}'$. The set $\tilde{E} = \varphi^{-1}\,\varphi(E)$ is not analytic in $\tilde{R}$, in general. But it is a countable union of local nowhere dense analytic subsets like $\varphi(E)$ in X is.

If $(x_1,x_2) \in \tilde{R}$ is a point there are many holomorphic maps $\psi : D \to \tilde{R}$ of the unit disc $D \subset \mathbb{C}$ around $0 \in \mathbb{C}$ with:

1) $\psi(0) = (x_1,x_2)$,

2) $\psi^{-1}(\tilde{E})$ is countable,

3) $\psi(D \setminus \{0\}) \subset \tilde{R}'$.

We consider the (set theoretic) fibred product $\tilde{R} \times_X D \subset \tilde{R} \times D$ and take the union Z of all irreducible components which do not lie over a single point of D, completely. All fibres $Z_t := Z \cap (\tilde{R} \times \{t\})$ have dimension $n_j - c$ in all points over $X_j \times X_i$. If t is generic we have $Z_t = \varphi^{-1}(\psi(t))$. So we call the Z_t *fibres in* $\tilde{R}$. The set of the fibres in $\tilde{R}$ is a *fibration in* $\tilde{R}$. We denote it by $\mathfrak{M}_\varphi$. Some of the fibres of $\mathfrak{M}_\varphi$ may cross. All $S \in \mathfrak{M}_\varphi$ are contained in a $\varphi^{-1}(x)$.

3. The set of images $(p_1 \circ \pi)(S)$, $S \in \mathfrak{M}_\varphi$ gives a *fibration* $\mathfrak{M}$ of X in pure c-codimensional analytic sets. This is considered to be the fibration given by the meromorphic equivalence relation R. We shall construct a quotient space of X by R. The points of this will be the fibres $S \in \mathfrak{M}$. In general, probably, it is only a weakly normal complex space.

But, probably, the quotient space will not exist without further assumption. We require a condition which can be verified before constructing the quotient space.

Definition: R *is called regular if for every compact subset* $K \subset X$ *a relatively compact open subset* $B \subset\subset \tilde{R}$ *exists, such that for all* $x \in K$ *all irreducible components* S' *of fibres* $S \in \mathfrak{M}_\varphi$ *enter in* B, *if* S' *is completely contained in* $\varphi^{-1}(x) \cap E$.

It follows directly that any S to a regular meromorphic equivalence relation R has finitely many irreducible components in E, only.

In order to obtain the quotient space X/R we have to construct a proper modification $\tilde{\pi} : \tilde{X} \to X$ of X, such that thereafter $\mathfrak{M}$ becomes a normal complex equivalence relation in $\tilde{X}$. By this we prove in this paper:

Main Theorem: *If* R *is a regular meromorphic equivalence relation of codimension* c *in* X *there is a unique proper modification* $\tilde{\pi} : \tilde{X} \to X$ *together with an open holomorphic map* $q : \tilde{X} \to Q$ *of* $\tilde{X}$ *onto a c-dimensional weakly normal complex space* Q *such that:*

1) $\tilde{\pi}$ *maps the fibres* $\tilde{S}$ *in* $\tilde{X}$ *topologically holomorphically onto* $S = \tilde{\pi}(\tilde{S}) \in \mathfrak{M}$.

2) *The map* $\tilde{S} \to S$ *is a bijection* $Q \to \mathfrak{M}$.

We put $Q = X/R$ and call Q the *(generalized) quotient space of* X *by* R.

In § 1 we prove a crucial lemma, in § 2 we prove an important proposition and in § 3 the Main theorem, in § 4 we prove the existence of complex quotient spaces of normal complex spaces by an analytically closed Lie group action. Finally, § 5 brings the proof of the result on normal complex equivalence relations stated in the first paragraph of the introduction.

The main result is a solution of a problem on the existence of m-bases stated by K.Stein in [St]. Group quotients were considered in the proper case already in [Li] and [Fu] and the Crucial Lemma of § 1, however for the proper case only, is contained in Horonakas Flattening Theorem (see [Hi]).

We use the notations of [STCER]. So $\hat{R}$ always stands for the simple complex equivalence relation to a complex equivalence relation R and R_F, R_φ etc. denotes the complex equivalence relation to holomorphic maps F, φ, etc.

§ 1. The Crucial Lemma

1. We prove that $\tilde{R}_\varphi$ leads to a proper (in general only weakly normal) modification of $\tilde{R}$ and the second factor X of $X \times X$. Since this proposition is local with respect to the second factor we may assume that this has pure dimension n and is an analytic subset of a domain $U \subset \mathbb{C}^m$. So we have the holomorphic map $F : \tilde{R} \xrightarrow{\varphi} X \to \mathbb{C}^m$. The codimension of the generic fibre is $n + c$ and the degeneration set of F is E.

We just may consider the following general situation: Y is a normal complex space, $F : Y \to \mathbb{C}^m$ is a holomorphic map of Y onto a pure n-dimensional normal analytic subset $A \subset U \subset \mathbb{C}^m$, the degeneration set $E := \{y \in Y : \operatorname{codim}_y F^{-1} F(y) < n\}$ is nowhere dense in Y. - We denote by $\mathfrak{M}_F$ the fibration of Y into pure n-codimensional analytic subsets given by R_F (like $\tilde{R}_\varphi$ gives $\mathfrak{M}_\varphi$). We prove:

Lemma (n): *If* $K \subset Y$ *is compact there is a finite family of commutative diagrams:*

$$\begin{array}{ccc} Y_\lambda & \xrightarrow{\pi_\lambda} & Y \\ {\scriptstyle q_\lambda}\downarrow & & \downarrow{\scriptstyle F} \\ Q_\lambda & \xrightarrow{\underline{F}_\lambda} & \mathbb{C}^m \end{array}$$

for $\lambda = 1,\dots,l$ *with the following properties:*

1) Y_λ, Q_λ *are normal complex spaces;* π_λ, q_λ, $\underline{F}_\lambda$ *holomorphic maps. The dimension of* Q_λ *is* n, *the generic fibre of* $\underline{F}_\lambda$ *is* 0*-dimensional.*

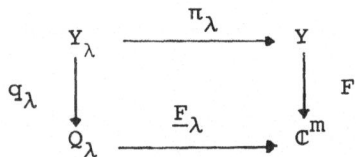

2) *All fibres* $Y_{\lambda t} := q_\lambda^{-1}(t) \subset Y_\lambda$ *with* $t \in Q_\lambda$ *have pure codimension* n . *They are mapped by* π_λ *finite open onto an open subset of a fibre* $S \in \mathfrak{M}_F$, *the subset containing* $S \cap K$.

3) *For* $S \in \mathfrak{M}_F$ *the intersection* $S \cap K$ *is in the image of a* $Y_{\lambda t}$.

4) *There are relatively compact open subsets* $Q'_\lambda \subset\subset Q_\lambda$ *such that after restriction to* Q'_λ *the property* 3) *still holds true.*

5) *If* $Q'_\lambda \subset\subset Q_\lambda$ *the map* $\pi_\lambda : q_\lambda^{-1}(\bar{Q}'_\lambda) \cap \pi_\lambda^{-1}(K) \to K$ *is proper.*

We shall deduce a proposition from Lemma (n) in § 2:

<u>Proposition (n)</u>: *Assume that* X, Y, Z *are normal complex spaces, that* X *is of pure dimension* n *and* $\varphi : Y \to X$ *is a surjective holomorphic map whose generic fibre has codimension* n, *such that* $\mathfrak{M}_\varphi$ *is regular, and that* $p : Y \to Z$ *is a finite holomorphic map. Moreover, we assume a holomorphic map* $\psi : Z \to X$ *with* $\varphi = \psi \circ p$. *Then there are unique proper modifications* $\hat{\pi} : \tilde{Y} \to Y$, $\tilde{\pi} : \tilde{X} \to X$, *which are biholomorphic outside* $\hat{\pi}^{-1}(E)$, $\tilde{\pi}^{-1}\varphi(E)$, *together with a holomorphic mapt* $\tilde{\varphi} : \tilde{Y} \to \tilde{X}$ *such that*

1) *the diagram*

$$\begin{array}{ccc} \tilde{Y} & \xrightarrow{\hat{\pi}} & Y \\ {\scriptstyle\tilde{\varphi}}\downarrow & & \downarrow{\scriptstyle\varphi} \\ \tilde{X} & \xrightarrow{\tilde{\pi}} & X \end{array}$$

is commutative,

2) *the fibres of* $\tilde{\varphi}$ *are of pure codimension* n,

3) *each fibre* $\tilde{S} = \tilde{\varphi}^{-1}(x) \in \mathfrak{M}_{\tilde{\varphi}}$ *is mapped by* $\hat{\pi}$ *topologically onto a fibre* $S \in \mathfrak{M}_{\varphi}$ *such that* $p \circ \hat{\pi}$ *is a bijection* $\mathfrak{M}_{\tilde{\varphi}} \to p\,\mathfrak{M}_{\varphi}$, *where* $p\,\mathfrak{M}_{\varphi}$ *is the set* $\{p(S)): S \in \mathfrak{M}_{\varphi}\}$ *(attention: the fibres are equipped with multiplicity of their irreducible components: see § 2.1).*

2. We prove the Lemmas (n) by an induction on n. In the case $n = 1$ the degeneration set E is empty. We simply put $Y_\lambda \to Q_\lambda$ equal $Y \to A$ with $A = F(Y) \subset U \subset \mathbb{C}^m$. $\mathfrak{M}_F$ coincides with the set of fibres of F. So nothing has to be proved.

We assume $n > 1$ and that Lemma (n-1) holds true. We take a linear projection $\mathbb{C}^m \to \mathbb{C}^n$ such that $A \to \mathbb{C}^m \to \mathbb{C}^n$ is discrete and compose this with a linear projection $\mathbb{C}^n \to \mathbb{C}^{n-1}$. The fibres in A of the composition $p : \mathbb{C}^m \to \mathbb{C}^{n-1}$ are of pure dimension 1. Hence, the image $V = p(A) \subset \mathbb{C}^{n-1}$ is open. We put $F^* = p \circ F$. The degeneration set E^* of F^* is contained in E and hence nowhere dense in Y. This means we are with F^* in the case of Lemma (n-1). We obtain a finite family of commutative diagrams:

$$\begin{array}{ccc} Y^*_\lambda & \xrightarrow{\ \pi^*_\lambda\ } & Y \\ {\scriptstyle q^*_\lambda}\downarrow & & \downarrow{\scriptstyle F^*} \\ Q^*_\lambda & \xrightarrow{\ \underline{F}^*_\lambda\ } & \mathbb{C}^{n-1} \end{array}$$

with $\lambda = 1,\dots,l^*$ and the properties 1) - 5). We may cut Q^*_λ into smaller pieces. Hence, we may assume that all Q^*_λ are analytic subsets of domains $U \subset \mathbb{C}^m$. We put $F^*_\lambda : Y^*_\lambda \xrightarrow{q^*_\lambda} Q^*_\lambda \xrightarrow{id} U$ and $F^+_\lambda = (f_\lambda, F^*_\lambda)$, where f_λ is the m-tupel of holomorphic functions defining the holomorphic map $F \circ \pi^*_\lambda$. The degeneration set E^*_λ of F^+_λ is the union of those irreducible components of the fibres of F^*_λ where f_λ is constant. It is contained in the inverse image of E and nowhere dense in Y^*_λ.

3. Now, we take an arbitrary point $t \in Q^*_\lambda$ and denote by $Y^*_{\lambda t\varkappa}$ the irreducible components of the fibre $Y^*_{\lambda t} = q^{*-1}_\lambda(t)$ and by $\mathring{Y}^*_{\lambda t\varkappa}$ the difference of $Y^*_{\lambda t\varkappa}$ and the other irreducible components. $\mathring{Y}^*_{\lambda t\varkappa}$ is an open subset of $Y^*_{\lambda t\varkappa}$. We take fixed points $y_\varkappa \in \mathring{Y}^*_{\lambda t\varkappa}$. We restrict ourselves to those irreducible components of $Y_{\lambda t}$, whose π^*_λ-images enter in K. This is a finite number.

We always represent a neighbourhood $W_\varkappa = W_\varkappa(y_\varkappa) \subset\subset Y^*_\lambda$ as an analytic covering of $\underline{W} \times V$ with a connected open $\underline{W} = \underline{W}_\varkappa \subset \mathbb{C}^{d+1}$, $d = d_\varkappa = \dim_{y_\varkappa} Y^*_\lambda - n+1$, and a fixed open $V(t) \subset\subset Q^*_\lambda$, such that $y_\varkappa$ is the only point over its image. We take a positive integer b such that the number of sheets $b_\varkappa$ of $W_\varkappa$ divides b, always. If we consider $W_\varkappa$ with multiplicity $b/b_\varkappa$, the coverings $W_\varkappa$ have b sheets always. If $z \in \underline{W}$, we denote by W_z the normalization of $W|z{\times}V$ (we have to observe multiplicity!) and obtain a b-sheeted normal analytic covering of V.

We denote by $W^!$, $W^!_z$ the symmetric power of W, W_z, which is a normal analytic covering of $\underline{W} \times V$ resp. V with $b!$ sheets. There are b canonical projections $W^! \to W$, $W^!_z \to W_z$.

We put $(z_\varkappa, t)$ for the image of $y_\varkappa$ in $\underline{W}_\varkappa \times V$, take a component f of f_λ and lift $f|q^{*-1}_\lambda(V)$ and $f|W_{z_\varkappa}$ to $Y^*_\lambda \times_V W_{z_\varkappa}$ and take the difference g. The normal complex space $Y^*_\lambda \times_V W_{z_\varkappa}$ is a b-sheeted analytic covering of $q^{*-1}_\lambda(V)$. If $y_\varkappa \in E^*_\lambda$ the function g vanishes on the fibre of $W_\varkappa \times_V W_{z_\varkappa}$ over t, identically. We lift g to functions $g_1,\ldots,g_b$ on $W_\varkappa^! \times_V W_{z_\varkappa}$ using the canonical projections $W_\varkappa^! \to W_\varkappa$. For $z \in \underline{W}_\varkappa$ each $W_z^! \times W_{z_\varkappa}$ is a normal analytic covering of V with $b \cdot (b!)$ sheets. We denote by $\mathcal{I}(z)$ the coherent ideal sheaf on $W_z^! \times_V W_{z_\varkappa}$ spanned by the inverse images of $g_1,\ldots,g_b$ under $W_z^! \times_V W_{z_\varkappa} \to W^! \times_V W_{z_\varkappa}$, which we have to take for all components f of $\mathfrak{f}_\lambda$. The $b \cdot (b!)$-symmetric functions of the direct image of $\mathcal{I}(z)$ on V span a coherent ideal sheaf $I(\varkappa,z)$ on V. After having replaced V by a relatively compact open neighbourhood of t we take several points $z_1,\ldots,z_r \in \underline{W}_\varkappa$ such that $I_\varkappa :=$ $I(\varkappa,z_1) + \ldots + I(\varkappa,z_r)$ is maximal on V.

If $y_\varkappa \notin E_\lambda^*$ we get $I_\varkappa = 1$ (if V is small enough) and nothing will happen later on. Therefore, we omit these irreducible components in our consideration.

<u>4.</u> We do the monoidal transformations of V by the ideal sheaves $I_1,\ldots,I_k$ defined to $y_\varkappa$ running through the irreducible components of $Y_{\lambda t}^*$ completely contained in E_λ^*. We obtain a proper modification $\tilde{V}$ of V. We lift everything to $\tilde{V}$ (to normal spaces!). Now, locally $I_\varkappa$ is spanned by one local cross-section! We denote by $\underline{g}_{\varkappa\nu}$ the b-th elementary symmetric function of g on Y_λ^* (= lifting of the old Y_λ^* to V), where g is the holomorphic function on $Y_\lambda^* \times_V W_{z_\varkappa}$ defined by the ν-th component f of $\mathfrak{f}_\lambda$. Since the ideals $I_\varkappa$ are locally principal in each point of $\tilde{V}$, the quotient $f_{\varkappa\nu} := \underline{g}_{\varkappa\nu}^{b!}/I_\varkappa$ can be locally along the fibres of $Y_\lambda^* \to \tilde{V}$ considered as a meromorphic function. If t stands for any

point out of $\tilde{V}$, which is mapped onto the old t, these functions are holomorphic in a neighbourhood of $\overset{\circ}{Y}{}^*_{\lambda t\varkappa}$, which is defined as the inverse image of the old object. But now $\overset{\circ}{Y}{}^*_{\lambda t\varkappa}$ will not be irreducible any longer in general. It can be seen rather easily that not all the $f_{\varkappa\nu}$ vanish identically on $\overset{\circ}{Y}{}^*_{\lambda t\varkappa}$.

5. Over a neighbourhood G of any point of $\tilde{V}$ we pass over to the graph of $(f_{\varkappa\nu} : \forall\ \varkappa,\nu)$ and obtain over G a proper modification $\tilde{Y}^*_\lambda$ of $Y^*_\lambda | G$. We put $\tilde{W}_\varkappa$ for the inverse image of $W_\varkappa$ in $\tilde{Y}^*_\lambda$. The generic fibre of $\tilde{Y}^*_\lambda \to G$ has codimension $n-1$, everywhere. However, there may be a degeneration set (over $q^*_\lambda(E^*_\lambda)$). By Proposition $(n-1)$ with $Z = \tilde{W}_\varkappa$ there are proper modifications $\tilde{W}^*_\varkappa \to \tilde{W}_\varkappa$, $\tilde{G}_\varkappa \to G$ and a holomorphic map $\tilde{\varphi} : \tilde{W}^*_\varkappa \to \tilde{G}_\varkappa$ such that the diagram

$$\begin{array}{ccc} \tilde{W}^*_\varkappa & \longrightarrow & \tilde{W}_\varkappa \\ \big\downarrow & & \big\downarrow \\ \tilde{G}_\varkappa & \longrightarrow & G \end{array}$$

commutes. All the fibres of $\tilde{W}^*_\varkappa \to \tilde{G}_\varkappa$ are of pure codimension $n-1$.

We apply this to all $\varkappa$ and obtain a proper modification $\hat{V}$ of G. We lift everything to $\hat{V}$, starting from $\tilde{Y}^*_\lambda$. The result is a map $\hat{q} : Y^*_\lambda \to \hat{V}$. The fibres of $\hat{q}$ have codimension $n-1$ everywhere now, and the meromorphic functions $f_{\varkappa\nu}$ are holomorphic maps $Y^*_\lambda \to \mathbb{P}_1$. Not all of these are constant on $Y^*_{\lambda t\varkappa}$ (here t any point over the old t).

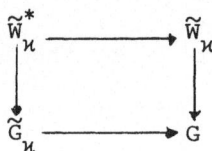

<u>6.</u> We wish to arrive at the case where not all $f_{\varkappa\nu}$ are constant on *each irreducible component* of $Y^*_{\lambda t\varkappa}$ So we have to decompose $Y^*_{\lambda t\varkappa}$ into irreducible components and to take a new enumeration of the irreducible components of $Y^*_{\lambda t}$ and do all of the construction again. But now the maximal number of sheets of the $W_\varkappa$ (without multiplicity) has decreased. So by an induction on this number we come to the desired case.

The generic fibre of the map $(f_{\varkappa\nu},\hat{q})$ has codimension n, now. The codimension is also n in each point of $Y^*_{\lambda t}$ over K. The degeneration set is closed. By making $\hat{V}(t)$ and Y^*_λ smaller we obtain that the codimension is n everywhere and, moreover, that for $Y^*_\lambda \to \hat{V}$ the properties 2) and 5) are satisfied (for the case of Lemma (n-1)).

<u>7.</u> The fibration $\mathfrak{M}_F$ of Y can be lifted to a fibration $\mathfrak{M}_\lambda$ of Y^*_λ, since all maps are constant on the fibres of $\mathfrak{M}_F$. However, we had to pass over to normalization several times. This may lead to analytic coverings. The holomorphic map $\pi_\lambda : Y_\lambda = Y^*_\lambda \to Y$ induces an analytic covering map of an open subset of each $\tilde{S} \in \mathfrak{M}_\lambda$ onto an open subset of a fibre $S \in \mathfrak{M}_F$ containing $S \cap K$.

$\mathfrak{M}_\lambda$ is finer than the fibration defined by $R_{(f_{\varkappa\nu},\hat{q})}$ and coarser than the simple equivalence relation to this fibration. Over a dense set of Y_λ the fibres of $\mathfrak{M}_\lambda$ coincide with those of $R_{F\circ\pi_\lambda}$ (by construction). Thus, over the inverse image of $Y \setminus E$ the fibres of $R_{F\circ\pi_\lambda}$ and $\mathfrak{M}_\lambda$ are the same. By taking the closure of $R_{F\circ\pi_\lambda} | Y_\lambda \setminus \pi_\lambda^{-1}(E)$ in $Y_\lambda \times Y_\lambda$ we obtain a complex equivalence relation R_λ in Y_λ whose fibre set is $\mathfrak{M}_\lambda$. The projections $R_\lambda \rightrightarrows Y_\lambda$ are open (since $A \subset U \subset \mathbb{C}^m$ was assumed to be locally irreducible). Hence, R_λ is a normal complex equivalence relation in Y_λ. We denote the normal quotient space of Y_λ by Ω_λ. It has pure dimension n . Since Y_λ is normal, by § 5. We have the quotient map $q_\lambda : Y_\lambda \to Q_\lambda$. We obtain $\underline{F}_\lambda$ since $F \circ \pi_\lambda$ is constant on the fibres of q_λ, which are those of π_λ.

8. We may replace the Q_λ^* we started with by a relatively compact open subset. Thus, we can cover Q_λ^* and the fibration over Q_λ^* and K by the images of a finite number of $Y_\lambda^* \to \hat{V}$. We even may shrink the $\hat{V}$. So we arrive at a finite family

$$\begin{array}{ccc} Y_\lambda & \xrightarrow{\pi_\lambda} & Y \\ \downarrow & & \downarrow \\ Q_\lambda & \longrightarrow & \mathbb{C}^m . \end{array}$$

The conditions 2), 3), 5) are trivial from construction. Since we may even shrink the $\hat{V}$, also 4) is satisfied. Thus, the Lemma (n) has been proved.

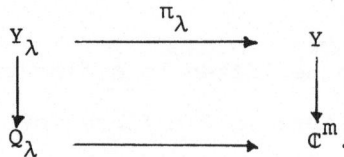

§ 2. Proof of the Proposition (n)

1. We prove the Proposition (n) from § 1.1. The map $\varphi : Y \to X$ is surjective and holomorphic, the normal complex space X is pure n-dimensional and the generic fibre of φ has codimension n. The fibration $\mathfrak{M}_\varphi$ is regular. We take a relatively compact open subset $G \subset\subset X$ and to the compact subset $\bar{G}$ a relatively compact open subset $B \subset Y$, $B = p^{-1}(\underline{B})$, $\underline{B} \subset\subset Z$, such that for $x \in \bar{G}$ all irreducible components S' of fibres $S \in \mathfrak{M}_\varphi$ enter in B, if S' is completely contained in $\varphi^{-1}(x) \cap E$. We apply Lemma (n) to $K = \bar{B}$. Since X can be locally embedded in an open subset $U \subset \mathbb{C}^m$ we obtain a finite family of commutative diagrams:

$$\begin{array}{ccc} Y_\lambda & \xrightarrow{\pi_\lambda} & Y \\ {\scriptstyle q_\lambda}\downarrow & & \downarrow{\scriptstyle \varphi} \\ Q_\lambda & \xrightarrow{\underline{F}_\lambda} & X \end{array}$$

for $\lambda = 1,\ldots,l$ with the properties 1) - 5). Any fibre $p(S)$, $S \in \mathfrak{M}_\varphi$ may be the image of fibres $Y_{\lambda_1 t_1}$, $Y_{\lambda_2 t_2}$ with $\lambda_1 \neq \lambda_2$ or $t_1 \neq t_2$, i.e. such that $p \circ \pi_{\lambda_1}(Y_{\lambda_1 t_1} \cap K) = p \circ \pi_{\lambda_2}(Y_{\lambda_2 t_2} \cap K)$. In such a case we call points $y_1 \in Y_{\lambda_1 t_1}$, $y_2 \in Y_{\lambda_2 t_2}$ equivalent if they are over the same point of Y. We also consider the points t_1, t_2 to be equivalent. By this we obtain an equivalence relation $\tilde{R}$ in $\hat{Y} := \dot{\bigcup}_{\lambda=1,\ldots,l} Y_\lambda$ and an equivalence relation R in $\hat{Q} = \dot{\bigcup} Q_\lambda$. We denote by $\hat{Y}^*$ the part of $\hat{Y}$ lying over $B^* = B \cap \varphi^{-1}(G)$, by $\hat{Q}^*$

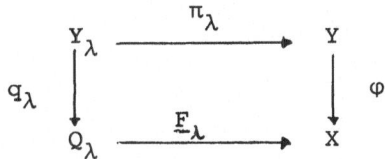

the part of $\hat{Q}$ over G and prove:

Proposition: $\tilde{R}^* = \tilde{R}|\hat{Y}^*$, $R^* = R|\hat{Q}^*$ *are complex, holomorphic and semiproper.*

Proof: If $S \in \mathfrak{M}_\varphi$ is a fibre over G, we take points $z_1,\ldots$ $\ldots, z_r \in p(S) \cap \underline{B}^*$, $\underline{B}^* := \underline{B} \cap \varphi^{-1}(G)$ on the p-image of the various images sets of those irreducible components of S which are completely contained in E. We take pure n-dimensional local complex subspaces $A_i \subset Z$, $i=1,\ldots,r$ with the following properties:

1) $A_i \subset U_i \subset\subset Z$ are n-dimensional complete intersections (relatively to U),

2) $\bar{A}_i \cap p(S) = \{z_i\}$.

A neighbourhood $\underline{V}(S) \subset \mathfrak{M}_\varphi$ is defined as the set of all $S_1 \in \mathfrak{M}_\varphi$ over G such that $S_1 \cap B^*$ can be connected with $S \cap B^*$ by a chain of holomorphic 1-parameter families $S_\varkappa(t) \cap B^*$, $|t| < 1$ over G with $p\, S_\varkappa(t) \cap \partial A_i \neq \emptyset$ and $\varkappa = 1,\ldots,k$.

We denote by $\hat{V}$ the inverse image of V in $\hat{Q} := \bigcup_{\lambda=1}^{1} Q_\lambda$. Clearly, $\hat{V}$ is open and its inverse image $\hat{U} \subset \hat{Y}$ is open again. By projection we obtain subsets $U \subset \hat{Y}^*/\tilde{R}^*$, $V \subset \hat{Q}^*/R^*$ which are open in the quotient topology.

There is a multiplicity for the irreducible components of the fibres $S \in \mathfrak{M}_\varphi$ coming from the generic fibres of a neighbourhood of S. If S is generic this multiplicity always is one. Otherwise it can be a higher integer. It might happen that a settheoretic S has various multiplicities coming from different neighbourhoods. Then these have to be considered as different fibres. Thus, a fibre is the set S equipped with multiplicity of the irreducible components. - The multiplicity carries over to $p(S)$. We take this multiplicity! Then each $p\,S_1$, $S_1 \in \underline{V}$ has the same intersection number with A_i.

We use the multiplicity to define the multiplicity for the intersection points $S_1 \cap A_i$, $S_1 \in \underline{V}$. If g is a holomorphic function on A_i, we take the elementary symmetric polynomials of the values of g in $S_1 \cap A_i$. These all lead to holomorphic functions on $\hat{V}$. We may assume that an embedding of A_i in a complex number space is given by finitely many holomorphic functions. We take these for g.

We define $\hat{V}$ so small that it is over an open subset of X, which is isomorphic to an analytic subset of a domain in the complex number space. We also take the holomorphic functions on $\hat{V}$ coming from the finite number of coordinate functions on this open subset.

Altogether, we obtain a finite set H of holomorphic functions on $\hat{V}$. The differences $f(t) - f(\tau)$ for $f \in H$, $(t,\tau) \in \hat{V} \times \hat{V}$ define a coherent ideal sheaf I on $\hat{V} \times \hat{V}$. We wish to have the I maximal.

We proceed as follows: Firstly, we take an infinite set of z_i on $p(S) \cap \underline{B}^* \cap p\,E$, which is somewhere dense on each p-image of any irreducible component of S, which is completely contained in E. We take the $A_i \subset Z$ such that $\underline{V}$ is still a neighbourhood of S. This is possible! We just have to use the parametrization of fibres given by Lemma (n) and that the inverse image of the "point" S is closed in $\hat{Q}$ and that we can make $\hat{Q}$ smaller. The set H is infinite now. But the ideal sheaf I is still coherent. The functions of H separate the different fibres of $\underline{V}$: We have to use that for $t \to t_o$ each irreducible component of $q_\lambda^{-1}(t)$ converges against some irreducible components of $q_\lambda(t_o)$: this is a well-known statement.

After having made $\hat{Q}$ and $\underline{V}$ smaller we find finitely many points among our infinite set $\{z_i\}$ such that the ideal I is spanned by the functions to these z_i already. Now H has become finite and the functions H separate the fibres out of $\underline{V}$.

We have $R^*|\hat{V} = R_H$ and $\tilde{R}^*|\hat{U} = R_{(H,\pi_\lambda)}$. So R^*, $\tilde{R}^*$ are holomorphic equivalence relations. - We may replace Q_λ by a relatively compact open subset $Q'_\lambda \subset\subset Q_\lambda$. Since $\bar{Q}'_\lambda$ is compact and the maps $\pi_\lambda : q_\lambda^{-1}(\bar{Q}'_\lambda) \cap \pi_\lambda^{-1}(B^*) \to B^*$ are proper, it follows immediately that R^* and $\tilde{R}^*$ are semiproper. So the Proposition is proved.

2. Hence $\tilde{Y}^* = \hat{Y}^*/\tilde{R}^*$, $\tilde{X}^* = \hat{Q}^*/R^*$ are complex spaces. We have holomorphic maps $\tilde{\varphi}^* : \tilde{Y}^* \to \tilde{X}^*$, $\hat{\pi}^* : \hat{Y}^* \to B^*$, $\tilde{\pi}^* : \tilde{X}^* \to G$ such that the diagram

$$\begin{array}{ccc} \tilde{Y}^* & \xrightarrow{\hat{\pi}^*} & B^* \\ \tilde{\varphi}^* \downarrow & & \downarrow \varphi \\ \tilde{X}^* & \xrightarrow{\tilde{\pi}^*} & G \end{array}$$

is commutative. All fibres of $\tilde{\varphi}^*$ are of pure codimension n. They are mapped by $\hat{\pi}^*$ finite open onto an $S \cap B^*$, $S \in \mathfrak{M}_\varphi$. By $\tilde{R}^*$ this map is bijective. Hence, it is topological. The composition $p \circ \hat{\pi}^*$ gives a bijection $\mathfrak{M}_{\tilde{\varphi}^*} \to p(\mathfrak{M}_{\varphi} \cap B^*)$. The maps $\hat{\pi}^*$, $\tilde{\pi}^*$ are proper. Since the inverse image of a generic point consists of one point only, they are proper modifications.

3. $\tilde{X}^*$ is defined as a set independently of the choice of B. Since every irreducible component S' of a fibre $S \in \mathfrak{M}_{\varphi}$ over G with $S' \subset E$ enters in B^* the set $\tilde{X}^*$ is just the set of pS, $S \in \mathfrak{M}_{\varphi}$ over G . But already the topology of $\tilde{X}^*$ might depend on the choice of B. By making B larger it might become finer. We would obtain a proper modification $\tilde{X}^{**} \to G$ instead of $\tilde{X}^* \to G$, but also a bijective holomorphic map $\delta : \tilde{X}^{**} \to \tilde{X}^*$ such that

$$\begin{array}{ccc} \tilde{X}^{**} & \searrow & \\ \delta \downarrow & & G \\ \tilde{X}^{*} & \nearrow & \end{array} \qquad \text{commutes.}$$

In general, $\tilde{X}^*$, $\tilde{X}^{**}$ will not be normal complex spaces. But they are *quasi normal*: any local continuous complex function which is holomorphic outside a nowhere dense analytic set, is holomorphic.

When we pass over to the normalization of $\tilde{X}^{**}$, $\tilde{X}^*$ the map δ is still bijective and holomorphic: This follows from the modification properties. By a well-known theorem (see [CAS]) it is biholomorphic then. This implies that already the old δ is biholomorphic. Thus, the proper modification $\tilde{X}^* \to G$ depends only on G .

The same holds true for $\tilde{Y}^* \to B^*$. If B^{**} is larger we have a commutative diagram

$$\begin{array}{ccc} \tilde{Y}^* & \longrightarrow & B^* \\ {\scriptstyle \hat{\delta}}\downarrow & & \downarrow{\scriptstyle \mathrm{id}} \\ \tilde{Y}^{**} & \longrightarrow & B^{**} \end{array}$$

where $\hat{\delta}$ is a biholomorphic map of $\tilde{Y}^*$ onto an open subset of $\tilde{Y}^{**}$.

4. Now we exhaust X by a sequence of relatively compact open subsets $G_\nu \subset\subset G_{\nu+1} \subset\subset X$ and Y by a similar sequence $B_\nu \subset\subset B_{\nu+1} \subset\subset Y$. We construct the commutative diagrams

$$\begin{array}{ccc} \tilde{Y}^*_\nu & \longrightarrow & B^*_\nu \\ \downarrow & & \downarrow \\ \tilde{X}^*_\nu & \longrightarrow & G^*_\nu \end{array}$$

and get isomorphisms $\delta_\nu : \tilde{X}^*_\nu \xrightarrow{\sim} \tilde{X}^*_{\nu+1} | G_\nu$, $\hat{\delta}_\nu : \tilde{Y}^*_\nu \xrightarrow{\sim} \tilde{Y}^*_{\nu+1} | B^*_\nu$. We glue together and obtain proper modifications $\hat{\pi} : \tilde{Y} \longrightarrow Y$, $\tilde{\pi} : \tilde{X} \longrightarrow X$ such that the diagram

$$\begin{array}{ccc} \tilde{Y} & \longrightarrow & Y \\ \downarrow & & \downarrow \\ \tilde{X} & \longrightarrow & X \end{array}$$

is commutative. The properties 2) and 3) of Proposition (n) are satisfied. Its proof is completed!

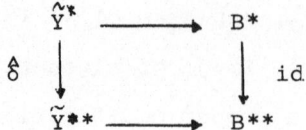

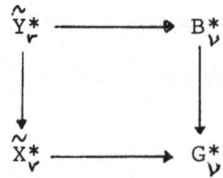

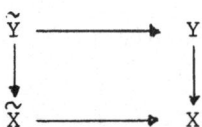

§ 3. The Proof of the Main Theorem

1. Assume now that X is a normal complex space and that R is a regular meromorphic equivalence relation in X of codimension c. We denote by $P \subset X$ a polar set. We take the normalization $\pi : \tilde{R} \to R$ and put $\varphi = p_2 \circ \pi : \tilde{R} \to X$. The set E is the degeneration set of φ. The space X decomposes into connected components X_i of dimension n_i. The codimension of the generic fibre of φ over X_i is $c + n_i$.

We apply Proposition $(c + n_i)$ with $Y \to Z$ equal $\pi : \tilde{R} \to X \times X$ and ψ equal $p_2 : X \times X \to X$. We obtain a commutative diagram

$$\begin{array}{ccc} \tilde{Y} & \xrightarrow{\hat{\pi}} & \tilde{R} \\ {\scriptstyle\hat{\varphi}}\downarrow & & \downarrow{\scriptstyle\varphi} \\ \hat{X} & \xrightarrow{\tilde{\pi}} & X \end{array}$$

where $\hat{X}$ is a proper modification of X. We take the inverse image of R in $\hat{X} \times \hat{X}$ under the map $\tilde{\pi} \times \tilde{\pi}$ and omit all irreducible components which are completely in $\hat{P} \times \hat{X}$ or in $\hat{X} \times \hat{P}$ with $\hat{P} = \tilde{\pi}^{-1}(P^*)$, $P^* = P \cup \varphi(E)$. We note: If a point of a fibre $S \in \mathfrak{M}_\varphi$ is contained in E then a full irreducible component through this point is contained in E. Since R is regular it follows that the *projection* $\varphi : E \to X$ *is semiproper*. So $\varphi(E) \subset X$ and $\hat{P} \subset \hat{X}$ are nowhere dense analytic sets.

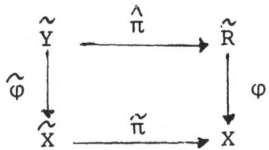

Thus, we obtain an analytic set $R^o \subset \hat{X} \times \hat{X}$, which contains the diagonal and is invariant under reflexion. The blowing-up took place in $\varphi(E)$ only. So we have $\hat{X} \setminus \hat{P} = X \setminus P^*$. Hence, the restriction $R^o | \hat{X} \setminus \hat{P}$ is a normal complex equivalence relation. R^O is simply the closure of $R \cap (X \setminus P^*) \times (X \setminus P^*))$ in $\tilde{X} \times \tilde{X}$.

By a generic fibre $S \in \mathfrak{M} := p_1 \mathfrak{M}_\varphi$ we understand a fibre X_x, $x \in X \setminus P^*$ such that $X_x \setminus P^*$ is dense in X_x. Clearly, each generic fibre has pure codimension c in X. To every point $x \in \hat{X}$ the fibre $S = p_1 \pi \hat{\pi} \tilde{\varphi}^{-1}(x)$ in X is attached. We call a map into $\tilde{X}$ *constant to fibres* if the composition with $x \to S$ is constant. For a generic $S \in \mathfrak{M}$ the map $\tilde{\pi}^{-1}$ is constant to fibres in $S \setminus P^*$. Since $\tilde{X} = \mathfrak{M}_\varphi$, it has a unique continuation to S which is also constant to fibres. We denote this by $\tilde{\pi}^{-1} : S \to \tilde{X}$.

The fibres of the projection $P^* \times X \to X$ have codimension $\leq n_i$ in P^*. We denote by $A^\cdot$ the set of points, where this codimension is $< n_i$ (degeneration set). The image $\underline{A} \subset X$ is nowhere dense. All fibres X_x, $x \in X \setminus P^* \setminus \underline{A}$ are generic, all other fibres $S \in \mathfrak{M}$ through an $x \in X \setminus P^*$ are approximated by such fibres. Thus, we also have a $\tilde{\pi}^{-1} : S \to \tilde{X}$ which is constant to fibres. Now, every $S \in \mathfrak{M}$ can be approximated by S_1 through an $x \in X \setminus P^*$. So we have a fibre constant $\tilde{\pi}^{-1} : S \to \tilde{X}$ always: We put $\underline{R} = (p_1 \pi \hat{\pi}, \tilde{\varphi})(\hat{Y}) \subset X \times \tilde{X}$. Then $\underline{R}_x \subset X$ is the fibre over x and $\underline{R}_{\tilde{\pi}^{-1}(x)}$ is constant on each $S \in \mathfrak{M}$.

Hence $\tilde{\pi}^{-1} : S \twoheadrightarrow \hat{S} \subset \tilde{X}$ is a topological map. Its inverse $\tilde{\pi}|\tilde{S} : \tilde{S} \to S$ is a finite open map. We obtain a fibration $\tilde{\mathfrak{M}}$ in $\tilde{X}$ into pure c-codimensional analytic sets $\tilde{S}$. By the construction of $\hat{X}$ two different $\tilde{S} \in \tilde{\mathfrak{M}}$ never intersect. Each point of $\hat{X}$ is contained in an $\tilde{S}$. So $\tilde{\mathfrak{M}}$ is an equivalence relation in $\tilde{X}$.

We wish to prove that $\tilde{\mathfrak{M}}$ is given by R^o. Then we also know that R^o is a complex equivalence relation in $\hat{X}$. If $x \in \tilde{X}$ we denote by $\tilde{S}_x \in \tilde{\mathfrak{M}}$ the fibre through x. For $x \to x_o$ the fibre $\tilde{S}_x$ always converges against S_{x_o}. We have $R^o | \tilde{X} \setminus \tilde{P} = \tilde{\mathfrak{M}} | \tilde{X} \setminus \tilde{P}$. We put $\mathfrak{M}' = p_1 \pi^o \mathfrak{M}_{\varphi^o}$, where $\mathfrak{M}_{\varphi^o}$ is the fibration in the normalization $\pi^o : \tilde{R}^o \to R^o$ given by the projection onto $\tilde{X}$. If $x_o \in \tilde{X}$ there is a sequence of generic points $x_\nu \in \tilde{X} \setminus \tilde{P}$ converging against x_o. Then we have for the fibres $S'_{x_\nu} \in \mathfrak{M}'$ the equation $S'_{x_\nu} = S_{x_\nu}$ and so the S'_{x_ν} always converge against S_{x_o}. Moreover, $R^o \cap (\tilde{P} \times \tilde{X}) \cup (\tilde{X} \times \tilde{P})$ is nowhere dense in R^o. This all implies that the degeneration set of $p_2 : R^o \to \tilde{X}$ has to be empty and we get $S_{x_o} = S'_{x_o}$ for all $x_o \in \tilde{X}$. That means $\tilde{\mathfrak{M}} = \mathfrak{M}' = R^o$.

Since the fibres are of pure codimension c and S'_x converges for $x \to x_o$ always against S'_{x_o}, $p_2 : R^o \to \tilde{X}$ has to be open. So R^o is normal and $Q = \tilde{X}/R^o$ a weakly normal complex space. We have to use § 5.6. In our case Q is a usual quotient and the fibres over Q are the fibres of $\tilde{\mathfrak{M}}$. Each fibre in $\tilde{X}$ is topologically mapped onto a fibre in X. The map is a bijection of fibres. That completes the proof of the Main Theorem.

§ 4. Quotients by Lie Groups. Meromorphic Reduction

1. We assume that X is an n-dimensional normal complex space and that L is a complex Lie group acting holomorphically on X. We assume that there is no smaller union of connected components of X, on which L acts. We denote the dimension of the generic orbit by d and put $c = n-d$. We define R_L as the graph of orbits:

$$R_L = \{(x_1,x_2) \in X \times X : x_1 \in L\, x_2\} .$$

In general, the closure $R = \bar{R}_L$ is not an analytic set. But it contains the diagonal and is invariant under reflexion $X \times X \overset{\sim}{\to} X \times X$.

Definition: L *acts analytically closed on* X *if* R *is an analytic set of dimension* $n+d$ *and* R *is regular in the sense of paragraph 3 of the introduction.*

In the case that an algebraic group acts algebraically on an algebraic space X, the set R is always analytic of dimension $n + d$. If, moreover, X is complete L acts analytically closed on X.

We prove:

Theorem: *If* L *acts analytically closed on* X, *the graph closure* R *is a meromorphic equivalence relation on* X *and hence* $X/L := X/R$ *is a weakly normal complex space of pure dimension* c.

We call X/L the quotient of X by the complex Lie group L.

2. Since R is the closure of R_L each fibre $R_x = R \cap (X \times x)$

is closed against the action of L on X . The set of points $x \in X$, where the vector space of infinitesimal transformations has dimension less than d , is a nowhere dense analytic subset $P \subset X$. In $X \setminus P$ all orbits have pure dimension d . so no fibre of $p_2 : R \to X$ has dimension less than d and the generic case is that the fibre dimension is d . We denote by E the degeneration set of $p_2 : R \to X$. The map $p_2 : R \setminus E \to X$ is open.

We define the fibration $\mathcal{M}_\varphi$ in the normalization $\tilde{R}$ of R , obtain by projection the fibration $\mathcal{M}_{p_2}$ in R and by p_1 the fibration $\mathcal{M}$ in X . All fibres $S \in \mathcal{M}$ are pure d-dimensional and invariant against group action. Hence, they are singularity-free in $X \setminus P$. But in $X \setminus P$ they may consist of an at most countable number of connected components. The group L acts on such a component always transitively. - We call the fibres $S \in \mathcal{M}$ generalized orbits of L.

<u>3. Proof of the Theorem</u>: Since R is regular the projection $p_2 : E \to X$ is semiproper (see § 3.1). So $\underline{E} = p_2(E) \subset X$ is a nowhere dense analytic set. We put $\check{X} = X \setminus P \setminus \underline{E}$.

We take a point $x_o \in \check{X}$ and a connected component R_1 of $R_{x_o} \cap \check{X}$. There are connected open neighbourhoods $V(x_o) \subset \check{X}$ and $U(y_o) \subset\subset \check{X}$ to $y_o \in R_1$, such that $R_x \cap (U \times V)$, $x \in V$ is a regular family of connected d-dimensional submanifolds A_t for t in an analytic covering B of V. Since $R = \bar{R}_L$ the set $B_1 \subset B$ of points $t \in B$ with $A_t \subset R_L$ is dense. Because of the continuity of the action of L it is open.

We denote by V_1 the open image of B_1 in V. The set $R_1 \cap (U \times x_o)$ is in the accumulation set of the orbits Lx, $x \in V_1$. Since L acts on R_1 transitively and acts on $\mathring{X} \times V$ this holds also true for the whole set R_1.

We take finitely many connected components $R_1, \ldots, R_l$ of $R_{x_o} \cap \mathring{X}$ and take a fixed V and put $V^* = V_1 \cap \ldots \cap V_l$. This set is dense and open in V. The set $R_1 \cup \ldots \cup R_l$ is in the accumulation set of the orbits Lx, $x \in V^*$. This implies $R_{x_o} \cap \mathring{X} = R_{x_1} \cap \mathring{X}$ if $x_1 \in R_{x_o} \cap \mathring{X}$. So the transitive law for the relation $R|\mathring{X}$ is valid, i.e., $R \mid \mathring{X}$ is a complex equivalence relation. Moreover $R|\mathring{X}$ is normal and $R \cap (\mathring{X} \times \mathring{X})$ is dense in R. That completes the proof of the Theorem.

<u>4.</u> We assume that X is a compact connected n-dimensional normal complex space. We denote by $M = M(X)$ the field of meromorphic functions on X. We wish to construct a reduction $\underline{X}$ of X, which is a Moishezon space and a biregular invariant of X, such that M is the field of meromorphic functions on $\underline{X}$.

We denote by c the degree of transcendency of M and take meromorphic functions $f_1, \ldots, f_c$ which are algebraically (and analytically) independent, and a nowhere dense analytic set $P \subset X$ which contains the polar sets of $f_1, \ldots, f_c$. The c-tupel $f = (f_1, \ldots, f_c)$ gives a meromorphic map $f : X \to \mathbb{P}^c = \mathbb{P} \times \ldots \times \mathbb{P}$, c-times. The graph $G \subset X \times \mathbb{P}^c$ is an n-dimensional analytic subset. The holomorphic projection $G \to \mathbb{P}^c$ defines a complex equivalence relation $\tilde{R} \subset G \times G$ which projects by $G \to X$ to a meromorphic equivalence relation $R \subset X \times X$. Since $G|X \setminus P \to X \setminus P$ is biholomorphic the restriction $R|X \setminus P$ is a complex equivalence relation.

From now on we assume that P is so large that it contains also the image of the degeneration set $G \to X$. For $R|X \setminus P$ there is the well defined simple equivalence relation $\widehat{R|X \setminus P}$ which is the finest complex equivalence relation, whose fibres are locally the same as those of $R|X \setminus P$. We denote by $\hat{R}$ the closure of $\widehat{R|X \setminus P}$ in R. Then $\hat{R}$ is a meromorphic equivalence relation in X with polar set P. The map $R \cap ((X \setminus P) \times (X \setminus P)) \to (X \setminus P)$ has fibres of pure codimension $c+n$ and is open. It is immediate that $\hat{R}$ is independent of the choice of P and moreover it does not depend on the choice of $f_1, \ldots, f_c$, it is an invariant of X.

The meromorphic equivalence relation R is trivially regular and defines the unique proper modification $\tilde{X}$ of X and the quotient space Q which is a weakly normal space of dimension c. We have the commutative diagram

$$\begin{array}{ccc} & \tilde{X} & \\ \swarrow & & \searrow \\ Q & & X \end{array}$$

Every meromorphic function $f \in M$ lifts to a meromorphic function $\tilde{f}$ on $\tilde{X}$ which is constant on the fibres $\tilde{X} \to Q$ and hence comes from a meromorphic function $\underline{f}$ on Q. The degree of transcendency of Q (and of its normalization) is c. Hence, Q is a Moishezon space.

<u>Theorem</u>: *For every compact connected complex space* X *there is a unique Moishezon space* Q *, connected with* X *by a commutative diagram*

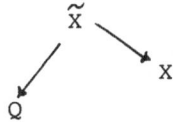

$$\begin{array}{ccccc} & & \tilde{X} & & \\ & {\scriptstyle q}\swarrow & & \searrow{\scriptstyle \pi} & \\ Q & & & & X \end{array}$$

, where π *is a proper modification and* q *a surjective fibre map with pure* $(n-d)$*-dimensional fibres, such that for the fields of meromorphic functions* $M(Q) = M(\tilde{X}) = M(X)$.

We call Q the meromorphic reduction of X .

§ 5. Appendix

1. Assume that Q is a normal complex space and that R is a semiproper complete equivalence relation in Q with the following properties:

1) If $R_1 \subset R$ is an irreducible component of R, which is different from the diagonal $D \subset Q \times Q$, then $p_2(R_1) \subset Q$ is nowhere dense.

2) R is discrete, i.e., $p_2 : R \to Q$ is a discrete map.

Since R is semiproper, for each point $t_o \in Q$ there is an open neighbourhood $U(t_o) \subset\subset Q$, which is mapped by the quotient map $Q \to Q/R$ onto an open neighbourhood $V(\underline{t}_o)$ of the image point of t_o. Since $p_2 : R \to Q$ is discrete, there are only finitely many points in U which are equivalent with t, for any point $t \in U$. If $t_1 \in Q \setminus \bar{U}$ is a point equivalent to t, then the quotient image of an open neighbourhood $W(t_1)$ is in V. But because of property 1) there would be points in W not equivalent to points of U. This is a contradiction. It follows:

$$R \cap p_2^{-1}(U) = R \cap (\bar{U} \times U)$$

Thus, the projection $p_2 : R \cap p_2^{-1}(U) \to U$ is proper. If B is the union of irreducible components of R different from the diagonal, the set $\underline{B} = p_2(B)$ is a nowhere dense analytic subset of Q.

The complex equivalence relation R sews Q together on the set $\underline{B}$: always finitely many points are identified. However, probably, the quotient Q/R will no longer be a complex space in general. There will be not enough local holomorphic functions. We define:

Definition: *A sutured complex space is a normal complex space together with a semiproper complex equivalence relation with the properties 1) and 2).*

2. We prove the following:

Theorem: *If X is a weakly normal complex space and R a normal complex equivalence relation in X then R is semi-proper and holomorphic. So X/R is a weakly normal complex space. If X is normal then also X/R is normal.*

Proof: We denote the codimension of the fibres to R with c. We take the normalization $\tau : \tilde{X} \to X$ of X and lift R to $\tilde{X}$. We obtain a complex equivalence relation $\tilde{R}$ in $\tilde{X}$ with pure c-codimensional fibres. We pass over to the simple equivalence relation $\hat{R}$ belonging to $\tilde{R}$. Then $Y = \tilde{X}/\hat{R}$ is a normal complex space of pure dimension c. The equivalence relation $\tilde{R}$ is obtained from a discrete complex equivalence relation $\underline{R}$ on Y by lifting.

We have $\tilde{X}/\tilde{R} = Y/\underline{R}$. But in general $\underline{R}$ will not be open. We shall define the open part R^* of $\underline{R}$. We denote by R^* just the union of those irreducible components of $\underline{R}$, which have dimension c. Then the projection $p_2 : R^* \to Y$ is open. The rest of $\underline{R}$ is mapped onto a nowhere dense subset of Y. The set R^* contains the diagonal, it is invariant under reflexion. We have to prove that it is an equivalence relation. For every $y_o \in Y$ we have

$$p_1 R^*_{y_o} = \lim_{\substack{y \to y_o \\ y = \text{generic}}} p_1 R^*_y$$

since $p_2 : R^* \to Y$ is open. If y is generic, we have $R^*_y = \underline{R}_y$ and since $\underline{R}$ is an equivalence relation, for each $y_1 \in p_1 R^*_y$ also $p_1 R^*_y = p_1 R^*_{y_1}$. So we get also for $y_1 \in p_1 R^*_{y_o}$ the equation $p_1 R^*_{y_o} = p_1 R^*_{y_1}$. That means the transitive law and that R^* is an equivalence relation.

3. We prove that R^* is holomorphic and semiproper. We denote by $y_o \in Y$ a point and by $W(y_o)$ an open neighbourhood. If $y_1 \in Y$ is a point equivalent with $y \in W$, we have $(y_1, y) \in R^*$. There are open neighbourhoods $U(y_1)$, $V(y) \subset W$ *such that* $R^* \cap (U \times V)$ is an analytic covering of V. There is an open neighbourhood $U'(y_1) \subset U$ such that $R^* \cap (U' \times V)$ is an analytic covering of U'. So the points of U' are equivalent with points of W: The set $\hat{W}$ of points in Y equivalent with points of W is open. By definition of the quotient topology $q(W) \subset Y/R^*$ is open.

We may assume $W \subset\subset Y$. Then $\bar{W}$ is compact and $q(W) \subset q(\bar{W})$. Thus, R^* is semiproper. - To prove that R^* is holomorphic, we consider the case $V = V(y_o) = U = U'$. We take a holomorphic function f in V, lift it to $R \cap (U \times V)$ and pass over to the b-th elementary symmetric function $\underline{f}$ on U. This $\underline{f}$ is constant on the fibres in U.

If V is isomorphic to an analytic set in a domain of holomorphy we can separate any two equivalence classes in U by an $\underline{f}$. There are finitely many $\underline{f}_1, \ldots, \underline{f}_l$ such that the fibres of $\underline{F} = (\underline{f}_1, \ldots, \underline{f}_l)$ have dimension 0. Now, $F = (\underline{F} \circ q^{-1}) \circ q$ is holomorphic on the open set $\hat{U} := q^{-1} q(U)$ and we have

$D = \hat{R}_F \subset R^* \subset R_F$. That means that R^* is holomorphic. So $Q = Y/R^*$ is a complex space.

We prove that Q is normal. If g is a bounded holomorphic function in $q(U) \setminus A$, where $A \subset q(U)$ is a nowhere dense analytic set, $g \circ q$ is holomorphic in $\hat{U} \setminus q^{-1}(A)$. The set $q^{-1}(A)$ is a nowhere dense analytic set in $\hat{U}$ because q is open. Since $\hat{U}$ is normal, the bounded holomorphic function $g \circ q$ has a unique analytic extension $\hat{g}$ to $\hat{U}$, which is fibre constant: there is a unique complex function $\underline{g}$ on U with $\underline{g} \circ q = \hat{g}$. We have $\underline{g}|q(U) \setminus A = g$. By definition of the complex structure on $q(U)$ this $\underline{g}$ is holomorphic. So the first Riemann Extension Theorem is valid: Q is a normal complex space.

<u>4.</u> If $X = \tilde{X}$ the projection $p_2 : \underline{R} \to Y$ is open, since $R \to X$, $X \to Y$ are open, then. So we have $R^* = \underline{R}$ and $Q = X/R$, i.e. Q is the usual quotient space of X . In the general case we have a holomorphic map $\tilde{X} \to Q$. The normalization map $\tau : \tilde{X} \to X$ gives a commutative diagram:

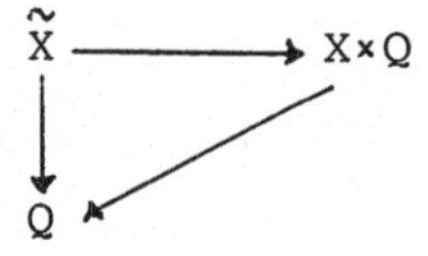

The map $\tilde{X} \to X \times Q$ is finite. Hence, the image X^* of $\tilde{X}$ in $X \times Q$ is an analytic subset and a reduced complex space. We have a projection $X^* \to Q$ and a holomorphic map $X^* \to X$. This is finite since we have the factorization $\tilde{X} \to X^* \to X$.

<u>5.</u> We can also prove that R is semiproper. We take a point $x_o \in X$, an open neighbourhood $V(x_o) \subset X$, a point $(x_1, x_o) \in R_{x_o}$ and an analytic covering $A \subset p_2^{-1}(V) \cap R$ with

$A \subset\subset R$, $(x_1,x_o) \in A$. Then $p_1(p_2^{-1}(V) \cap R)$ is an open neighbourhood of S with $S = p_1(R_{x_o})$. It is saturated by fibres. So it is the inverse image of an open neighbourhood W of the point $q(S) \in X/R$. The image $K:=p_1(\bar{A}) \subset X$ is compact and we have $q(K) \supset W$. This proves the statement.

We get immediately, that $\underline{R}$ is semiproper. $\underline{R}$ is obtained by lifting a semiproper equivalence relation R' from Q to Y. We have $\underline{R} = R^* \cup A$, where A is the union of irreducible components of dimension less than c. The image B of A in $Q \times Q$ under $Y \times Y \to Q \times Q$ is an analytic set of dimension less than c, since locally $Y \to Q$ is an analytic covering. The dimension of B is less than $c = \dim Q$. We have $R' = D \cup B$, R' is discrete and $\underline{B}:=p_2(B) \subset Q$ nowhere dense. That proves that (Q,R') is a sutured complex space.

The generic fibre of $\tilde{X} \to Q$ is mapped by $\tilde{X} \to X$ onto a fibre of X. Since $p_2 : R \to X$ is open, this holds true for all fibres of $\tilde{X} \to Q$, hence it is true for the fibres $X^* \to Q$, but these are analytic subsets of X. So the map $\pi^*:X^* \to X$ restricted to fibres is biholomorphic. Outside $\underline{B}$ the map π^* is bijective.

Since π^* is finite it is topological there. If f is a local holomorphic function there, $f \circ \pi^{*-1}$ is a local continuous function which is holomorphic outside a nowhere dense analytic set. Because X is weakly normal, it is holomorphic. So $\pi^* : X^*|Q\setminus\underline{B} \xrightarrow{\sim} \pi^*(X^*|Q\setminus\underline{B}))$ is a biholomorphic map. The rest is a nowhere dense analytic set. That means that π^* is a proper modification.

We call $\pi^* : X^* \to Q$ the *normalization* of R in X .

<u>6.</u> In the case of the Main Theorem each fibre in X (there denoted by $\tilde{X}$) has a well defined multiplicity on its irreducible components. We can use like in § 2.1 complete intersections $\bar{A}_i$ to construct local holomorphic functions in Q/R' which separate the fibres of X . This proves that R' is holomorphic and that Q/R' is a weakly normal complex space. So, in our case we can divide by R' and the sutured complex spaces are not necessary: The quotient X/R is a weakly normal complex space. But this is also like this in general. It can be seen that a multiplicity can be locally resp. X/R defined in all of the cases of our Theorem. So X/R always is a weakly normal complex space. This proves our Theorem.

Bibliography

[Fu] Fujiki,A.: On Automorphism Groups of Compact Kähler Manifolds. Invent.Math.44, 225-258 (1978).

[STCER] Grauert,H.: Set Theoretic Complex Equivalence Relations. Math.Annalen 265, 137-148 (1983)

[CAS] Grauert,H. and R.Remmert: Coherent Analytic Sheaves. Springer Heidelberg 1984.

[Hi] Hironaka,H.: Flattening Theorem in Complex-Analytic Geometry Amer.J.Math.97, 503-547 (1975).

[Li] Lieberman: Compactness of the Chow Scheme: Fonctions de Plusieurs Variables Complexes III (Séminaire Norguet). Lecture Notes in Mathematics 670 (1978).

[Re] Remmert,R.: Holomorphe und meromorphe Abbildungen komplexer Räume. Math.Ann.133, 328-370 (1957).

[St] Stein,K.: Maximale holomorphe und meromorphe Abbildungen I,II. Ann.J.85, 298-315 (1963); 86, 823-869 (1964).

Recent Developments in Homogeneous CR-Hypersurfaces

A. Huckleberry and W. Richthofer

Dedicated to Wilhelm Stoll*

1. INTRODUCTION

Let $\hat{G}$ be a connected complex Lie group and $X = \hat{G}/\hat{H}$ a complex manifold homogeneous under a holomorphic $\hat{G}$-action. In order to understand X, e.g. how it fits into a fine classification, details of its function theory, etc., one should use as much Lie theoretic information about $\hat{G}$ as is possible. In particular it is often useful to study the orbit structure of real subgroups of $\hat{G}$. Such orbits are usually not complex submanifolds of X.

Conversely, if G is a connected real Lie group and one wishes to understand G and its representations, then one naturally looks for orbits M = G/H where strong analytic or algebraic tools are at hand. A very interesting setting is where X is a complex manifold where G is acting as a group of holomorphic transformations and M = G/H is a G-orbit in X. If $\mathfrak{g}$ is the Lie algebra of G, then the complex Lie algebra $\hat{\mathfrak{g}} = \mathfrak{g} + i\mathfrak{g}$ (not necessarily a direct sum) is represented as an algebra of holomorphic vector fields on X. Ideally the vector fields in $\hat{\mathfrak{g}}$ are integrable so that we have an induced holomorphic action of the associated complex Lie group $\hat{G}$. In this situation we have the inclusion of orbits $G/H \hookrightarrow \hat{G}/\hat{H}$ and it is possible to derive information about M from the complex homogeneous space $\hat{G}/\hat{H}$.

If M = G/H is an orbit in X as above, then we refer to X as a $\mathfrak{g}$-complexification of M. If in addition the Lie algebra of vector fields induces a G-action, then we refer to X as a G-complexification of M. In either case M inherits a G-invariant Cauchy-Riemann structure from X (see [G],[AHR] for generalities). With this in mind we can refine our line of

* Stoll's beautiful results on parabolic spaces were one of the strong motivating forces for the research described in the present paper.

questioning: Given an orbit $M = G/H$ of a real Lie group, what are the possible G-invariant CR-structures on M? What are the possible $\mathfrak{g}$-complexifications X so that $M \hookrightarrow X$ is CR-embedded? What are the possible G-complexifications? Is it possible to find $X = \hat{G}/\hat{H}$ complex homogeneous so that $M \hookrightarrow X$ is CR-embedded as a G-orbit?

If we reach the ideal situation, i.e. $X = \hat{G}/\hat{H}$ and $M = G/H$ is CR-embedded as a G-orbit, then there is so much structure at hand that it is often possible to make very strong statements. These statements can go both ways: The structure of M yields information about X or vice versa.

Example 1: Let $V := \mathbb{P}_1 \times \mathbb{P}_n$ be embedded via the Segre embedding in $\mathbb{P}_m$, $m = 2m-1$. This is equivariant with respect to the usual $G = SL_2(\mathbb{C}) \times SL_{n+1}(\mathbb{C})$ action on V. Note that G acts transitively on the complement $\mathbb{P}_m \setminus V =: X = G/H$. Let $K = SU_2 \times SU_{n+1}$. It turns out that the CR-structure of the minimal K-orbit M in X reflects the fact that V is not a complete intersection in $\mathbb{P}_m$ (see [BFS]). □

Example 2: Let $\hat{G}$ be a complex semi-simple group and $\hat{H}$ a parabolic subgroup, i.e. $\hat{G}/\hat{H}$ is a homogeneous rational manifold. Let G be a non-compact real form of $\hat{G}$. Then the generic orbit of G is open and G has a unique orbit M of minimal dimension ([WO]). Theorems from complex analysis lead one to consider certain CR-cohomology spaces on M. They are interesting G-moduls, because they have stable Hilbert subspaces which realize interesting representation theory for G (see [RP] for a special case). □

We began studying the interplay between CR-orbits and homogeneous spaces from the point of view of the homogeneous space, i.e. given $X = \hat{G}/\hat{H}$, what influence do the natural CR-orbits $M \hookrightarrow X$ have on the classification theory for X? A most striking application arose in the classification of homogeneous surfaces, where Tomilieri's classification of invariant CR-structures on Heisenberg groups ([TO]), the Andreotti-Fredricks embedding theorem ([AF]), and Tanaka's extension theorem for CR-maps of real quadrics ([TA]) provided a fundamental step in the case of solvable groups (see [OR], [H]).

On the other hand, while working in the complex homogeneous situation, we realized that the CR-manifolds $M = G/H$ are important in their own right and therefore began an organized study of the subject. The purpose of the present paper is to outline the recent developments in this direction so that the reader has a guide to the somewhat technical details in [AHR and R]. Up to this point most of the fine classification results have been proved for compact homogeneous CR-hypersurfaces. However, the basic tools are available for the higher codimensional case.

As far as we know, the first results in the classification theory for compact homogeneous CR-hypersurfaces are due to Morimoto and Nagano ([MN]). They considered the situation where $M = G/H$ is CR-embedded in a Stein manifold X. In this case one immediately sees that $M = \partial\Omega$, where Ω is a strongly pseudo-convex open subset of X. In particular the Levi-form of M is everywhere positive dfinite. Since Ω can be recovered as the envelope of holomorphy of the CR-functions on M, it follows that G acts as a group of holomorphic transformations on Ω. In order to go ahead with their classification theory, Morimoto and Nagano needed the theory of compact groups. Thus they assumed that $\pi_1(M)$ is finite and as a result any maximal compact subgroup K of G acts transitively on M ([S]). Under these assumptions they proved that either $\Omega \cong \mathbb{B}_n$, the ball in $\mathbb{C}^n$, and M is CR-equivalent to S^{2n-1} with its induced structure, or M is a sphere bundle in the tangent bundle of a compact symmetric space of rank 1 and Ω is the tube around the 0-section with this sphere bundle as boundary. Since the symmetric spaces of rank 1 are completely classified, i.e. spheres, projective spaces over $\mathbb{R}$, $\mathbb{C}$, or $\mathbb{H}$, and the Cayley projective plane, this is certainly a <u>fine</u> classification theorem.

Rossi ([R1]) extended the work of Morimoto and Nagano in two ways. First, he used the theory of "filling in holes" ([R2]) to show that if $M = G/H$ is an abstract strongly pseudo-convex compact CR-hypersurface with $\dim_{\mathbb{R}} M \geq 5$, then M is G-equivariantly the boundary of a domain Ω in a Stein <u>space</u> X. For the same reason as above G acts as a group of holomorphic

transformations on Ω. Rossi's methods also require that a maximal compact subgroup K of G acts transitively on M. Thus he also assumed $\pi_1(M)$ to be finite. His classification goes as follows. If Ω is smooth, then the results of Morimoto and Nagano may be applied. If not, then it is easy to see that it has exactly one singular point x_0. Taking a K-equivariant minimal desingularization $\pi: \tilde{\Omega} \to \Omega$, it follows that $Q := \pi^{-1}(x_0)$ is a K-orbit and $\tilde{\Omega}$ can be identified with a K-invariant tube neighborhood of Q in this normal bundle. Ampleness criteria show that Q is homogeneous rational and the normal bundle is very ample. Thus the singular cases in the classification arise as follows:
Let G be a semi-simple complex Lie group and P a parabolic subgroup. Let $Q := G/P$ and recall that any very ample principal $\mathbb{C}^*$-bundle over Q is G-homogeneous, i.e. the bundle can be described by a homogeneous fibration $G/H \xrightarrow{\mathbb{C}^*} G/P$. Attaching the ∞-section and blowing it down to a point x_0, we obtain an affine cone C with vertex x_0 so that $C \setminus \{x_0\} = \hat{G}/\hat{H}$. Let K be a maximal compact subgroup of G. Then any K-orbit $M = K/L$ in $C \setminus \{x_0\}$ is a strongly pseudoconvex CR-hypersurface. Any two K-orbits are equivalent under the right $\mathbb{C}^*$-action and any two maximal compact subgroups are conjugate. Thus the construction only depends on the ample bundle. The only possibility for a non-singular vertex is when the bundle is the hyperplane section bundle over $\mathbb{P}_n$, i.e. $C = \mathbb{C}^{n+1}$ ([HO]).

Using non-trivial analytic methods and a complicated check of cases in the spherical case, Burns and Shnider ([BS]) showed that $\pi_1(M)$ <u>is</u> finite if M is a strongly pseudoconvex compact homogeneous CR-hypersurface with $\dim_{\mathbb{R}} M \geq 5$. Thus, except for possibilities in the 3-dimensional case, the fine classification of Morimoto-Nagano-Rossi is indeed complete. □

In [AHR] we began a study of abstract homogeneous CR-manifolds and applied our methods to the classification problem for compact homogeneous CR-hypersurfaces. One should note that if X is an arbitrary compact complex homogeneous manifold, then $M = S^1 \times X$ is a homogeneous CR-hypersurface. Thus a classification of homogeneous CR-hypersurfaces would naturally contain a classification of homogeneous compact complex manifolds. This is of course not possible at the present time, and thus it is necessary to make restrictions. The general theory in fact shows that problems with Levi-flatness, e.g. $S^1 \times X$, are critical. Thus we began by considering homogeneous compact CR-hypersurfaces with non-degenrate Levi-form. Of course this contains the strongly pseudoconvex case discussed above. A fine classification is proved in ([AHR]) and described in § 4 of the present paper. There are no restrictions on dimension. Since one can't always "fill in the holes", there are certain cases where there are no G-complexifications (see § 3 for examples). However we give an exact description of these cases.

The fine classification of compact homogeneous CR-hypersurfaces M with non-degenerate Levi-form is given in terms of a root theoretic description of a canonical fibration of M which has a strongly pseudoconvex fiber and a compact complex homogeneous rational base. The simplest case of this fibration is the S^1-fibration of M induced by the cone fibration $\hat{G}/\hat{H} \underset{\mathbb{C}^*}{\to} \hat{G}/\hat{P}$ in the case discussed above. The Levi-form of M, which is itself an interesting invariant of the group theory, can be explicitely calculated ([AZ]). In fact the resulting characteristic polynomials have a very simple form.

In § 5 of the present paper we give a complete description of the compact homogeneous CR-hypersurfaces M = G/H which possess a Kähler structure, i.e. there is a CR-embedding $M \hookrightarrow X$ in a Kähler manifold. On the one hand, this may be thought of as a continuation of the classification theory of Borel-Remmert ([BR]) and Matsushima ([M]) in the complex Kähler case. On the other hand, we hope that this is the beginning of a project which will significantly aid in understanding complex homogeneous manifolds. For example, if $X = \hat{G}/\hat{H}$ is Kähler, e.g. Stein or quasi-projective and M = G/H is a compact orbit of some real subgroup of $\hat{G}$, then M inherits this Kähler structure. In fact one only needs for some neighborhood of M to be Kähler. For example this is guaranteed if there are sufficiently many holomorphic or meromorphic functions on X. The results in the hypersurface case indicate that the existence of a Kähler structure on M is very restrictive. This then imposes strong funtion-theoretic conditions on the ambient space.

□

2. SOME BASIC METHODS

Let M = G/H be a homogeneous real analytic generic CR-submanifold of a complex manifold X with complex structure J. If R is the maximal J-invariant subbundle of TM, then we refer to (R,J) as the induced CR-structure on M. We wish to study how properties of X affect M and vice versa. The "vice versa" needs to be clarified, because only the germ of X along M can be determined by M. However, in many situations involving group actions there is a global relationship, e.g. if $X = \hat{G}/\hat{H}$ is complex homogeneous and G is a subgroup of $\hat{G}$ having compact hypersurface orbits in X.

If X is not complex homogeneous as above, it is often possible to construct an embedding

$$M = G/H \hookrightarrow X \hookrightarrow \hat{G}/\hat{H} = \hat{M} ,$$

where $M \hookrightarrow \hat{M}$ is equivariant. In this case $\hat{M}$ is called a G-complexification of M.

There are at least two different ways of approaching the classification of (M,X): 1) Use complex analytic methods and concentrate on X(extrinsic); 2) Forget X and study M as a real manifold with some additional structure (intrinsic). One of the basic ingredients of the following is the interaction between intrinsic and extrinsic methods.

We begin by giving an intrinsic characterization of a homogeneous CR-manifold.

Proposition 1: Let M = G/H be a real-homogneneous manifold. The G-invariant CR-structures (R,J) on M are in 1-1 correspondence with the pairs $(\tilde{R},\tilde{J})$ which satisfy the following conditions:

(0) $\tilde{R}$ is a subspace of $\mathfrak{g}$ with $\mathfrak{h} \subset \tilde{R} \subset \mathfrak{g}$ and $\tilde{J} : \tilde{R} \to \tilde{R}$ is an endomorphism;

(1) $\tilde{J}X = 0$ *if and only if* $X \in \mathfrak{h}$;

(2) $\tilde{J}^2X + X \in \mathfrak{h}$ *for all* $\tilde{X} \in \tilde{R}$;

(3) $Ad_g X \in \tilde{R}$ *and* $\tilde{J}Ad_g X - Ad_g \tilde{J}X \in \mathfrak{h}$ *for all* $g \in H$, $X \in \tilde{R}$;

(4) $[X,Y] - [\tilde{J}X,\tilde{J}Y] \in \tilde{R}$ *and*
$\tilde{J}([X,Y] - [\tilde{J}X,\tilde{J}Y]) - [\tilde{J}X,Y] - [X,\tilde{J}Y] \in \mathfrak{h}$ *for all* $X,Y \in \tilde{R}$.

Two pairs $(\tilde{R},\tilde{J})$ *and* $(\tilde{R}',\tilde{J}')$ *are equivalent if and only if* $\tilde{R} = \tilde{R}'$ *and* $\tilde{J}X - \tilde{J}'X \in \mathfrak{h}$ *for all* $X \in \tilde{R} = \tilde{R}'$. *If* H *is connected, then* (3) *may be replaced by*

(3)' $[X,Y] \in \tilde{R}$ *and* $\tilde{J}[X,Y] - [X,\tilde{J}Y] \in \mathfrak{h}$
for all $X \in \mathfrak{h}$, $Y \in \tilde{R}$. □

The proof of Prop. 1 is a straightforward consequence of the definitions (see [R], p. 17).

Corollary 2: *Every* G-*invariant* CR-*structure on* G/H *is analytic*.

The following results of Andreotti-Fredricks provide the first steps toward connecting the intrinsic and extrinsic points of view.

Theorem 3 ([AF]). *Let* M *be an analytic* CR-*manifold of type* (m,ℓ). *Then* M *admits a generic complexification* $(\hat{M},\tau)$. *Given a generic complexification* $(\tilde{M},\tilde{\tau})$ *of* M, *there are open neighborhoods* U *of* $\tau(M)$ *in* $\hat{M}$ *and* $\tilde{U}$ *of* $\tilde{\tau}(M)$ *in* $\tilde{M}$ *and a biholomorphic map* $f : U \to \tilde{U}$ *such that* $f \cdot \tau = \tilde{\tau}$.

Remark: If $\hat{M}$ is a generic complexification of M and M is of type (m,ℓ) then $\dim_{\mathbb{C}}\hat{M} = m - \ell$.

Theorem 4 ([AF]). *Let* $f : M \to M'$ *be an analytic* CR-*map between two analytic generic* CR-*manifolds* $M \subset \hat{M}$ *and* $M' \subset \hat{M}'$. *Then there are open neighborhoods* U *of* M *in* $\hat{M}$ *and* U' *of* M' *in* $\hat{M}'$ *and a holomorphic map* $\hat{f} : U \to U'$ *such that* $\hat{f}|M = f$.

As a result we have the following simple but useful

Corollary 5: *Let* M *be an analytic generic* CR-*submanifold of a complex manifold* $\hat{M}$. *Then for every analytic* CR-*vector field*

$X \in \Gamma_{CR}(M,TM)$ there is an open neighborhood U of M in $\hat{M}$ and a holomorphic vector field Z on U such that

$$X = (Z + \bar{Z})|M.$$

Given a homogeneous CR-manifold $M = G/H$, Prop.1-Cor.5 show that M is an analytic generic CR-submanifold of a complex manifold $\hat{M}$. For every $g \in G$ there are open neighborhoods U_g, V_g of M in $\hat{M}$ and a biholomorphic map $f_g : U_g \to V_g$ such that $f_g|M = g$. Taking $\hat{M}$ to be smaller if necessary, we may assume that every $\mathfrak{g}$-vector field on M is the "restriction" of a holomorphic vector field on $\hat{M}$ as in Cor. 5. If G acts almost effectively on M, i.e. the ineffectivity of the G-action is discrete, we have an embedding $\mathfrak{g} \to \Gamma_{\mathcal{O}}(\hat{M},T\hat{M})$. The complex hull of $\mathfrak{g}$ with respect to this embedding is denoted by $\hat{\mathfrak{g}}$ and is called the $\hat{M}$-complexification of $\mathfrak{g}$. Any such complex manifold $\hat{M}$ is refered to a a $\mathfrak{g}$-complexification of $M = G/H$.

One of the main methods for studying M is to find G-equivariant fibrations $M = G/H \to G/I$ such that G/I and I/H are known. We begin by considering the normalizer fibration $G/H \to G/N_G(H)$, where $N_G(H) := \{g \in G | gHg^{-1} = H\}$. In some sense this fibration factors out the group manifolds which are equivariantly contained in G/H. Of course we must take the CR-structure into consideration. This is done at the Lie algebra level and therefore the fiber is only locally a group.

With Prop.1 in mind it is natural to define $N_{CR}(H)$, the CR-normalizer of H, as

$$N_{CR}(H) := \{g \in G | Ad_g X \in \tilde{R} \text{ and } Ad_g \tilde{J}X - \tilde{J}Ad_g X \in \mathfrak{h} \text{ for all } X \in \tilde{R}\}.$$

Using Prop.1, a direct calculation shows that $N_{CR}(H)$ consists of the $g \in N_G(H^\circ)$ such that the right-translation $aH^\circ \to agH^\circ$, $G/H^\circ \to G/H^\circ$, are CR-mappings with respect to the CR-structure on G/H° which is induced by the covering map $G/H^\circ \to G/H$.

We are now in a position to give intrinsically defined fibrations of G/H which are reasonable in the category of CR-manifolds.

Theorem 6. *Let $M = G/H$ be a homogeneous CR-manifold and let $L \subset N_{CR}(H)$ be a closed subgroup such that $H \subset L$ and $\tilde{J}(\mathfrak{l} \cap \tilde{R}) \subset \mathfrak{l} \cap \tilde{R}$. Then there is a unique G-invariant CR-structure on G/L such that $G/H \to G/L$ is a CR-submersion.*

□

Using the properties of $\tilde{J}$ described in Prop.1, it follows that $\tilde{J}(\mathfrak{n}_{CR} \cap \tilde{R}) \subset \mathfrak{n}_{CR} \cap \tilde{R}$ and thus we have

Corollary 7. *There is a unique G-invariant CR-structure on $G/N_{CR}(H)$ such that $G/H \to G/N_{CR}(H)$ is a CR-submersion.*

□

If we look at $N_{CR}(H)/H$ as a submanifold of G/H with the induced CR-structure, then $N_{CR}(H)$ "normalizes" the CR-structure. As a result we have

Proposition 8. *The fibers of the fibration $G/H \to G/N_{CR}(H)$ are Levi-flat. Moreover the distribution generated by the maximal complex subspaces tangent to the fibers of $G/H \to G/N_{CR}(H)$ is contained in the Levi kernel of M.*

□

We now assume M is embedded in a $\mathfrak{g}$-complexification $\hat{M}$. Let $\pi : G \to G/H$ denote the projection and $0 := \pi(e)$. It is useful to note that the isotropy Lie algebra $\hat{\mathfrak{h}} := \{Z \in \hat{\mathfrak{g}} \mid Z(0) = 0\}$ is directly connected with the CR-structure of G/H. To see this recall that $\hat{\mathfrak{g}} = \mathfrak{g} + J\mathfrak{g}$, where J is the complex structure on $\hat{M}$ and $\mathfrak{g}$ is regarded as a real subalgebra of $\Gamma_0(\hat{M}, T\hat{M})$. Hence $Z = X + JY \in \hat{\mathfrak{h}}$ if and only if $JX(0) = Y(0)$. This means that, considered as elements in TG_e, X and Y are contained in $\tilde{R}$ and $\tilde{J}X = Y \pmod{\mathfrak{h}}$. Elementary calculations then show that

$$N_{CR}(H) = \{ g \in G \mid \mathrm{Ad}_g\, \hat{\mathfrak{h}} = \hat{\mathfrak{h}} \},$$

where $\mathrm{Ad} : G \to \mathrm{Aut}(\hat{\mathfrak{g}})$ denotes the adjoint action of G on $\hat{\mathfrak{h}}$ which is induced from the usual adjoint action of G on $\mathfrak{g}$. This (extrinsic) argument shows that $G/N_{CR}(H)$ naturally inherits a G-invariant CR-structure from the Grassmannian defined by the complex subspaces of $\hat{\mathfrak{h}}$ which have the same dimension as $\hat{\mathfrak{h}}$: $N_{CR}(H)$ is just the isotropy N at the "point" $\hat{\mathfrak{h}}$. Using Plücker-coordinates we therefore have a G-equivariant map

$G/H \to G/N \to \mathbb{P}_k(\mathbb{C})$.

This map is also given by the holomorphic sections of the anti-canonical bundle of $\hat{M}$ which are generated by the $\hat{\mathfrak{g}}$-sections and is therefore a CR-map. We refer to it as the $\mathfrak{g}$-*anticanonical* fibration of G/H. Since $N = N_{CR}(H)$ and $G/H \to G/H_{CR}(H)$ is a CR-submersion, the map $G/N_{CR}(H) \to G/N$ is CR, but in general *its inverse is not*. However, if G/H is a CR-hypersurface, is totally real, or is a complex manifold, the map $G/N_{CR}(H) \to G/N$ is a CR-isomorphism.

The equivariant embedding $G/N \hookrightarrow \mathbb{P}_k(\mathbb{C})$ yields a representation of G in $PSL_{k+1}(\mathbb{C})$. Let $\tilde{G}$ denote the smallest complex Lie subgroup which contains the image of G (Recall that G is connected!), and let $\tilde{N}$ be the $\tilde{G}$-isotropy group at the point N. Consequently we have

$$G/H \to G/N \hookrightarrow \tilde{G}/\tilde{N} \hookrightarrow \mathbb{P}_k(\mathbb{C}),$$

where G/N is a generic CR-submanifold of the complex homogeneous manifold $\tilde{G}/\tilde{N}$.

Since $N \subset N_G(H^0)$, the fiber N/H^0 of the $\mathfrak{g}$-anticanonical fibration can be written as the quotient A/Γ, where $A = N/H^0$ and $\Gamma := H/H^0$ is a discrete subgroup in A. Moreover the $\mathfrak{g}$-anticanonical fibration of N/H is degenerate, i.e. the base is a point. We call such a manifold *flat*. Indeed *flat* implies *Levi-flat*, but not vice versa.

From now on we focus our attention on homogeneous compact CR-hypersurfaces. In this case *either G/N is projective rational and N/H is a compact flat CR-hypersurface or G/N is a compact hypersurface in $\tilde{G}/\tilde{N}$ and N/H is a compact parallelizable complex manifold*. In the latter case there is not always a G-complexification of G/H (see § 3). However, in the first case we have

Theorem 9. *Let G/H be a compact homogeneous CR-hypersurface where G is acting effectively. If the base G/N of the $\mathfrak{g}$-anticanonical fibration of G/H is projective rational, then there is a G-complexification $\hat{G}/\hat{H}$ of G/H and a commutative equivariant diagram of CR-maps*

$$\begin{array}{ccc} G/H & \longrightarrow & \hat{G}/\hat{H} \\ \downarrow & & \downarrow \\ G/N & \xrightarrow{\sim} & \hat{G}/\hat{N} \end{array},$$

where $\hat{N} = N_{\hat{G}}(\hat{H}^0)$. □

For details of the results in this section see [R].

3. THE BASE OF THE $\mathfrak{g}$-ANTICANONICAL FIBRATION OF A COMPACT HOMOGENEOUS CR-HYPERSURFACE.

Recall that if $M = G/H$ is a homogeneous compact CR-hypersurface, then we have the $\mathfrak{g}$-anticanonical fibration

$$G/H \to G/N \to \mathbb{P}_k.$$

In this section we make some detailed remarks about the base. Thus we assume that G is a real group of linear transformations acting on $\mathbb{P}_k$, $\hat{G}$ is the smallest complex Lie group in $PSL_{k+1}(\mathbb{C})$ which contains G, and that the G-orbit $M = G(x) = G/H$ is a real hypersurface in the $\hat{G}$-orbit $X = \hat{G}(x) = \hat{G}/\hat{H}$. Without loss of generality we may assume that G is acting almost effectively on M. Note that the base of the $\mathfrak{g}$-anticanonical fibration either satisfies these conditions or is a compact homogeneous rational complex manifold. The latter situation is well-understood.

The following is the fundamental for the classification of the linear situation described above.

<u>Proposition 1</u>. <u>Either M is the equivariant product $S^1 \times Q$ of a circle and a complex homogeneous rational manifold or the semi-simple part K_{ss} of any maximal compact subgroup K of G acts transitively on M. In the latter case $\pi_1(M)$ is finite.</u>

The proof of this fact goes by induction on dimension and uses fibrations of the complex homogeneous space $X = \hat{G}/\hat{H}$.

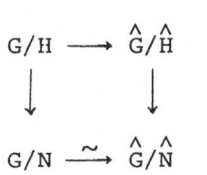

We assume that the Levi-flat case $M = S^1 \times Q$ is well-understood and only consider the latter case in the above proposition. Hence we assume that $G = K$ is a semi-simple compact linear group. Of course $\mathrm{Aut}_{CR}(M)$ may be much larger e.g. $M = \partial \mathbb{B}_n$, but going to the possibly smaller compact group has distinct advantages.

In the usual way we have

$$M = K/L \hookrightarrow S/H =: X \hookrightarrow \mathbb{P}_k ,$$

where M is a hypersurface in X. Since linear semi-simple groups are algebraic, H is an algebraic subgroup of S, and X is Zariski open in its closure in $\mathbb{P}_k$. It is therefore enough to understand *algebraic homogeneous spaces S/H of complex semi-simple linear groups where the general orbit of a maximal compact subgroup is a real hypersurface*.

If X = S/H is Stein, then we are in the case handled by Morimoto and Nagano which was described in § 1. In fact X is then the tangent bundle of a symmetric space of rank 1. If X is not Stein, i.e. H is not reductive, then H is contained in a proper parabolic subgroup of S. Let P be a minimal such subgroup. Since K_P has hypersurface orbits in the fiber P/H of S/H → S/P, we are in a good inductive situation and it is relatively easy to prove

Proposition 2. *Let S be a linear semi-simple complex group and H an algebraic subgroup. Suppose that the generic orbit of a maximal compact subgroup K of S in the homogeneous space X = S/H is a real hypersurface. Let P be a minimal parabolic subgroup of S containing H (P = S is allowed). Then P/H is Stein.*

This result had already been used in the classification of certain almost homogeneous spaces ([HS], [A]). In [AHR] it is pointed out that the "Stein-Rational" fibration S/H → S/P is essentially unique and exactly reflects the structure of M = K/L.

The only cases where P is not unique arise from building the following type of example inside of S. Therefore all possible such minimal parabolic groups can be described (see [AHR]).

Example. Let $S = SL_{n+1}(\mathbb{C})$ act on $\mathbb{P}_n \times \mathbb{P}_n$ by $A(p,q) = (A\bar{p}^{-t}, Aq)$. There are exactly two S-orbits in $\mathbb{P}_n \times \mathbb{P}_n$, i.e. an open one X = S/H which is the complement of a divisor orbit defined by $\{(v,w) \mid v^t \cdot w = 0\}$. There are exactly two minimal parabolic groups, P_1 and P_2, which contain H. They just arise by projecting on the respective factors of $\mathbb{P}_n \times \mathbb{P}_n$:

$$S/H \overset{\mathbb{C}^n}{\to} S/P_i = \mathbb{P}_n \text{ , } i = 1, 2 \text{ .} \qquad \square$$

Given that M = K/L is described by its complexification S/H and that S/H breaks into Stein and rational parts, i.e. P/H and S/P, it only remains to describe the possible pairs (H;P) coming from a given S. The situation is so concrete that in principle one could calculate virtually anything one needs, e.g. analytic cohomology as in [BFS]. So far calculations have been carried out in two directions: 1) A detailed root-theoretic description of all pairs (H,P) ([AHR]); 2) An explicit calculation of the Levi form of M ([AZ]). Since these results are a bit technical, we only discuss here an important case of 1).

Consider the class of algebraic manifolds X = S/H as above. Assume further that a Stein-rational fibration has the form $S/H \to S/P$ with $P/H = \mathbb{C}^n$, i.e. a homogeneous affine bundle where the generic K_p-orbit in $\mathbb{C}^n$ is a hypersurface. Under these assumptions it is easy to see that P is represented in the complex affine group, that $R_u(P)$ is represented as the full group of translations $\mathbb{C}^n$, and the reductive part of P is represented as $GL_n(\mathbb{C})$, $SL_n(\mathbb{C})$ or Sp_n in the standard way. The latter case can only occur when n is even.

Since K_p has a fixed point in P/H, we may take the reductive parts of P and H in a Levi-decomposition to be the same i.e. $P = L \ltimes R_u(P)$ and $H = L \ltimes R_u(H)$, and $R_u(H)$ is obtained from $R_u(P)$ by removing n 1-dimensional root groups. Now L is determined by a set π of simple roots, i.e. the simple factors in L are determined by orthogonal connected chains in the Dynkin diagram for S. We write π as a disjoint union $\pi_1 \cup \pi_2$, where the simple factors coming from π_1 yields the factor which is either represented as SL_n or Sp_n.

It is not difficult to show that P contains exactly one <u>simple</u> root group which is not in H. Let a be the root corresponding to this group. Then

$$R_u(P) = R_u(P_{\pi \cup \{a\}}) \; R_u(P_{\pi_2 \cup \{a\}})$$

and

$$R_u(H) = R_u(P_{\pi \cup \{a\}})\, R_u(P_{\pi_2 \cup \{a\}})',$$

where P_σ is the parabolic group corresponding to the set σ of simple roots. Of course π_1 must be orthogonal to $\pi_2 \cup \{a\}$.

Given the above information one can write down the possible diagram. Since π_1 is orthogonal to the connected chain $\pi_2 \cup \{a\}$, it is enough to describe the latter. The following is a list of all possibilities where the "white" circles represent the roots in π_2.

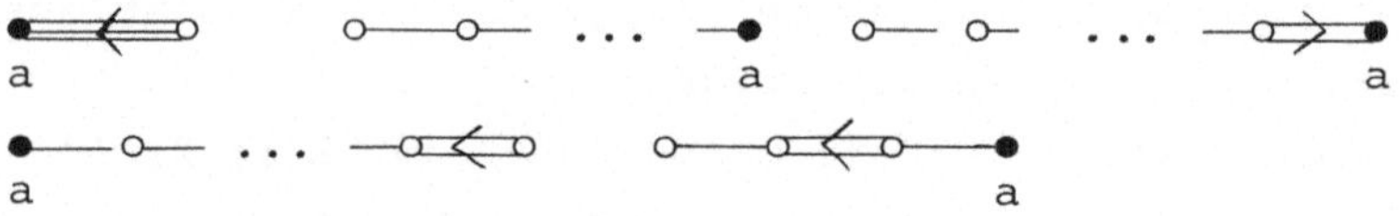

An Example. Here we discuss G-invariant CR-structures on S^3. The first such structure which comes to mind is $M = \partial\mathbb{B}_2$, the boundary of the ball in $\mathbb{C}^2$ with the induced structure. Let $K = SU_2$ act linearly on $\mathbb{C}^2$ as usual. In this way K acts freely and transitively on M and therefore inherits a left-invariant CR-structure. Since $\hat{K} = SL_2(\mathbb{C})$, one easily finds a G-complexification,

$$S^3 = M = K/L \to \hat{K}/\hat{L} = \mathbb{C}^2 \setminus \{(0,0)\},$$

and the k-anticanonical fibration is just the restriction of the Hopf fibration,

$$\mathbb{C}^2 \setminus \{(0,0)\} = \hat{K}/\hat{L} \xrightarrow[\mathbb{C}^*]{} \hat{K}/B = \mathbb{P}_1,$$

to M.

Now K is not the full group of CR-automorphisms of M. However, this group G is easy to describe: By analytic continuation

$$G = Aut_{CR}(\partial\mathbb{B}_2) \cong Aut_{\mathcal{O}}(\mathbb{B}_2) = PSU(3,1).$$

It should be remarked that the induced Hopf fibration of M is not G-equivariant. The -anticanonical fibration is just the embedding in $\mathbb{P}_2$ induced by the standard inclusion $\bar{\mathbb{B}}_2 \to \mathbb{P}_2$. In fact $\hat{G} = PSL_3(\mathbb{C})$ and we again have a G-complexification,

$$\partial\mathbb{B}_2 = M = G/H \hookrightarrow \hat{G}/\hat{H} = \mathbb{P}_2.$$

We now construct a family $\{M_r \mid r > 0\}$ of SU_2-invariant CR-structures on S^3 which are pairwise inequivalent and which are not equivalent to the above structure M. For this let $X := Q_{(2)}$ be the affine quadric which is $SO_3(\mathbb{C})$-homogeneous via the standard representation of $SO_3(\mathbb{C})$. Recall that $SL_2(\mathbb{C}) =: \hat{G}$ is the universal cover of $SO_3(\mathbb{C})$. Hence we can write $X = Q_{(2)} = \hat{G}/\hat{H}$, where $\hat{H}$ is the group of diagonal matrices in $\hat{G}$.

Let $g_r := \begin{pmatrix} 1 & r \\ 0 & 1 \end{pmatrix}$, $r \geq 0$, and N_r be the K-orbit of the point $g_r\hat{H}g_r^{-1}$ in $\hat{G}/\hat{H}$. Then $\{N_r \mid r \geq 0\}$ is a parameterization of the orbits of $K = SU_2$. $N_0 = S^2$ is embedded as totally real submanifold, and $N_r \cong \mathbb{P}_3(\mathbb{R})$ as a real analytic manifold. For $r > 0$ let M_r be the universal cover of N_r. Equip M_r with the CR-structure coming from the covering $M_r \to N_r$ and lift the K-action. It follows that M_r is K-homogeneous, is $\mathbb{R}$-analytically equivalent to S^3, and K acts freely and transitively. Thus, for each $r > 0$ we have a left-invariant CR-structure on K.

Identifying M_r with $K = SU_2$, for every $g \in K$ we have the $\mathbb{R}$-analytic automorphism $\operatorname{int}(g)$, $h \to ghg^{-1}$, of M_r.

Lemma 1. *$\operatorname{Int}(g) : M_r \to M_s$ is a CR-map if and only if $r = s$. If $\operatorname{int}(g)$ is a CR-map and g is not the identity, then $r = s = 1$ and* $g \in \{\begin{pmatrix} 1 & 0 \\ 0 & 1 \end{pmatrix}, \begin{pmatrix} -1 & 0 \\ 0 & -1 \end{pmatrix}, \begin{pmatrix} 1 & 1 \\ -1 & 1 \end{pmatrix}, \begin{pmatrix} -1 & -1 \\ 1 & -1 \end{pmatrix}\}$.

Proof. Let $g = \{\begin{pmatrix} a & -\bar{b} \\ b & \bar{a} \end{pmatrix}, |a|^2 + |b|^2 = 1\}$, and $\varphi := \operatorname{Int}(g)$. Then φ is a CR-map if and only if $\varphi_* : T_e^{CR}(M_r) \xrightarrow{\sim} T_e^{CR}(M_s)$ and $\varphi_* \circ J_r = J_s \circ \varphi_*$. In this case $\varphi_* = \operatorname{Ad}(g)$,

$$T_e^{CR}(M_r) = ((\begin{pmatrix} 0 & -1 \\ 1 & 0 \end{pmatrix}, \begin{pmatrix} -ir^{-1} & i \\ i & ir^{-1} \end{pmatrix}))_{\mathbb{R}} =: ((v, w_r))_{\mathbb{R}},$$

and $J_r(v) = w_r$.

The condition $\varphi_* : T_e^{CR}(M_r) \xrightarrow{\sim} T_e^{CR}(M_s)$ can be formulated as follows:

$$(1) \quad 2s\, \operatorname{Im}\{ab\} = -\operatorname{Im}\{b^2 + \bar{a}^2\}$$

so that

$$\varphi_*(v) = \operatorname{Re}\{z\} \cdot v + \operatorname{Im}\{z\}\ w_s,$$

where $z = b^2 + \bar{a}^2$,

and

$$(2) s \cdot r^{-1}(|a|^2 - |b|^2 + 2r\operatorname{Re}\{ab\}) = \operatorname{Re}\{\bar{a}^2 - b^2 - 2r^{-1}\bar{a}b\}$$

so that

$$\varphi_*(w_r) = -\operatorname{Im}\{w\} \cdot v + \operatorname{Re}\{w\}\ w_s,$$

where $w = \bar{a}^2 - b^2 - 2r^{-1}\bar{a}b$.

The condition that φ_* commutes with the J-operators can be spelled out as follows:

$$-\operatorname{Im}\{z\}\cdot v + \operatorname{Re}\{z\}\cdot w_s = -\operatorname{Im}\{w\}\cdot v + \operatorname{Re}\{w\}\cdot w_s.$$

Equivalently, $w = z$ or $b^2 \cdot r = -\bar{a}b$. Hence $b = 0$ or $rb = -\bar{a}$. If $b = 0$, then (1) implies that $\operatorname{Im}\{\bar{a}^2\} = 0$. Since $|a|^2+|b|^2 = 1$, it follows that $a = \pm 1$. Consequently $g = \pm I$ and $r = s$.

If $rb = -\bar{a}$, then (1) again implies that $\operatorname{Im}\{\bar{a}^2\} = 0$. Thus $a, b \in \mathbb{R}$. Applying (1) and (2), we see that $\varphi_*(v) = v$ and $\varphi_*(w_r) = w_s$. Now $\varphi_* = \operatorname{Ad}(g)$ and therefore application of φ_* amounts to conjugation with the matrix $g = \begin{pmatrix} a & -\bar{b} \\ b & \bar{a} \end{pmatrix}$. In particular, $\det(w_r) = \det(w_s)$, i.e. $r = s$.

Finally, a concrete calculation of centralizers shows that unless $\operatorname{Int}(g) = I$ we have $r = 1$ and g is in the group of order 4 in the statement of the proposition. □

Since a CR-isomorphism $\varphi\colon M_r \to M_s$ induces an automorphism of $\operatorname{Aut}_{CR}(M_r)$, it is useful to note

<u>Lemma 2</u>. <u>Let</u> SU_2 <u>act on</u> M_r <u>as above</u>. <u>Then</u> $\operatorname{Aut}_{CR}(M_r)^0 = SU_2$.

Proof. Set $G := \operatorname{Aut}_{CR}(M_r)^0$ and consider the $\mathfrak{g}$-anticanonical fibration

$$M = G/H \to G/N.$$

The fiber is infinitesimally defined by the isotropy algebra $\hat{\mathfrak{n}} := \mathfrak{n}_{\hat{\mathfrak{g}}}(\hat{\mathfrak{h}})$. Since $\hat{\mathfrak{g}} \supset \mathfrak{sl}_2(\mathbb{C})$ and the anticanonical fibration for the affine quadric is finite (in fact 2-1), the fibration $G/H \to G/N$ is likewise finite. Thus $\hat{G}/\hat{N}$ is a homogeneous surface on which $SL_2(C)$ acts and has an open orbit. For the same reason as above, the $\mathfrak{sl}_2(\mathbb{C})$-anticanonical bundle of this orbit is finite. The classification of homogeneous surfaces then shows that $\hat{G} = SL_2(C)$ (see [HL], [OR], [H]). Thus, since $G \supset SU_2$ and G is a real form of $\hat{G}$ (M_r is strongly pseudoconvex!), it follows that $G = SU_2$. □

<u>Remarks</u> (1) Since SU_2 acts transitively on M_r, no other Lie group of CR-transformations acts transitively.

(2) Even though M_r is strongly pseudoconvex, one can not "fill in the hole" (see [AHR]).

(3) There is no G-complexification of M_r, because $M_r = G/H$ would be contained in $\hat{G}/\hat{H}$, $\hat{G} = SL_2(\mathbb{C})$ and $\dim_{\mathbb{C}}\hat{H} = 1$. A simple

check shows that, no matter which 1-parameter group $\hat{H}$ is chosen, the generic G-orbit is not simply-connected. However, $\pi_1(M_r) = 1$. □

<u>Proposition 3</u>. <u>Let</u> N <u>be</u> <u>a</u> <u>homogeneous</u> CR-<u>hypersurface</u> <u>which</u> <u>is</u> $\mathbb{R}$-<u>analytically</u> <u>the</u> 3-<u>sphere</u> S^3. <u>Let</u> $G = Aut_{CR}(N)^0$. <u>Then</u> <u>either</u> $N = M = \partial\mathbb{B}_2$ <u>and</u> $G = PSL(3,1)$ <u>or</u> $N = M_r$ <u>for</u> <u>some</u> $r > 0$. <u>In</u> <u>the</u> <u>latter</u> <u>case</u> $G = SU_2$. <u>The</u> CR-<u>manifolds</u> M, M_r, $r > 0$, <u>are</u> <u>pair-wise</u> <u>different</u>.

<u>Proof</u>. Since the su_2-anticanonical fibration of M has positive dimensional fiber, $M \neq M_r$, $r > 0$. Analytic continuation arguments show that $Aut_{CR}(M)^0 = Aut_{\mathcal{O}}(\mathbb{B}_2)$. Thus it remains to discuss the manifolds M_r.

Suppose that $\varphi : M_r \to M_s$ is a CR-isomorphism. We may assume that $\varphi(e) = e$, where M_r and M_s are $\mathbb{R}$-analytically identified with $K = SU_2$ and e is the identity in K. Define $\Psi \in Aut(K)$ by $g \to \varphi g \varphi^{-1}$.

Now $Aut(SU_2) = Int(SU_2) \rtimes \mathbb{Z}_2$, where $\mathbb{Z}_2$ is generated by complex conjugation b. Thus $\Psi = int(h)$ or $\Psi = int(h) \circ b$ for some $h \in K$. Note that b stabilizes $T_e^{CR}(M_r)$ for all r. Since $\Psi(g) \cdot f = \varphi(g \cdot \varphi^{-1}(f))$ for all $f \in K$ and $\varphi(e) = e$, it follows that $\Psi(g) = \Psi(g) \cdot e = \varphi(g)$ for all $g \in K$. We have already shown that $\varphi = int(h)$ implies that $r = s$ (Lemma 1). Thus it remains to handle the case where $\varphi(g) = int(h)(\bar{g})$. We do this by carrying out essentially the same calculations as in the proof of Lemma 1.

Using the fact that $\varphi_*: T_e^{CR}(M_r) \xrightarrow{\sim} T_e^{CR}(M_s)$ we obtain exactly the same equations as in Lemma 1, i.e. (1) and (2). Further, $\varphi_* J_r = J_s \varphi_*$ yields $z = -w$. Thus in this case $r\bar{a}^{2} = \bar{a}b$ and arguing along the same lines as in Lemma 1 yields $r = s$. □

4. THE FIBER OF THE $\mathfrak{g}$-ANTICANONICAL FIBRATION; CLASSIFICATION AND THE CASE OF NON-DEGENERATE LEVI FORM.

The aim of this section is to understand the fiber N/H of the $\mathfrak{g}$-anticanonical fibration $M = G/H \to G/N$. Without imposing further conditions it is impossible to say more than we already know, i.e. N/H is either complex parallelizable or a flat hypersurface. Thus we will impose two kinds of further conditions which we believe to be quite natural from the complex analytic viewpoint: We first consider the case where the Levi form of $M = G/H$ is non-degenerate. Secondly, without any Levi condition, we assume that G/H admits a Kählerian complexification, i.e. a Kähler manifold X which contains G/H as a real analytic compact hypersurface. In this case we refer to $M = G/H$ as a <u>Kählerian homogeneous</u> CR-<u>hypersurface</u>. Kählerian structures on CR-manifolds have been primarily studied by methods coming from Riemannian geometry (see [KY]). It is however our aim to handle this situation by purely group theoretic and complex analytic techniques.

The first part of this section is devoted to the study of a compact homogeneous CR-hypersurface $M = G/H$ where the Levi form of M is non-degenerate, i.e. no zero eigenvalues. The basic tools for this are Prop. 2. 8, Thm. 2. 9 and the discussion just prior to Thm. 2. 9.

So let us have a look at the $\mathfrak{g}$-anticanonical fibration $G/H \to G/N$. If G/N is a CR-hypersurface, then N/H is a complex manifold and 3. 8 (the "moreover" part) implies that N/H is finite. If G/N is a complex manifold, then there is a G-complexification $\hat{G}/\hat{H}$ of G/H as described in 2. 9. In this case N/H is a flat CR-hypersurface. Again by 2. 8 it follows that $N/H = S^1$ (Note that N is connected, because G/N is simply-connected.). From the proof of 2. 9 (see [AHR] for details) it is clear that $\hat{G},\hat{H}$ can be chosen such that $\hat{N}/\hat{H} = \mathbb{C}^*$.

Summing up we have the following

<u>Theorem 1</u>. <u>Let</u> G/H <u>be a compact homogeneous</u> CR-<u>hypersurface with non-degenerate Levi form</u>. <u>Then either</u> G/N <u>is projective</u>

rational and $G/H \to G/N$ is an S^1-CR-principal bundle which is equivariantly embedded in a homogeneous non-trivial $\mathbb{C}^*$-principal bundle over G/N, or $G/H \to G/N$ is finite. □

Remark. Together with the results in 3., Theorem 1 gives a complete fine classification of homogeneous compact CR-hypersurfaces with non-degenerate Levi form. □

Now assume that $M = G/H$ is a Kählerian homogeneous CR-hypersurface. If the base G/N of the $\mathfrak{g}$-anticanonical Fibration of G/H is a hypersurface, then N/H is obviously a compact complex homogeneous parallelizable manifold, i.e. every connected component of N/H is a compact torus [W].

If the base G/N is not a CR-hypersurface, then it is projective rational. In this case by 2. 9 we have a G-complexification G/H so that

$$\begin{array}{ccc} G/H & \longrightarrow & \hat{G}/\hat{H} \\ \downarrow & \circlearrowleft & \downarrow \\ G/N & \xrightarrow[\sim]{} & \hat{G}/\hat{N} \end{array} ,$$

where N/H is a flat CR-hypersurface in $\hat{N}/\hat{H}$. Since G/H lives in a Kähler tube X and since the germ of a generic complexification is unique, there is an open neighborhood U of G/H in $\hat{G}/\hat{H}$ which is Kählerian. Thus the same is true for $N/H \to \hat{N}/\hat{H}$. Now $\hat{N}/\hat{H} = \hat{A}/\hat{\Gamma}$, where $\hat{A} = \hat{N}/\hat{H}^0$ and $\hat{\Gamma} = \hat{H}/\hat{H}^0$. Hence we must understand the following situation:

Let $\hat{G}$ be a complex Lie group, $\hat{\Gamma} < \hat{G}$ a discrete subgroup and G a Lie subgroup of $\hat{G}$ such that the G-orbit G/Γ of $\hat{\Gamma}$ in $\hat{G}/\hat{\Gamma}$ is a compact hypersurface admitting a Kählerian neighborhood in $\hat{G}/\hat{\Gamma}$.

Before stating the result in this setting, we want to discuss two special cases. First, assume that $\mathcal{O}(\hat{G}/\hat{\Gamma}) \neq \mathbb{C}$. Let $\mathfrak{m} := \mathfrak{g} \cap i\mathfrak{g}$ be the maximal complex subspace of $\mathfrak{g}$ in $\hat{\mathfrak{g}}$ Now $\mathfrak{m} \triangleleft \hat{\mathfrak{m}}$ and, since G is a hypersurface in $\hat{G}$, $\mathfrak{m}$ is 1-codimensional in $\hat{\mathfrak{g}}$. Let $G_{\mathfrak{m}}$ denote the normal complex subgroup of $\hat{G}$ which is generated by m. Since $G_{\mathfrak{m}} < G$, the $G_{\mathfrak{m}}$-orbit of $\hat{\Gamma}$ is contained in the compact hypersurface G/Γ . Let $f \in \mathcal{O}(\hat{G}/\hat{H})$ be non-constant and note that $f \mid G/\Gamma$ is bounded. Consequently the restriction of f to a $G_{\mathfrak{m}}$-orbit in G/Γ is constant, and, since the fibers of f and the $G_{\mathfrak{m}}$-orbits are

1-codimensional, it follows that the $G_{\mathfrak{m}}$-orbits are closed. Hence if $\mathcal{O}(\hat{G}/\hat{H}) \neq \mathbb{C}$, then we have

$$\begin{array}{ccc} G/\Gamma & \longrightarrow & \hat{G}/\hat{\Gamma} \\ \downarrow & \circlearrowleft & \downarrow \\ G/G_{\mathfrak{m}}\Gamma & \longrightarrow & \hat{G}/\hat{G}_{\mathfrak{m}}\hat{\Gamma} \end{array} .$$

It follows that $G/G_{\mathfrak{m}}\Gamma = S^1 \to \hat{G}/\hat{G}_{\mathfrak{m}}\hat{\Gamma} = \mathbb{C}^*$. Thus G/Γ is a torus bundle over S^1 which is equivariantly embedded in a torus bundle over $\mathbb{C}^*$. In particular $\hat{G}$ is abelian or 1-step solvable.

The second special case is where G is simply-connected and abelian. In this case $\hat{G} = \mathbb{C}^n$ and $\hat{\Gamma}$ is a discrete subgroup of rank $2n$ or $2n-1$ and Γ has rank $2n-1$. We then have

$$G/\Gamma = \mathbb{R}^{2n-1}/\Gamma_{2n-1} \to \mathbb{C}^n/\Gamma_{2n-1} = G/\Gamma$$

and there is always a fibration

$$\begin{array}{ccc} G/\Gamma & \to & \hat{G}/\Gamma \\ \downarrow & & \downarrow \\ G/L & \xrightarrow{\sim} & \hat{G}/\hat{L} \end{array} ,$$

where $S^1 = L/\Gamma \to \hat{L}/\Gamma = \mathbb{C}^*$ and $\hat{G}/\hat{L} = G/L$ is a torus (see e.g. [V]). Thus if $\hat{G}$ is abelian, we always have an S^1-CR-fibration of G/Γ onto a compact torus. Note that if $\hat{\Gamma}$ is of rank $2n$ and $\hat{G}/\hat{\Gamma}$ is an irreducible torus, then this fibration is not induced by a fibration of $\hat{G}/\hat{\Gamma}$.

If $\hat{G}$ is abelian and simply-connected and $\mathcal{O}(\hat{G}/\hat{\Gamma}) \neq \mathbb{C}$, then rank $\hat{\Gamma} = 2n - 1$ and the above shows that

$$G/\Gamma = S^1 \times G/L \to \mathbb{C}^* \times \hat{G}/\hat{L} = \hat{G}/\hat{\Gamma}.$$

If $\mathcal{O}(\hat{G}/\Gamma) = \mathbb{C}$, then $\hat{G}/\Gamma$ is called a Cousin group.

With these preparations we can now state the main results in the flat Kähler case. For this it is convenient to refer to $(\hat{G}, G, \hat{\Gamma}, \Gamma)$, $G/\Gamma \to \hat{G}/\hat{\Gamma}$ as above, as the data of a "flat Kähler CR-hypersurface, FKH".

Theorem 2. *Let $(\hat{G}, G, \hat{\Gamma}, \Gamma)$ be the data of FKH. Then $\hat{G}$ is solvable and there is a closed normal complex subgroup $\hat{I} \subset \hat{G}$ such that $\hat{I}^\circ$ is contained in G, is abelian, with $\hat{I} = \hat{I}^\circ\Gamma$ and $\hat{G}'\hat{\Gamma} \subset \hat{I}$. If G/Γ admits a non-constant CR-function, then one can choose $\hat{I} = G_{\mathfrak{m}}\hat{\Gamma}$.*

□

We now state the main structure theorem for flat Kähler CR-hypersurfaces.

Theorem 3. The following statements are equivalent:

i) $(\hat{G}, G, \hat{\Gamma}, \Gamma)$ is the data of FKH;

ii) $G_{\mathfrak{n}}$ is abelian and either $\hat{G}$ is abelian or the $G_{\mathfrak{n}}$-orbits in G/Γ are closed and the image of the representation of $\rho : \Gamma \to \mathrm{Aut}(G_{\mathfrak{n}})$, $\rho(\gamma)(g) := \gamma g \gamma^{-1}$, is finite.

Remarks: (1) If $(\hat{G}, G, \hat{\Gamma}, \Gamma)$ is the data of a FKH and $\mathcal{O}(\hat{G}/\Gamma) = \mathbb{C}$, then Theorem 3 shows that $\hat{G}$ is abelian.

(2) If $\hat{G}$ is non-abelian, then $\mathcal{O}(\hat{G}/\Gamma) \neq \mathbb{C}$. The holomorphic reduction $\hat{G}/\Gamma \to \hat{G}/G_{\mathfrak{n}}\Gamma$ is a torus bundle over $\mathbb{C}^*$ which admits a finite covering which is trivial. Conversely, given a homogeneous torus bundle which is trivial after going to a finite cover, the preimage of S^1 is a FKH.

□

Sketch of the Proof of Theorem 3. Using Theorem 2 and the fact that there is a G-invariant measure on G/Γ, standard integration arguments yield a neighborhood of G in $\hat{G}$ which is right G-invariant and admits a right G-invariant Kähler metric. Explicit calculations show that $\hat{G}$ is abelian or $G_{\mathfrak{n}}$ is the nilradical of $\hat{G}$ and G. In the latter case the $G_{\mathfrak{n}}$-orbits in G/Γ are closed ([MOS]). Furthermore, direct calculations show that $\mathrm{ad}(\mathfrak{y})$ has purely imaginary spectrum and, since $\rho(\Gamma)$ stabilizes a full lattice, it consists of torsion elements and is therefore finite.

□

The proof of Theorem 2 is carried out by induction on $\dim_{\mathbb{R}} G/H$ (see [R]). To give a taste of what is done, we give a sketch of the proof of one step in the solvable case.

Proposition 4. Let $\hat{G}$, G, $\hat{\Gamma}$ and Γ be given as in Theorem 2. Assume the following:

1) $\hat{G}$ and G are solvable and simply-connected;

2) $\mathcal{O}(\hat{G}/\hat{\Gamma}) = \mathbb{C}$;

3) The center of $\hat{G}$ is discrete;

4) There is an abelian connected complex normal subgroup $\hat{N}$ of $\hat{G}$ such that $\hat{N}\hat{\Gamma}$ is closed in $\hat{G}$ and $\hat{G}' \subset \hat{N}$.

Then there is a closed complex subgroup $\hat{I} \subset \hat{G}$ such that $\hat{\Gamma} \subset \hat{I}$,

$\hat{I}/\hat{\Gamma}$ is connected and $\hat{I}^\circ$ is a non-trivial complex Lie subgroup of $G_{\mathfrak{m}}$.

Sketch of the Proof of Proposition 4. Since $\hat{N}$ is an abelian normal subgroup of a simply-connected solvable group, it can be identified with its Lie algebra. Consider the map $\mathrm{int} : \hat{G} \to \mathrm{Aut}(\hat{N}) \cong GL(\hat{\mathfrak{n}})$, $\mathrm{int}(g)(n) = gng^{-1}$. Since $\hat{N}$ is abelian this factors through the abelian quotient $\hat{G}/\hat{N}$ and induces a representation

$$\rho : \hat{G}/\hat{N} \to \mathrm{Aut}(\hat{N}).$$

If $\hat{N} \subset G$, then there is nothing to prove. Thus we assume that $\hat{N} \not\subset G$. In this case $N = \hat{N} \cap G$ is an abelian normal subgroup of $\hat{G}$ (Note that $\hat{G}' \subset G_{\mathfrak{m}} \subset G$!) and $G/N = \hat{G}/\hat{N}$. Let $H := N\Gamma/N$. Since H is a lattice in $G/N = \hat{G}/\hat{N}$, it follows from 3) that $\rho(H) \subset \mathrm{Aut}(\hat{N})$ is not unipotent. Hence there exists $h \in H$ such that $\mathrm{ad}_h = \rho(h) - \mathrm{id}_{\hat{N}}$ is not nilpotent.

Now consider the following chains of subspaces in $\hat{N}$:

$$\hat{N}^\circ = \hat{N}, \quad \hat{N}^{k+1} = \mathrm{ad}_h(\hat{N}^k), \quad k \geq 0$$

$$N^\circ = N, \quad N^{k+1} = \mathrm{ad}_h(N^k), \quad k \geq 0$$

$$N^\circ_{\mathfrak{m}} = N_{\mathfrak{m}} = N \cap G_{\mathfrak{m}}, \quad N^{k+1}_{\mathfrak{m}} = \mathrm{ad}_h(N^k_{\mathfrak{m}}), \quad k \geq 0.$$

It follows that

$$N^{k+1} \subset \hat{N}^{k+1} \subset N^k_{\mathfrak{m}} \subset N^k \subset \ldots..$$

Now ad_h is not nilpotent. Thus the chain $(\hat{N}^k)_{k \in \mathbb{N}}$ becomes stationary for some k_0. Therefore M^{k_0} is a complex Lie subgroup of $G_{\mathfrak{m}}$ and one can show that $\hat{I} = N^{k_0} \cdot \hat{\Gamma}$ does the job.

□

The proof of Prop. 4 shows one of the main difficulties in studying homogeneous CR-manifolds: One can rather easily find fibrations by real groups, but it is much more difficult to find the appropriate complex subgroups.

□

5. CLASSIFICATION OF KÄHLERIAN CR-HYPERSURFACES

In this section we describe a fine classification of homogeneous compact Kählerian CR-hypersurfaces $M = G/H$. Let $\hat{M}$ be a Kählerian tube with $M \to \hat{M}$ and look at the $\mathfrak{g}$-anticanonical fibration $G/H \to G/N$ of G/H. We first consider the case where G/N is projective rational. In this case it follows from Thm. 2.9 that there is a G-complexification $\hat{G}/\hat{H}$ with an open Kählerian neighborhood U of G/H in $\hat{G}/\hat{H}$ and we have the diagram

$$\begin{array}{ccc} G/H & \longrightarrow & \hat{G}/\hat{H} \\ \downarrow & & \downarrow \\ G/N & \xrightarrow{\sim} & \hat{G}/\hat{N} \end{array}$$

where $\hat{N} = N_{\hat{G}}(\hat{H}^{\circ})$. By Thm. 4.2 we know that $\hat{N}/\hat{H}^{\circ}$ is solvable and that we have fibrations

$$\begin{array}{ccc} N/H & \longrightarrow & \hat{N}/\hat{H} \\ \downarrow & & \downarrow \\ N/I & \longrightarrow & \hat{N}/\hat{I} \\ \downarrow S^1 & & \\ N/L & & \end{array}$$

where $N/I = \hat{N}/\hat{I}$ is a torus, $\hat{I}/\hat{H}$ is an abelian group, and N/L is a torus. Thus we have the following picture:

$$\begin{array}{ccc} G/H & \longrightarrow & \hat{G}/\hat{H} \\ \downarrow & & \downarrow \\ G/I & \longrightarrow & \hat{G}/\hat{I} \\ \downarrow S^1 & & \\ G/L & & \\ \downarrow & & \\ G/N & \xrightarrow{\sim} & \hat{G}/\hat{N}. \end{array}$$

The most striking question at this point is whether or not G/L is Kähler. A theorem of Blanchard ([BL]) implies that G/I is a Kählerian hypersurface in $\hat{G}/\hat{I}$. So we have an S^1-CR-principal fibration of G/I onto G/L which extends to a holomorphic submersion on some Kählerian open neighborhood U of G/I in $\hat{G}/\hat{I}$. This does not in general imply that the base of the S^1-CR-principal bundle is Kähler.

Example (F. Lescure). Take $\alpha \in \mathbb{C}^*$ with $|\alpha| < 1$ and let

$$\Gamma := \left\{ \begin{pmatrix} \alpha^n & & 0 \\ & \alpha^n & \\ 0 & & \alpha^n \end{pmatrix} \in GL_3(\mathbb{C}) \mid n \in \mathbb{Z} \right\}.$$ Consider the following

fibrations:
$$\mathbb{C}^3 \smallsetminus \{0\} = GL_3(\mathbb{C})/\hat{H} \to GL_3(\mathbb{C})/\Gamma\hat{H} =: X$$
$$\downarrow \mathbb{C}^* \qquad\qquad T$$
$$\mathbb{P}_2(\mathbb{C}) = GL_3(\mathbb{C})/\hat{N} \quad,$$

where

$$\hat{H} = \left\{ \begin{pmatrix} 1 & a & b \\ 0 & & \\ 0 & & A \end{pmatrix} \mid a,b \in \mathbb{C}, \ A \in GL_2(\mathbb{C}) \right\}$$

and

$$\hat{N} = \left\{ \begin{pmatrix} \lambda & a & b \\ 0 & & \\ 0 & & A \end{pmatrix} \mid \lambda \in \mathbb{C}^*, \ a,b \in \mathbb{C}, \ A \in GL_2(\mathbb{C}) \right\}.$$

Obviously $\Gamma\hat{H}/\hat{H} \cong \mathbb{Z}$ and $T = \hat{N}/\Gamma\hat{H} = \mathbb{C}^*/\mathbb{Z}$ is a compact complex torus. Now let $E = \mathbb{C}^3 \smallsetminus \{0\} \times_{\mathbb{P}_2(\mathbb{C})} X$ denote the fiber product of the bundles $\mathbb{C}^3 \smallsetminus \{0\} \to \mathbb{P}_2(\mathbb{C})$ and $X \to \mathbb{P}_2(\mathbb{C})$. Thus we have the following diagram of holomorphic fiber bundles:

$$
\begin{array}{ccccc}
 & & E & & \\
 & T \swarrow & | & \searrow \mathbb{C}^* & \\
\mathbb{C}^3 \smallsetminus \{0\} & & | T \times \mathbb{C}^* & & X \\
 & \mathbb{C}^* \searrow & \downarrow & \swarrow T & \\
 & & \mathbb{P}_2(\mathbb{C}) & &
\end{array}
$$

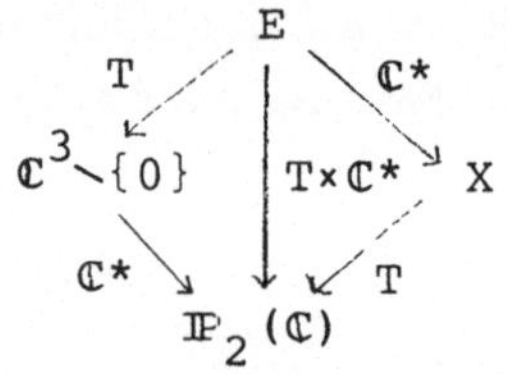

Since $H^1(\mathbb{C}^3 \smallsetminus \{0\}, \mathcal{O}) = H^2(\mathbb{C}^3 \smallsetminus \{0\}, \mathbb{Z}) = 0$, the bundle $E \xrightarrow{T} \mathbb{C}^3 \smallsetminus \{0\}$ is trivial. Thus E is Kählerian. The restriction of $E \xrightarrow{\mathbb{C}^*} T$ to $T \times S^5 \subseteq E$ is an S^1-CR-principal bundle over X. But X is not Kähler. □

The critical point in the above example is that the S^1 is fibered out in the wrong direction. In order to make this precise we need to introduce some notation.

Let X be a Kählerian complex manifold containing a compact hypersurface M. Let J be the complex structure on X, g a J-invariant (i.e. Hermitian) Riemannian metric, and $\omega(X,Y) = g(X,JY)$ its associated Kähler-form. For $p \in M$
$$E_p := \{X \in TM_p \mid \omega(X,Y) = 0 \ \forall \ Y \in TM_p\}$$

is 1-dimensional and transversal to RM_p, i.e. $TM_p = RM_p \oplus E_p$. Elementary differential geometric techniques lead to the following result.

<u>Proposition 1</u>. <u>Let</u> M <u>be a real hypersurface in a Kähler manifold</u> X. <u>Let</u> $\pi : X \to Y$ <u>be a holomorphic map onto a complex manifold</u> Y, <u>and assume that</u> $\pi|M$ <u>is an</u> S^1-CR-<u>principal bundle</u>. <u>If one of the following conditions hold, then</u> X <u>is Kähler</u>:

1) M <u>is Levi-flat</u>;
2) $\ker(\pi|M)_* = E_p$ <u>for all</u> $p \in M$.

□

The above shows that in order to show that G/L is Kähler we need to find an S^1-fibration in the "E_p direction". By induction this is reduced to the case where $\hat{N}/\hat{H}°$ is abelian. In this situation one has the N/H-CR(resp. $\hat{N}/\hat{H}$-holomorphic)-principal bundle $G/H \to G/N$(resp. $\hat{G}/\hat{H} \to \hat{G}/\hat{N}$). Since G/N is projective rational, a maximal semi-simple compact subgroup K of G acts transitively on G/N. Thus the group $A = N/H \times K$ acts transitively as a group of CR-transformations on G/H and we may write G/H = A/D. Since A is compact and also acts on $\hat{G}/\hat{H}$, we may assume that the Kählerian neighborhood U and the Kähler-metric ω are A-invariant.

Let $\pi : A \to A/D$ denote the projection and let $0 := \pi(e)$. For a subspace $V \subset T(A/D)_0$ let $\tilde{V}$ denote the space $\pi_*^{-1}(V) \subset \mathfrak{a}$. Since the Kähler-form ω is A-invariant, it determines an element $F \in H^2(\mathfrak{a}, \mathbb{R})$ given by $F = \pi^*(\omega|A/D)|\mathfrak{a}$. Note that C = N/H is the connected component of the center of A, and thus the Lie algebra splits, $\mathfrak{a} = \mathfrak{c} \oplus \mathfrak{k}$. Let $\rho_1 : \mathfrak{a} \to \mathfrak{c}$ and $\rho_2 : \mathfrak{a} \to \mathfrak{k}$ denote the projections. Now $[\mathfrak{k},\mathfrak{k}] = \mathfrak{k}$, $H^2(\mathfrak{k},\mathbb{R}) = 0$, and $d\varphi = 0$ for all $\varphi \in C^2(\mathfrak{c},\mathbb{R})$. Hence, elementary calculations show that $F = \rho_1^*(F_1) + \rho_2^*(dF_2)$, where $F_1 \in H^2(\mathfrak{c},\mathbb{R})$ and $F_2 \in C^1(\mathfrak{k},\mathbb{R})$. By the non-degeneracy of the Killing form B on $\mathfrak{k}$, we may write $F_2(X) = B(X,W)$ for some $W \in \mathfrak{k}$. Thus we have

$$F(X,Y) = F_1(\rho_1 X, \rho_1 Y) + B([\rho_2 X, \rho_2 Y], W).$$

It follows that $\tilde{E}_0 = \rho_1(\tilde{E}_0) + \rho_2(\tilde{E}_0)$. Furthermore, since the C-orbits in A/D carry a CR-hypersurface structure we obtain $\tilde{E}_0 = \rho_1(\tilde{E}_0) + \mathfrak{d}$. The above formula also implies that $\tilde{E}_0 \cap \mathfrak{k}$ is the centralizer of $W \in \mathfrak{a}$ and D is contained in the centralizer of the torus in K which is generated by W.

Now, excluding the Levi-flat case in which we show $A/D \underset{CR}{\cong} N/H \times G/N$ and in which G/L is Kähler (Prop.1), we show that $\tilde{E}_0 = \rho_1(\mathfrak{d}) + \mathfrak{d}$. From this and Prop.1 it follows that the fibration with the group $\langle \exp(\tilde{E}_0 \cap k) \rangle \cdot D$ yields an S^1-CR-principal fibration $G/H \to G/L$ where G/L is Kählerian.

It is now possible to state the first result in the classification of G/H.

<u>Theorem 2</u>. <u>Let G/H be a homogeneous compact Kählerian CR-hypersurface, where G is connected and the base G/N of the $\mathcal{Y}$-anticanonical fibration of G/H is projective rational. Then there exists a complex Lie group $\hat{G}$ with $G \subset \hat{G}$, N/H° is solvable, and one has</u>

$$\begin{array}{ccc} G/H & \longrightarrow & \hat{G}/\hat{H} \\ \downarrow & & \downarrow \\ G/N & \xrightarrow{\sim} & \hat{G}/\hat{N} \end{array}$$

<u>as in Thm. 2.9. Moreover there is a closed complex Lie group $\hat{I} \subset \hat{N}$ such that $\hat{I}^\circ \subset N \cdot \hat{H}^\circ$, $\hat{N}' \cdot \hat{H} \subset \hat{I}$, $\hat{I}^\circ/\hat{H}^\circ$ is abelian, $\hat{I}/\hat{H}$ is a compact torus, and there is a closed subgroup $L \subset G$ such that $H \subset G \cap \hat{I} \subset L \subset N$ and $G/G \cap \hat{I} \to G/L$ is an S^1-CR-principal bundle over the homogeneous compact Kähler manifold G/L. If $\hat{J} \subset \hat{G}$ is a closed complex subgroup such that $\hat{I} \subset \hat{J}$, $\hat{J}/\hat{H} \subset G/H$, and $\hat{J} = \hat{J}^\circ \cdot \hat{I}'$, then $\hat{J} \subset \hat{N}$ and $\hat{J}$ has the same properties as $\hat{I}$.</u>

□

The remaining case is that in which G/N is a CR-hypersurface. In this situation we know that either $G/N = S^1 \times Q$ (Q projective rational) or a maximal compact semi-simple subgroup K of G acts transitively on G/N (see Prop.3.1). In the latter case $\pi_1(G/N) = 1, \mathbb{Z}_2$ (see [AHR]). In both cases the components of N/H are compact complex tori. The fine classification in this case is handled by methods similar to those in the proof of Thm. 2. For this consider the bundle $G/N^\circ \cap H \to G/N^\circ$, i.e. a torus principal bundle. Again K acts transitively on G/N° and $A := N^\circ/N^\circ \cap H \times K$ acts transitively on $G/N^\circ \cap H$. Using F as in the proof of Thm. 2 and observing that the orbits of

$C = N°/N° \cap H$ are complex, one shows that

$$G/N° \cap H \underset{CR}{=} N°/N° \cap H \times G/N°.$$

We close by stating the main result on the bundle structure of a compact homogeneous Kählerian CR-hypersurface.

Theorem 3. *Let M be a compact homogeneous connected Kählerian CR-hypersurface. Then either*

1) *M is a torus bundle over an S^1-CR-principal bundle over a homogeneous complex Kähler manifold $T \times Q$ (T a complex torus, Q projective rational),*

or

2) *either M or a 2-1 covering of M is a CR-product $T \times M'$, where T is a compact compact torus and M^1 is a simply-connected compact homogeneous Kählerian CR-hypersurface which either lies in $\mathbb{P}_n(\mathbb{C})$ or is a 2-1 covering of a CR-hypersurface in $\mathbb{P}_n(\mathbb{C})$.*

□

Remarks. (1) The hypersurfaces M^1 in (2) above are classified via the Stein-Rational fibration (see § 3).
(2) The reader should note that by using Thm.4.3 one can say much more about the structure of M in case (1) above. In particular the fiber of the $\mathfrak{g}$-anticanonical fibration $M \to Q$ is either an S^1-bundle over a torus or vice versa.

For detailed proofs of the results in this section see [R].

References

[AHR] Azad,H., Huckleberry,A., Richthofer,W.: Homogeneous CR-Manifolds, Crelles J. (to appear)

[AF] Andreotti,A.,Fredricks,G.A.: Embeddability of real analytic Cauchy-Riemann Manifolds, Ann. Scuola Norm. Pisa 6, 285-304 (1979)

[AZ] Azad,H.: Levi-Curvature of Manifolds with a Stein-Rational Fibration, Manuscripta Math. (1985)

[BFS] Buchner,M., Fritsche,K., Sakai,T.: Geometry and cohomology of certain domains in the complex projective space, J.reine ang. Math. 323, 1-52 (1981)

[BL] Blanchard,A.: Sur les variétes analytiques complexes, Ann. Sci. Ec. Norm. Sup. 73, 157-202 (1956)

[BR] Borel,A., Remmert,R.: Über kompakte homogene Kählersche Mannigfaltigkeiten, Math. Ann. 145, 429-439 (1962)

[BS] Burns,D., Shnider,S.: Spherical hypersurfaces in complex manifolds, Inv. Math. 33, 223-246 (1976)

[G] Greenfield, S.J.: Cauchy-Riemann equations in several complex variables, Ann. Scuola Norm. Pisa, 257-314 (1968)

[H] Huckleberry, A.: Homogeneous Surfaces, (to appear)

[HL] Huckleberry, A.T., Livorni, E.L.: A classification of homogeneous surfaces, Can. J. Math., Vol.XXXIII, No.5, 1097-1110 (1981)

[HO] Huckleberry, A.T., Oeljeklaus, E.: A characterization of complex homogeneous cones, Math. Z. 170, 181-194 (1980)

[HS] Huckleberry, A.T., Snow, D.: Almost-homogeneous Kähler manifolds with hypersurface orbits, Osaka J. Math. 19, 763-786 (1982)

[KY] Kon, Masahiro, Yano, Kentaro: CR Submanifolds of Kählerian and Sasakian Manifolds, Progress in Math. v.30, Birkhäuser, (1983)

[M] Matsushima, Y.: Sur les espaces homogènes kählériens d'un groupe de Lie réductif, Nagoya Math. J. 11, 53-60 (1957)

[MN] Morimoto. Y., Nagano, T.: On pseudo-conformal transformation of hypersurfaces, J. Math. Soc. Japan 15, 289-300 (1963)

[MOS] Mostow, G.D.: Some applications of representative functions to solv-manifolds, Am. J. Math. 93, 11-32 (1971)

[OR] Oeljeklaus, K., Richthofer, W.: Homogeneous Complex Surfaces, Math. Ann. 268, 273-292 (1984)

[R] Richthofer, W.: Homogene CR-Mannigfaltigkeiten, Dissertation, Ruhr-Universität Bochum (1985)

[R1] Rossi, H.: Homogeneous strongly pseudoconvex hypersurfaces, Rice Studies 59 (3), 131-145 (1973)

[R2] Rossi, H.: Attaching analytic spaces to an analytic space along a smooth pseudoconcave boundary, Proc. of the Conf. on Complex Analysis, Minneapolis 1964, Springer Verlag (1965)

[RP] Rossi, H., Patton, C.H.: Unitary Structures on Cohomology (to appear)

[S] Samelson, H.: A class of complex-analytic manifolds, Portugaliae Math. 12, 129-132 (1953)

[TA] Tanaka, N.: On the pseudoconformal geometry of hypersurfaces of the space of n complex variables, J. Math. Soc. Japan 14, 397-429 (1962)

[TO] Tolimieri, R.: Heisenberg manifolds and theta functions, Trans. Am. Math. Soc. 239, 293-319 (1978)

[V] Vogt, C.: Line bundles on toroidal group, J.reine ang. Math. 335, 197-215 (1982)

[W] Wang, H.C.: Complex parallelisable manifolds. Proc. Am. Math. Soc. 5, 771-776 (1954)

[WO] Wolf, J.A.: The action of a real semi-simple group on a complex flag manifold. I: Orbit structure and holomorphic arc components. Bull. Amer. Math. Soc. 6, 1121-1237 (1969)

A. Huckleberry and W. Richthofer
Fakultät und Institut für Mathematik
Ruhr-Universität Bochum
Universitätsstraße 150
D-4630 Bochum 1, FRG

Problems of Value Distribution in Complex Analysis for Several Variables

P. Lelong
Université de Paris VI
Mathématiques
4, Place Jussieu F-75005 Paris

dedicated to Wilhelm Stoll

INTRODUCTION

Value distribution theory for holomorphic functions and holomorphic mappings is an important part of complex analysis - For a long time, the theory was established only for mappings $\mathbb{C} \to \bar{\mathbb{C}}$, i.e. , for the values of entire or meromorphic functions defined in the complex plane $\mathbb{C}$, with ranges in $\mathbb{C}$ or $\bar{\mathbb{C}}$ (identified with the Riemann sphere).

We are indebted to Wilhelm Stoll for giving beautiful extensions of the theory to complex manifolds, opening ways to new problems of value distribution theory, and giving to the theory a larger domain of research.

How does one generalize to C^n the classical results obtained formely for holomorphic mappings $f : \mathbb{C} \to \bar{\mathbb{C}}$? If f is not a rational function, the image $f(C)$ gives a covering of $\bar{\mathbb{C}}$ of unbounded degree. To consider the general situation of $f : X \to Y$, we must assume that an exhaustion $\{X_r\}_{r>o}$ of X is given , $X_r \supset X_{r'}$, for $r > r'$, and $\bigcup_{r>o} X_r = X$. We suppose X_r is a relatively compact domain in X , with image $f(X_r) \subset Y$ of bounded degree $n(r)$, defined by

$$n(r,a) = \text{card } [X_r \cap f^{-1}(a)] \qquad \text{for} \quad a \in Y ,$$

and

$$n(r) = \sup_{a \in Y} n(r,a) < \infty \qquad \text{for} \quad 0 < r < \infty .$$

and the problem is to discover situations in which $n(r,a)$ has the same "asymptotic growth" as $r \to +\infty$, for almost all $a \in Y$. Then the theory will consist in giving connections between types of asymptotic growth and classes of "exceptional sets". Historically the theory was built after the famous theorem of E.Picard (1883), to obtain a more comprehensive view of Picard's result ; the importance of the growth indicators in this problem appeared in the works of E.Borel, R.Nevanlinna (see [12]). On the other hand much of the theory was proved to be a consequence of particular properties of conformal mappings (or more generally of some classes of quasi-conformal mappings $\mathbb{C} \to \bar{\mathbb{C}}$) according to the paper of L.Ahfors [1].

The general problem was to define a class Γ of manifolds and a class φ of mappings $f : X \to Y$, for X and Y in Γ , with the following properties :

(I) - The fiber $f^{-1}(a)$ is discrete or of the same pure dimension for "almost all" $a \in Y$.

(II) - There exists an exhaustion $\{X_r\}$ of X by relatively compact subsets of X such that $n(r,a)$ or some 'area" of $f^{-1}(a) \cap X_r$ has the same asymptotic growth for "almost all" $a \in Y$.

Results for this very general problem were given by M.H.Schwartz (see [14] and [15]) for differentiable manifolds in R^n and C^n . She extended Nevanlinna's theory using the Gauss-Bonnet formula in the same way that Ahlfors in [1] used the Euler-Poincaré characteristic of the Riemann sphere $Y = \bar{C}$.

In the following we suppose X and Y are complex analytic subvarieties in C^n of (complex) pure dimension p for X and q for Y , $q \leqslant p \leqslant n$; $f : X \to Y$ will be a holomorphic mapping $X \to Y$ with dense image in Y . The fiber $f^{-1}(a)$ is an analytic subvariety in X , but it is not of pure dimension $p-q$ for all $a \in Y$. Thus, the class of the "exceptional sets" we have to introduce in the theory must contain the class of the analytic subvarieties of Y .

Comparison with results of the classical case gives another indication on how one should define this class. For $X = \mathbb{C}$ and $Y = \mathbb{C}$, and $f : X \to Y$ an entire function, we have a general result. Let be $M(r,a) = \frac{1}{2\pi}\int_0^{2\pi} \log|f(re^{i\theta}) - a|d\theta$, such that $n(r,a) = \frac{\partial}{\partial \log r} M(r,a)$;

$T(r) = \frac{1}{2\pi}\int_0^{2\pi} M(r,e^{i\varphi})d\varphi$ is an indicator of growth of f for $r \to +\infty$. Then (see [12] , p. 264) : $\lim_{r=\infty} \frac{M(r,a)}{T(r)} = 1$ for all a with the possible exception of a polar set (i.e. a set of null capacity) in C. This result is not the deepest of the theory, but it proves that the set A of the Picard's or Nevanlinna's exceptionnal values (we have card $A \leqslant 1$ in the first case, A is at most countable in the second case) belongs to the class of polar sets in C . Indeed the preceding result is a consequence of the property of that non-polar subsets of $\mathbb{C}$ are control sets (see later the definition) in the following situations : (I) for negative subharmonic functions in a domain $G \supset A$ - (II) for the class of subharmonic functions $V(z)$ in $\mathbb{C}$ satisfying the estimate $V(z) \leqslant C_V + \log[1 + |z|]$. Obviously $M(r,a)$ is in the second class for $r > 0$ fixed and for $a \in \mathbb{C}$.

In C^n , the sets which are not pluripolar will be control sets for plurisubharmonic functions. For the general problem of holomorphic mappings

$$F : X \to Y$$

with X , Y complex analytic subvarieties in C^n , of pure dimension p and q , the exhaustion will be given by the balls $B(r) = [z \in C^n ; \|z\| < r]$ and $X_r = B(r) \cap X$. A general result of L.GRUMAN [7] gives a bound of the area $\sigma_X(a,r)$ of the fiber $X_r \cap F^{-1}(a)$, with the possible exception of a pluripolar subset of Y. Thus, for a wide class of problems in value distribution theory in C^n , or in complex spaces, we will probably find for exceptional sets classes Γ of sets satisfying both of the properties : (I) Γ is a subclass of the class of pluripolar sets. (II) - Γ contains the class of analytic subvarieties.

Most results given here will appear in the forthcoming book [10]. First we recall the property that non-pluripolar sets are control sets (see [9b]) and consequences of the works of E.Bedford and B.A.Taylor [2] and of J.Siciak [16]. In the second part, a short survey of the results of L.Gruman [7] is given. In the last part of this lecture, I will consider a particular problem of value distribution theory with F of finite order. Then it is possible to obtain (using elementary properties of holomorphic functions) classes of analytic sets as exceptional sets. The result is a generalization of a result [9a] , I have given former. It contains results given recently in [15] by N.Sibony and Pitt Man Wong .

I. PLURIPOLAR SETS AND CONTROL SETS (see [9b])

I recall : a set A in a complex manifold X is called pluripolar in X if there exists a function V , plurisubharmonic in X , and $A \subset A' = [z \in X ; V(z) = -\infty]$. We denote by $PSH(X)$ the convex cone on R_+ of plurisubharmonic functions defined in X . If X is a domain of C^n or a Stein manifold, locally pluripolar sets in X are pluripolar in X . We denote by $g^*(z) = \limsup_{y \to z} g(y)$ the upper regularization of g . If $g = \limsup_n f_n$, where $f_n \in PSH(X)$ is a sequence locally bounded from above, we have $g^* \in PSH$ and the set $[z \in X ; g(z) < g^*(z)]$ is known to be locally pluripolar (see [2]). Given a set A in a domain G of $\mathbb{C}^n$ we denote by $PSH(A,G)$ the convex cone of functions $V \in PSH(G)$ with conditions $V(z) \leq 0$ for $z \in G$ and $V(z) \leq -1$ for $z \in A$.

Definition. A set A in a bounded domain G of C^n is called a control set for the class $PSH(G)$ if

$$g_{A,G}(z) = [\sup_V V(z) \; ; \; V \in PSH(A,G)]$$

has a strictly negative regularization $g^*_{A,G}(z)$ in G.

Indeed, $g^*_{A,G}(z) = \limsup_{y \to z} g_{A,G}(y) \in PSH(G)$ and $g^*_{A,G}(z) \leq 0$. Then we have only two possibilities : (I) $g^*_{A,G}(z) \equiv 0$, or (II) $g^*_{A,G}(z) < 0$ for all $z \in G$, by the maximum principle. Then :

Proposition 1. A set A in a bounded domain G of C^n is a control set (for $PSH(G)$) if and only if it is not contained in the $-\infty$ of a negative function $V \in PSH(G)$.

If $A \subset A' = [z \in G \; ; \; V(z) = -\infty]$ and $V(z) \leq 0$, $V \in PSH(G)$, there exists $z_o \in G$, such that $V(z_o) > -\infty$. Then for the sequence $V_q = \sup(-1, \frac{1}{q} V) \in PSH(G)$, we have $V_q \in PSH(A,G)$, $\lim V_q(z_o) = 0$ and $g^*_{A,G}(z_o) = g_{A,G}(z_o) = 0$. Then $g^*_{A,G}(z) \equiv 0$; therefore A is not a control set. Conversely, if A is not a control set, then $g^*_{A,G}(z) \equiv 0$. The set $g_A(z) = g^*_A(z)$ is a dense subset in G ; then there exists $z_o \in G$ such that $g_{A,G}(z_o) = g^*_{A,G}(z_o) = 0$. We consider a sequence $V_q \in PSH(A,G)$ such $\lim V_q(z_o) = 0$. We can suppose, taking a subsequence, that $\sum_q |V_q(z_o)| < \infty$. Then $S(z) = \sum V_q(z) \in PSH(G)$ has value $-\infty$ on A, which proves that A is contained in the $-\infty$ of the negative plurisubharmonic function $S(z)$.

We call $g^*_{A,G}(z)$ the extremal function of A with respect to G ; $g^*_{A,G}(z) = [\sup \; V(z) \; , \; V \in PSH(A,G)]^*$.

Proposition 2 (Inverse function theorem for plurisubharmonic functions) (see [9b] and [10]).

Given a domain $G \subset C^n$ and $V(z,u) \in PSH(G \times C)$, $z \in G$, $u \in \mathbb{C}$, we consider

$$M(z,r) = \sup V(z,u) \quad \text{for} \quad |u| \leq r, \; u \in \mathbb{C} \; ,$$

and for $m > M(z,1)$ (respectively for $\zeta \in \mathbb{C}$, $|\zeta| < e^{-M(z,1)}$), we define

$$\delta(z,m) = [\sup r \ , \ r>0, \ M(z,r)<m] \ ; \ U_1(z,m) = -\log\delta(z,m)$$

(1) $$\delta'(z,\zeta) = \delta(z, -\log|\zeta|) = [\sup r, \ M(z,r) + \log|\zeta| < 0]$$

$$U_2(z,\zeta) = -\log\delta'(z,\zeta) \ .$$

Then U_1 is a negative plurisubharmonic function defined for $z \in G_q = [z \in G \ ; \ M(z,1) < q]$ and $m > q$; $U_2(z,\zeta)$ is a negative plurisubharmonic function in $\Delta = [z \in G \ , \ |\zeta| < e^{-M(z,1)}$ which remains plurisubharmonic for $\zeta = 0$. The theorem replaces an unbounded family of functions $z \to M(z,r)$, for $r \to +\infty$, by a family of negative plurisubharmonic functions. For the proof see [9b] and [10].

Remarks. (I) If a set A is pluripolar in a manifold X , it is pluripolar in each subdomain $G \subset X$. Hence, if A is a control set in G it is a control set in $G' \supset G$ for the class $PSH(G')$ with condition $V(z) \leq 0$ in G' .

(II) If G is an unbounded domain in C^n , a set $A \subset G$, is not pluripolar if and only if there exists a compact $K \subset A$ and a bounded domain $G' \subset G$ such that K is not pluripolar in G' . For $G = C^n$, A is not pluripolar if and only if it is a control set for the class of the functions $V \in PSH(C^n)$ with condition

(2) $$V(z) \leq C_V + \log(1 + \|z\|) \quad , \quad V \in PSH(C^n)$$

A is not pluripolar if and only if $\gamma(z) = [\sup_V V(z)]^*$, for V with conditions (2) is plurisubharmonic in C^n and not the constant $+\infty$ (see [2]) .

Proposition 3. (I) Given a bounded domain G in C^n and a compact $A \subset G$, not pluripolar, and $\psi(t)$, a function of t defined for $t \geq 0$, which is increasing and convex of t , with $\lim_{t=\infty} \psi(t) = \infty$, we consider in $G \times \mathbb{C}$, for $x \in G$, $u \in \mathbb{C}$, a function $V \in PSH(G \times \mathbb{C})$, and $M(x,r) = \sup_{|u|\leq r} V(x,u)$ for $x \in G$, $r > 0$.

We suppose a bound is given

$$M(x,r) \leq \psi(\log r) \qquad \text{for} \quad x \in A \ , \ r > r_0 > 0 \ .$$

Then if $M(x,1) \leq m_0$ is bounded in G , there exists a control function $\sigma_{A,G}(x) \in PSH(G)$, not depending on V but only on A, G and $\sigma_{A,G}(z) \geq 1$,

<u>such that we have the bound</u>

(3) $$M(x,r) \leq \psi[\sigma_{A,G}(x) \log r] \quad \underline{\text{for}} \quad x \in G \ , \ r > r_1$$

<u>with</u> $r_1 = \sup [r_o, \psi^{-1}(m_o)]$.

(II) <u>If we do not suppose</u> $M(x,1)$ <u>bounded in</u> G, <u>by the same assumptions on</u> G, A , <u>we have</u>

$$\limsup_{r=\infty} [\psi(\sigma_{A,G}(x) \log r)]^{-1} M(x,r) \leq 1 .$$

For the proof of (I) , see [9b] or [10] : the function $\sigma_{A,G}(x)$ is defined by $\sigma_{A,G}(x) = [- g^*_{A,G}(x)]^{-1} \in PSH(G)$.

For (II) we take an exhaustion of G by domains G_q where G_q is an open component of the set $M(x,1) < q$, $q \in \mathbb{N}$, $q \to +\infty$.
For $q > q_o$, the compact set A is contained in G_q and by (I) we obtain a bound

(4) $$M(x,r) \leq \psi[\sigma_q(x) \log r] \text{ for } x \in G_q \ , \ r > r_q$$

$r_q = \sup [r_o , \psi^{-1}(q)]$. For $q \to +\infty$, x fixed, the sequence of the control functions g^*_{A,G_q} which is defined for $q > q(x)$ is decreasing and has a limit $g^*_{A,G} \in PSH(G)$. Then on a compact set $K \subset G$, $\sigma_{A,G_q}(x)$ is defined and decreasing in $q > q(K)$; $\sigma_{A,G_q}(x)$ has a limit $\sigma_{A,G}(x)$ uniformly on compact subsets of G .

<u>Remark</u>. If A is not a pluripolar compact subset of G , there exists $x_o \in A$, such that $g^*_{A,G}(x_o) = -1$, and $\sigma_{A,G}(x_o) = 1$. The open sets $\sigma_{A,G}(x) < 1 + \varepsilon$ are not empty for $\varepsilon > 0$ (see [2]) .

Now we consider the order $\rho(x)$ in x of $u \to V(x,u)$ and we can define for holomorphic functions $F(x,u)$ and for $V(x,u) \in PSH(G \times \mathbb{C})$ the class of functions of <u>finite order</u> with respect to u . For $V \in PSH(G \times \mathbb{C})$, we define $M'(x,r) = \sup [1, M(x,r)]$ and

(5) $$\rho(x) = \limsup_{r=\infty} (\log r)^{-1} \log M'(x,r)$$

or equivalently, we consider $V' = \sup(V,0)$.
For $F \in H(G \times G)$, an holomorphic function, we define the order $\rho(x)$ as the order of $V = \log |F|$ in $x \in G$.

Proposition 4. <u>Given</u> $V \in PSH(G \times \mathbb{C})$, <u>there exists only the following possibilities for the ordre</u> $\rho(x)$ <u>of</u> $u \to V(x,u)$:

(I) - $\rho(x) \equiv 0$

(II) - $\rho^*(x) \equiv +\infty$ and $\rho(x) < \infty$ is a pluripolar set

(III) - $\rho^*(x) \in PSH(G)$ and the set $\rho(x) < \rho^*(x)$ is pluripolar.

Moreover if $V(x,u) \in PSH(C^n \times C)$ is of finite total order in $C^n \times C$, then in (III), $\rho^*(x)$ is a constant ρ.

For the proof, one applies Proposition 2 with $\delta(x,m) = [\sup r, r > 0, M(x,r) < m]$ and

$$-\frac{1}{\rho(x)} = \limsup_{p=\infty} \frac{-\log \delta(x,m)}{\log m} \leqslant 0 .$$

We obtain $W(x) = -\frac{1}{\rho^*(x)} \in PSH(G)$ or $W(x) \equiv -\infty$. If $W(x) = -\infty$, then $\rho^*(x) = \rho(x) = 0$. If $W(x) \in PSH(G)$, we have (III) if $W(x) < 0$, and (II) if $W(x) \equiv 0$.

Thus, we can give a definition :

Definition. <u>A function</u> $V(x,u) \in PSH(G,C)$, $x \in G$, $u \in \mathbb{C}$ <u>is called of finite ordre with respect to</u> u <u>if the order</u> $\rho(x)$ <u>of</u> $u \to V(x,u)$ <u>is finite for</u> $x \in A$, <u>and</u> A <u>a not pluripolar subset in</u> G.

Proposition 5. <u>If</u> $V(x,u)$, <u>a plurisubharmonic function of</u> $(x,u) \in G \times C$, $x \in G$, $u \in \mathbb{C}$, <u>is of finite order with respect to</u> u, <u>its order</u> $\rho(x)$ <u>is bounded on each compact in</u> G, <u>and</u> $\rho(x) \leqslant \rho^*(x)$, <u>with</u> $\rho^* \in PSH(G)$ <u>and the set</u> $\rho(x) < \rho^*(x)$ <u>is pluripolar in</u> G.

It is a consequence of Proposition 4 and of the property that a "negligible" set (i.e. , the set of all points $x \in G$ such that $\rho(x) < \rho^*(x)$) is pluripolar (see [2], [3])) .

II. THE GENERAL RESULTS OF L.GRUMAN (see [7]) .

A - A first result of L.Gruman is for the holomorphic mappings

$$F : X \to Y$$

where X and Y are analytic subvarieties in C^n, with pure (complex) dimension p for X and q for Y, $q \leqslant p \leqslant n$. The exhaustion in X is given by balls $B(r) = [\, z \in C^n ; \|z\| \leqslant r \,]$.

Theorem 1. <u>Given</u> $\varepsilon > 0$, $\alpha > 1$, <u>there exists a pluripolar set</u> $E \subset Y$, <u>with following properties</u> : (I) - $F^{-1}(a)$ <u>is of complex dimension</u> $p-q$ <u>for</u> $a \in Y$, $a \notin E$. (II) <u>If</u> $\sigma_X(a,r)$ <u>is the area of</u> $B(r) \cap F^{-1}(a) \cap X$ <u>and</u> $\sigma_X(r)$ <u>the</u> (p <u>dimensionnal)</u> <u>area of</u> $B(r) \cap X$, <u>and</u> $M_F(r) = \sup. \|F(z)\|$ <u>for</u> $\|z\| \leqslant r$, <u>then</u>

for $a \in Y$, $a \notin E$,

(6) $$\lim_{r=o} \frac{r^{2q}\ \sigma_X(a,r)}{(\log r)^{\alpha}\ \sigma_X(r+\varepsilon r)[\log M_F(r+\varepsilon r)]^q} = 0 .$$

An important step in the proof of (8) is the following : let be $V \in PSH(C^n)$, $V > 0$, $M_V(r) = \sup_{z<r} V(z)$ and $[X]$ the current of integration on the sub-variety X . Given γ , $0 < \gamma < 1$, there exists a bound

(7) $$\int_{B'(r)} [X] \wedge \beta^{p-q} \wedge (i\partial\bar\partial V) \leqslant c(\gamma,n)\,[M_V(r)]^q\ \sigma_X(a,r)r^{2q}$$

in a ball $B'(r) = B(\gamma^q r)$, for a constant $c(\gamma,n)$ depending only on n and γ; $\beta = i\ \partial\bar\partial\|z\|^2$. The proof uses the Monge-Ampère operator defined for $V \in PSH \cap L^\infty_{loc}$ by E.Bedford and B.A.Taylor (see [2]) and the property that the extremal function of a non-pluripolar set is a control function. The denominator of (6) can be replaced by

$(\log r)^{\alpha}\int_{B(r+\varepsilon r)} i\partial\bar\partial \log[1 + |F|^2]^q \wedge [X] \wedge \beta_{p-q}$ (β_s is the volume élément in C^s) .

B - A second result of L.Gruman in [7] gives an upper bound and a lower bound of the area $X \cap \ell \cap B(r)$ for "almost" all linear subspaces ℓ of C^n .

Let be $X \subset C^n$ an analytic subvariety of pure dimension p , and $q \geqslant n-p$. We denote by $G_q(C^n)$ the Grassmann manifold of the subspaces of dimension q in C^n , and by Θ_ℓ the current of integration on $\ell \cap X$. Let be

$$\sigma_X(\ell,r) = \int_{B(r)} \Theta_\ell \wedge \beta_{p+q-n}$$

For $z \in C^n$, we denote $L_z = [\,\ell \in G_q(C^n)\ ;\ z \in \ell\,]$. Then if Y is an irreducible analytic variety in a domain $\Omega \subset G_q(C^n)$, such that for each $z \in C^n$, we have $\dim(Y \cap L_z) = \dim Y - (n-q)$ or $Y \cap L_z = \phi$, we say that Y is non-degenerate.

Theorem 2. If Y is a non-degenerate irreducible subvariety in a domain $\Omega \subset G_q(C^n)$ and $E \subset Y$ is a non-pluripolar set in Y , and if X is an analytic subvariety of pure dimension $n-q$ in C^n , then :

(I) - For $\varepsilon > 0$, $\beta > 1$, there exists $\ell \in E$ such that

$$\lim_{r=\infty} [(\log r)^{\beta}\ \sigma_X(r+\varepsilon 1)]^{-1}\ \sigma_X(\ell,r)r^{2(n-q)} = 0 .$$

(II) Given an increasing sequence r_m , $\lim r_m = +\infty$, there exists

$t > 0$ and $\ell \in E$ such that

(8) $$\lim_{m\to\infty} \inf[\sigma_X(tr_m)]^{-1} \quad \sigma_X(\ell,r_m)r_m^{2(n-q)} = 0 .$$

Such general results have important consequences for the development of value distribution theory : future researchs may focus attention on exhibiting situations -perhaps very particular- in which we can assert that the "exceptional" set has more precise properties, for example that it is an analytic set or an algebraic set. In the following we consider an example where the mappings is of finite order.

III. SPECIAL PROPERTIES OF A MAPPING WITH FIBER OF FINITE ORDER.

Let be $\Omega = G \times \mathbb{C}$, $\Omega \subset C^n$ and G a pseudo-convex domain in C^{n-1} and $F(x,u) \in H(\ )$, holomorphic of (x,u) , $x \in G$, $u \in \mathbb{C}$.

Theorem. Let $F \in H(\Omega)$ be of finite order with respect to u for $x \in G$. We denote by $E_s \subset G$ the set of the points $x \in G$, such that $u \to F(x,u)$ has at most $s \geqslant 0$ zeros (with multiplicity). Let be E_∞ the analytic set in G of the x such $u \to F(x,u) \equiv 0$. Then

$$E_s' = E_s \cup E_\infty$$

is a closed set in G , and $E_s' = G$ or there exists an analytic subvariety $M_s \subset G$ such $E_s' \subset M_s$. Moreover $\widetilde{M}_s = M_s \times C$ is defined in Ω equation $\varphi_\nu(x,u) = 0$, where φ_ν is a polynomial with constant coefficients of a finite number of the variables $\xi_m = \dfrac{\partial^m F}{\partial u^m}(x,u)$

First we prove that $G \setminus E_s'$ is open. Let $x_o \in G$, $x_o \notin E_s'$: the entire function $\varphi : u \to F(x_o,u)$ has $s' > s$ isolated zeros , and there exists $r > 0$ such that the disk $|u| < r$ contains s' zeros of φ and $|F(x_o,u)| \geqslant c > 0$ for $|u| = r$. By continuity, there exists $\alpha > 0$ such that $|F(x,u)| \geqslant \frac{c}{2}$ and

$$n(x,r) = \frac{1}{2\pi i} \int_{|a|=r} \frac{\partial F}{\partial u}(x,u) F^{-1}(x,u)du = s'$$

for $\|x-x_o\| < \alpha$; the ball $B(x_o,\alpha)$ does not belong to E_s' ; E_s' is closed.

Now we prove that E_s' is contained in an analytic subvariety M in a compact ball $B \subset G$, or (equivalently) that $\widetilde{E}_s' = E_s \times C$ is contained in an analytic subvariety $\widetilde{M}$ of $\widetilde{B}$. We give first the equations of $\widetilde{M}$

in a domain $\Delta \subset \widetilde{B}$, $\Delta : \|x - x_o\| < \alpha'$, $|u - u_o| < r_o$, such that $\Delta \cap W = \phi$. Let be $x_o \in E_s$: there exists $u_o \in C$ such that $F(x_o,u_o) \neq 0$, and we define Δ of center (x_o,u_o) such that $\Delta \subset \widetilde{B}$, $\Delta \cap W = \phi$. Then

$$G(x,u) = \frac{\partial F}{\partial u}(x,u)\, F^{-1}(x,u)$$

is holomorphic in Δ , and for $x \in E_s \cap [\|x - x_o\| < \alpha'] = \eta_s$:

$$F(x,u) = \prod_1^s (u - u_j) \exp P_x(u)$$

with degreee $P_x(u) \leqslant \rho = \sup_{x \in B} \rho(x)$. Then

(9) $$G(x,u) = \frac{\partial P_x}{\partial u} [\sum_1^s \frac{1}{u-u_j}] = \frac{\partial P_x}{\partial u} R_x(u)\, \varphi_x^{-1}(u)$$

with $\varphi_x(u)$ of degree s in u .

Let be $G(x,u) = \sum_o^\infty a_q(x,u_o)\,(u-u_o)^q$ for $(x,u) \in \Delta$, with $a_q = a_q(x,u_o) = \frac{1}{q!} \frac{\partial^q}{\partial u^q} [F^{-1} \frac{\partial F}{\partial u}]_{u=u_o}$.

The coefficients $a_q(x,u_o)$ are holomorphic of (x,u_o) in $\Omega \setminus W$. By an easy calculation we obtain a polynomial $\widetilde{P}_q(\xi_o, \dots, \xi_{q+1})$ such that :

(10) $$a_q(x,u_o) = F^{-q-1}(x,u_o)\, a'_q(x,u_o)$$

and $a'_q(x,u_o) = \widetilde{P}_q [F(z,u), \dots \frac{\partial^{q+1}}{\partial u^{q+1}} F(x,u_o)]_{u=u_o}$ is a holomorphic function in all Ω . From (9) we deduce , for $\|x - x_o\| < \alpha'$, $x \in E_s$

(11) $$G(x,u)\, \varphi_x(u) = \frac{\partial P_x}{\partial u} R_x(u) .$$

Then $\frac{\partial P_x}{\partial u} R_x(u)$ is a polynomial of u of degree at most $\rho+s-1$. We write $\varphi_x(u) = \sum_o^s b_j (u-u_o)^j$, with b_j depending on x for $x \in E_s \cap [\|x - x_o\| < \alpha'] = \eta_s$, and

(12) $$G(x,u) = \sum_o^\infty a_j(x,u_o)\,(u-u_o)^q$$

which is convergent for $|u - u_o| < r_o$. Then for $x \in \eta_s$, we obtain from (9), (11) and (12) the equations

(13) $$\sum_{p=o}^s a_{\nu-p}(x,u_o)\, b_p = 0 \quad \text{for all } \nu \geqslant s+\rho \text{ and } x \in \eta_s .$$

The polynom $\varphi(u)$ is not the constant null, the coefficients b_p are not all zero, then the determinant

(14) $$D_\nu(x, u) = \begin{pmatrix} a_\nu & \dots , & a_{\nu-s} \\ & & \\ a_{\nu+s} & \dots , & a_\nu \end{pmatrix}$$

must vanish for $x \in \eta_s$ and $u = u_o$ and $\nu \geqslant s+\rho$. The same conclusion remains if we replace u_o by u'_o such $|u_o - u'_o| < \frac{r_o}{2}$; then

$$G(x,u) = \sum_{o}^{\infty} G_q(x,u'_o)(u-u'_o)^q$$

is convergent for $\|x - x_o\| \leqslant \alpha'$, $|u - u'_o| < \frac{r}{2}$, and

$a_q(x,u'_o) = \frac{1}{q!} \frac{\partial q}{\partial u^q} [F^{-1} \frac{\partial F}{\partial u}]_{u=u'_o}$. As a consequence the determinants $D_\nu(x,u)$ are holomorphic functions of $(x,u) \in \Delta$, which vanish if $\nu \geqslant s+\rho$ for $(x,u) \in (E_s \times \mathbb{C}) \cap \Delta$. If we replace $a_q(x,u)$ by $a'_q(x,u)$, we obtain

$$D_\nu(x,u) = F^{-\sigma}(x,u) D'_\nu(x,u)$$

with $\sigma = (s+1)(\nu+1)$ and $D'_\nu(x,y)$ is obtained writing $a'_{\nu-j}$ instead $a_{\nu-j}$ in (13) . Then $D'_\nu(x,u)$ is holomorphic of $(x,u) \in \Omega$ and is a polynomial of $\xi_o = F(x,u)$, ... $\xi_{\nu+s+1} = \frac{\partial F}{\partial u^{\nu+s+1}}(x,u)$ with constant coefficients.

a/ We suppose that the equations $D'_\nu(x,u) = 0$, for $\nu \geqslant s+\rho$, define an analytic subvariety $\widetilde{M}_s$ in $B \times C$. Obviously, $\widetilde{M}_s$ is invariant by the translations $(x,u) \to (x,u+v)$ for $v \in \mathbb{C}$. Moreover $\widetilde{M}_s$ contains $E_\infty \times \mathbb{C}$; E'_s in contained in B in the analytic subvariety $\widetilde{M}_s \cap [u=0]$. Then E'_s is contained in an analytic subvariety of G .

b/ If $D'_\nu(x,u) = 0$ in Ω for $\nu \geqslant s+\rho$, and if $F(x,u) \not\equiv 0$, both W and $E_\infty \times C$ are analytic subvarieties in Ω ; for $(x_o,u_o) \notin W$, we construct a neighboorhood $\Delta : \|x - x_o\| < \alpha''$, $|u - u_o| < r'_o$, such that $\Delta \cap W = \phi$; then there exists solutions $b_o, \ldots, b_s$, not all zeros, and $Q_x(u) = b_o + b_1(u-u_o), \ldots, + b_s(u-u_o)^s$ is a polynom of degree at most s and from (14) we deduce that $G(x_o,u)$ is a rational function of u with at most s poles (with multiplicity). Then $u \to F(x_o,u)$ has at most s zeros. The conclusion holds for all $x \in G$ such that we can choose u_o with $F(x_o,u_o) \neq 0$, i.e. for each $x \in G$ which is not in E_∞ . Then $(E_\infty \times C) \cup (E_s \times C) = \Omega$.

Corollary. 1/ _If_ $\rho^*(x)$ _is a constant (in particular for_ $\Omega = C^{n-1}$), _then the system of equations_ $D'_\nu(x,u) = 0$ _for_ $\nu \geqslant s+\rho$ _definies an analytic subvariety_ M_s _in_ $\Omega \times C$, _and_ $(E_s \cup E_\infty) = M_s$.

2/ If X is Cousin data of zeros of finite order ρ in C^n such that the origin is not in $|X|$ (the support of $|X|$) , the set E_s of the complex line through the origin such that $\text{card}(\ell \cap |X|) \leqslant s$ is a closed set and if it is not contained in the zeros of a homogeneous polynomial $\varphi(z) = 0$ in C^n , $|X|$ is an algebraic set .

For the proof consider $F(z)$ such that $X = [z \in C^n , F(z) = 0]$ an $F_1(z,u) = F(z_1 u ,\ldots, z_n u)$ for $z = (z_1,\ldots,z_n)$; we take for Δ a neighboorhood of the origin $z = 0$, $u = 0$ in which $F_1(z,u) \neq 0$; the $a'_q(z)$ are homogeneous polynomials of z . The set E_∞ is empty , and E_s is an analytic subvariety in the projective space $P(C^n)$ or $D'(z,u) \equiv 0$ for $\nu \geqslant s+\rho$. Then $E_s = P(C^n)$ and the indicatrix $\nu_X(r)$ of X is bounded (see [10]), X is algebraic giving the result of [15] .

BIBLIOGRAPHY

[1] L.AHLFORS : Zur Theorie der uberlagerungsflächen . Acta Math., t.65, (1935), p. 157-191.

[2] E.BEDFORD et B.A.TAYLOR : A new capacity for plurisubharmonic functions. Acta Math., t. 149, (1982), p. 1-40.

[3] E.BEDFORD : The operator $(dd^c)^n$ on complex spaces (preprint).

[4] J.CARLSON : A moving lemma for the transcendental Bezout problem. Ann. of Math., t. 103, (1976), p. 305-330.

[5] L.GRUMAN : The area of analytic subvarieties in C^n . Math. Scand., t. 41, p. 365-397.

[6] L.GRUMAN : La géométrie globale des ensembles analytiques dans C^n . Lecture Notes n° 822, Springer, 1980, p. 90-99.

[7] L.GRUMAN : Ensembles exceptionnels pour les applications holomorphes dans C^n . Lecture Notes Springer, n° 1028, (1983), p.125-162.

[8] C.O.KISELMAN : The growth of restrictions of plurisubharmonic functions (preprint).

[9a] P.LELONG : Sur les valeurs lacunaires d'une relation à deux variables. Bull. Sc. Math., t. 56, p. 103-112.

[9b] P.LELONG : A class of Fréchet spaces in which the bounded sets are C-polar sets. Holomorphy and approximation theory : Math. Studies, North-Holland, vol. 71, p. 253-272.

[10] P.LELONG et L.GRUMAN : Entire functions of several complex variables. Grundlehren der Math., n° 282, Springer, 1985.

[11] R.MOLZON, B.SCHIFFMAN et N.SIBONY - Average growth - estimates for hyperplane sections of entire analytic sets . Math. Ann. t. 257, 1981, p. 43-59.

[12] R.NEVANLINNA : Eindeutige analytische Funktionen. Grundlehren der Math., 46, Springer, (1936).

[13] M.-H.SCHWARTZ : Formules apparentées à la formule de Gauss - Bonnet. Acta Math., t. 91, (1954), p. 189-244.

[14] M.-H.SCHWARTZ : Formules apparentées à celles de Nevanlinna -Ahlfors pour certaines applications d'une variété à n-dimensions dans une autre. Bull. S.M.F., t. 84, (1954), p. 317-359.

[15] N.SIBONY et PITT MAN WONG : Some results on global analytic sets. Lecture Notes n° 822, p. 221-238.

[16] J.SICIAK : Extremal plurisubharmonic functions in C^n . Proceedings of the first Finnish. Polish Summer School in complex Analysis, p. 115-152.

17 J.A.TAYLOR : An estimate for an extremal plurisubharmonic function on C^n . Lecture Notes n° 1028, p. 318-328.

On the Boundary Behavior of Holomorphic Mappings

László Lempert[1]

Eötvös Lóránd University, Dept. of. Math., Budapest, Múzeum Krt. 6-8, 1088 Hungary

0. Introduction

Let D_1 , D_2 be bounded domains in $\mathbb{C}^n$ and let ϕ be a biholomorphic mapping between the two. Under what conditions is it possible to conclude that ϕ extends (with some regularity) to the closure $\bar{D}_1$ of D_1?

It is natural to require that the boundaries ∂D , ∂D_2 be regular to some extent; and based on the negative evidence of the lack of counterexamples, it is generally believed that no further assumptions are needed. However, the existing theorems do use further assumptions, partly of geometric, partly of analytic nature. Rather than listing all these theorems, we shall restrict ourselves to describe those that are strongest in the sense that they contain the others as special cases. Moreover, we shall always assume that the domains in question have at least C^∞-boundaries.

If both domains are such that their Bergman projections map C^∞ functions into C^∞ functions (this is the so-called condition R), then S. Bell and E. Ligocka proved that ϕ extends to a C^∞-diffeomorphism between the closures $\bar{D}_1$, $\bar{D}_2$, see [B-L]. D. Catlin proved that pseudoconvex domains of "finite type" satisfy condition R , see [C]. These domains include strictly pseudoconvex domains and also pseudoconvex domains with analytic boundaries.

When the domains are analytically bounded and the Bergman projection satisfies a "condition Q ", analogous to condition R but pertaining to analytic functions, then ϕ is known to extend to a biholomorphism between neighborhoods of $\bar{D}_1$, $\bar{D}_2$ (see S. Bell [Be]). The inconvenient feature of this theorem is, however, that apparently the only general geometric condition which implies condition Q is strict pseudoconvexity

[1]Supported in part by NSF grant MSF grant MCS 82-13077.

(D. Tartakoff [Ta], G. Komatsu [Ko], M. Derridj - D. Tartakoff [D-T]. On the other hand, for strictly pseudoconvex domains a simpler approach also yields the analytic continuation of biholomorphic mappings (H. Lewy [Le], S. Pinchuk [P]).

In [D-F2] K. Diederich and J.E. Fornaess consider analytically bounded domains in $\mathbb{C}^2$ such that (on both boundaries) the set of pseudoconvex points is separated from the set of pseudoconcave points by a totally real surfaces. Under that condition they show that ϕ extends to a homeomorphism between the closures.

Further results treat exceptional situations such as Reinhardt domains (see W. Kaup [Ka], S. Bell and H. Boas [B-B]), or, slightly more generally, domains with many symmetries (D. Barrett [Ba1]), and analytically bounded Hartogs domains in $\mathbb{C}^2$ (K. Diederich - J.E. Fornaess [D-F3]). The aim of this paper is to show that a geometric condition totally different from pseudoconvexity (in fact, one bearing on the first derivative of a defining function rather than on the second) also ensures the extendibility of a biholomorphism to the boundaries. More precisely, we shall prove

<u>Theorem 0.1</u> Let D_1 , D_2 be strictly starshaped, bounded domains in $\mathbb{C}^n$ having real analytic boundaries, ϕ a biholomorphic mapping from D_1 onto D_2 . Then ϕ extends to a Hölder-continuous homeomorphism between $\bar{D}_1$ and $\bar{D}_2$.

Strictly starshaped here means that there is a point O_i in D_i such that D_i is starshaped with respect to O_i and for any $z \in \partial D_i$ the angle between the radius vector $O_i z$ and the tangent hyperplane to ∂D_i in z is nonzero.

On certain parts of the boundaries the mapping can be shown to have better regularity than just Hölder-continuity. Indeed, S. Bell observed that once the continuous extension of ϕ is demonstrated the ideas displayed in [Be] prove that:

1) Points on ∂D_1 near which ∂D_1 is pseudoconvex are mapped to points on ∂D_2 with the same property, and near these points the extension of

ϕ is C^∞-smooth; and 2) Strictly pseudoconvex points on ∂D_1 are mapped to strictly pseudoconvex points on ∂D_2 and the mapping ϕ analytically extends across such points.

One of the important ingredients of our proof is a theorem by D. Barrett about a weak regularity property of the Bergman projection in strictly starshaped domains ([Ba3]).In the same paper he proves stronger regularity results, too, provided the domain in question satisfies more restringent conditions. It should be pointed out that when these regularity results become sufficiently strong, they can be coupled with ideas from [B-L], to prove better smoothness of biholomorphic maps. The results thus obtained have the following flavor. Suppose D_1 , D_2 are smoothly bounded and very strictly starshaped in the following sense: If O_i is the center of D_i , then for any $z \in \partial D_i$ the angle between the radius $O_i z$ and the tangent hyperplane to ∂D_i in z should be greater than 89° . Then any biholomorphic mapping $\phi : D_1 \to D_2$ extends to a C^1 diffeomorphism between the closures, provided the dimension n is low. If one wants to have the same regularity in somewhat higher dimensions, or get better regularity in low dimensions, it suffices to replace 89° in the assumptions by $89^o 59'$, and so on. However, in the case treated in this paper, only the weakest regularity result of D. Barrett is available and this requires and approach entirely different from that of [B-L], as will be explained in the next chapter.

Acknowledgement. A part of this work was done while the author was a visiting lecturer at Princeton University. I wish to express my gratitude to this institution. Thanks are also due to D. Barrett who kindly showed me a version of his manuscript [Ba3].

1. On the proof of Theorem 0.1

Our proof consists of two parts. In the first we prove that our biholomorphic mapping does not change the distance to the boundaries very radically. More precisely, there are positive constants c , η such that

$$\text{(1.1)} \qquad c\, \mathrm{dist}(z,\partial D_1)^{\eta} < \mathrm{dist}(\phi(z),\partial D_2) < c^{-1}\mathrm{dist}(z,\partial D_1)^{1/\eta}$$

for $z \in D_1$. This is achieved as follows. (1.1) is easily seen to be equivalent to a lower estimate on the Jacobian determinant of ϕ (and on its inverse) by a high power of the boundary distance. Now a lower estimate of $\det \phi'$ is equivalent to an upper estimate for $\det(\phi^{-1})'$. A suitable upper estimate is obtained by using D. Barrett's theorem in [Ba3], whence (1.1) will follow.

Before proceeding to the second part of the proof, we note that if r is a defining function of D_2 , (1.1) implies that $r \circ \phi$ is Hölder-continuous. Then we face the problem whether the regularity of a composed function implies the regularity of the inner function. When this question is considered in the greatest possible generality, the answer is clearly no. Indeed, ϕ may map into a level set of r and then $r \circ \phi$ will be smooth no matter how badly ϕ behaves. It is a most remarkable fact that, nevertheless, when ϕ is assumed to be holomorphic and r is supposed to have certain properties to be specified in a moment, then the Hölder-continuity of $r \circ \phi$ does imply that of ϕ .

The simplest instance of this principle is when r is supposed to be strictly plurisubharmonic and smooth. Then the α-Hölder-continuity of $r \circ \phi$ implies the $\alpha/2$-Hölder-continuity of ϕ $(\alpha \notin \mathbf{Z})$. For $\alpha = \infty$ this was first observed by N. Kerzman (see [Ke]).

Another instance of the above principle is when r is supposed to be real analytic. To get a precise theorem some additional conditions have to be imposed on r and ϕ ; however, these conditions are automatically met in the concrete situation considered. Thus in the second part of our proof we start with the Hölder-continuity of $r \circ \phi$ and conclude the Hölder-continuity of ϕ . This part does not use the fact that ϕ is biholomorphic.

The idea here is to show that the situation which gave rise to the counterexample above (namely ϕ mapping into a level set of r) cannot happen, not even approximately. On the contrary, ϕ has to be transversal in some sense and to a certain extent to the level sets of r . More exactly if ϕ is restricted to complex curves, the images of these curves cannot have arbitrarily large order of contact with the level hypersurfaces of r . This part of the work is based on results by K. Diederich and J.E. Fornaess, and J. D'Angelo. A quantitative statement about the contact between holomorphic curves and level hypersurfaces will then prepare the ground for the final step of the proof, which is the estimation of ϕ' . The estimate we thus obtain is equivalent to the Hölder-continuity of ϕ .

Finally a word about the notation. $c, c', \ldots$ will stand for positive constants (depending only on D_1 , D_2 and ϕ , unless stated otherwise). However, the c's will not necessarily mean the same constants on each occurrence.

2. Estimating the boundary distance

Let D_1 , D_2 and Φ be as in Theorem 0.1. The aim of this chapter is to prove

Lemma 2.1: There are positive constants η , c , such that

$$\text{(2.1)} \qquad c\cdot \operatorname{dist}(z,\partial D_1)^{\eta} < \operatorname{dist}(\Phi(z),\partial D_2) < c^{-1}\operatorname{dist}(z,\partial D_2)^{1/\eta}$$

for all $z \in D_1$.

Here $\operatorname{dist}(\cdot,\partial D_1)$ stands for distance to the boundary ∂D_1 . The statement of Lemma 2.1 is equivalent to

Lemma 2.2: There are positive constants c , η such that we have the following estimate for the Jacobian determinants of ϕ and $\psi = \phi^{-1}$:

$$\text{(2.2)} \qquad |\det \phi'(z)| > c \operatorname{dist}(z,\partial D_1)^{\eta} ,$$

$$\text{(2.3)} \qquad |\det \psi'(w)| > c \operatorname{dist}(w,\partial D_2)^{\eta} .$$

We shall first derive Lemma 2.2 and then show its equivalence with Lemma 2.1.

Proof of Lemma 2.2: Let P_i denote the Bergman projection in D_i . That is, P_i is the orthogonal projection of $L^2(D_i)$ to its subspace consisting of holomorphic functions. In [Be] S. Bell showed that there is a compactly supported $\varphi \in C_0^{\infty}(D_2)$ such that

$$\int_{D_2} \bar{h} = \int_{D_2} \varphi\bar{h}$$

for all holomorphic $h \in L^2$. In other words, $1 = P_2 1 = P_2\varphi$. Using the transformation formula for the Bergman projection, this implies $\det \phi' = P_1(\det \phi') = P_1((\det \phi')(\varphi \circ \phi))$.

We now invoke D. Barrett's theorem (see [Ba3]); according to which on smooth strictly starshaped domains the Bergman projection maps the Sobolev space $W^{1/2}$ into itself. Since $(\det \phi')(\varphi \circ \phi)$ is in $C_0^\infty(D_1)$, P_1 maps it into $W^{1/2}(D_1)$. In particular, there is an $\varepsilon > 0$ such that $\det \phi' = P_1((\det \phi')\varphi \circ \phi) \in L^{2+\varepsilon}$.

Hence

$$\infty > \int_{D_1} |\det \phi'|^{2+\varepsilon} = \int_{D_2} |\det \psi'|^{-\varepsilon} .$$

Applying now the submean value property to the subharmonic function $|\det \psi'|^{-\varepsilon}$ on balls centered at $w \in D_2$ and of radius $\operatorname{dist}(w, \partial D_2)$, the second inequality in Lemma 2.2 is obtained. Since the role of D_1 and D_2 is symmetric, the first inequality holds as well.

Proof of the equivalence of Lemmas 2.1 and 2.2: Suppose first that (2.1) holds. Then using the Cauchy estimates and the boundedness of Φ and ψ, we have

$$|\det \phi'(z)| = |\det \psi'(\phi(z))|^{-1} >$$

$$> c' \operatorname{dist}(\phi(z), \partial D_2)^n > c' c^n \operatorname{dist}(z, \partial D_1)^{n\eta} .$$

Conversely, suppose that (2.2) and (2.3) hold. We first want to estimate the length of $\phi'(z)v$, where v is an arbitrary unit vector in $\mathbb{C}^n$. Of course we have an upper estimate

$$(2.4) \qquad |\phi'(z)v| < c' \operatorname{dist}(z, \partial D_1)^{-1}$$

but we shall need an estimate from below. To this end, let $v_1 = v, v_2, \ldots, v_n$ be an orthonormal basis in $\mathbb{C}^n$. Then by Cramer's rule and (2.4) (applied to $v = v_2, v_3, \ldots$)

$$c \operatorname{dist}(z, \partial D_1)^\eta < |\det \phi'(z)| < n! \prod_{j=1}^{n} |\phi'(z)v_j| <$$

$$< n! |\phi'(z)v_1| \cdot (c')^{n-1} \operatorname{dist}(z, \partial D_1)^{1-n} ,$$

so that with some $M > 0$

$$(2.5) \qquad |\phi'(z)v| > c \operatorname{dist}(z,\partial D_1)^M \ .$$

Let now $\phi(z) = w$ and $w' \in \partial D_2$ the point which lies the nearest to w . Choose a unit vector $v \in \mathbb{C}^n$ so that $\phi'(z)v$ have the same direction as the vector ww' . Put $f(t) = \phi(z+tv)$. Then for $0 < t < \operatorname{dist}(z,\partial D_1)/2$ we have

$$(2.6) \qquad |f(t) - f(0) - f'(0)t| = |\int_0^t f''(u)(t-u)du| < c't^2 \operatorname{dist}(z,\partial D_1)^{-2} \ .$$

The point $f(0) + f'(0)t = w + \phi'(z)vt$ lies on the half line ww' , at distance $> ct \operatorname{dist}(z,\partial D_1)^M$ from w . Choose $t > 0$ so that this distance be $2|w-w'| = 2 \operatorname{dist}(w,\partial D_2)$; then

$$(2.7) \qquad t < 2c^{-1}|w-w'|\operatorname{dist}(z,\partial D_1)^{-M} ,$$

and $f(0) + f'(0)t$ lies in the complement of D_2 . Moreover, the distance of $f(0) + f'(0)$ to D_2 will be $|w-w'|$ (if z , and therefore w , is sufficiently near to the boundary ∂D_1 res. ∂D_2).

Now there are two possibilities. Either the value of t selected above satisfies $t < \operatorname{dist}(z,\partial D_1)/2$, or $t \geq \operatorname{dist}(z,\partial D_1)/2$. In the first case (2.6) shows that the distance of $f(0) + f'(0)t$ to $f(t) = \phi(z+tv) \in D_2$ is less than $c't^2\operatorname{dist}(z,\partial D_1)^{-2}$, hence in view of (2.7)

$|w-w'| < 4c'c^{-2}|w-w'|^2\operatorname{dist}(z,\partial D_1)^{-2M-2}$. This implies the first inequality in (2.1) since $|w-w'| = \operatorname{dist}(w,\partial D_2) = \operatorname{dist}(\phi(z),\partial D_2)$.

If, however, the second case arises, then, again by (2.7), we have

$$\operatorname{dist}(z,\partial D_1)/2 < 2c^{-1}|w-w'|\operatorname{dist}(z,\partial D_1)^{-M}$$

whence the first inequality on (2.1) again follows.

The second inequality in (2.1) is obtained by interchanging the roles of D_1, D_2 .

A consequence of Lemma 2.1 is the following. Let $r : \mathbb{C}^n \to \mathbb{R}$ be a real analytic defining function of D_2, i.e., $D_2 = \{w : r(w) < 0\}$ and $dr \neq 0$ on ∂D_2.

Lemma 2.3: The function $r \circ \phi$ is Hölder continuous on D_1.

Proof: Choose two points $z_1, z_2 \in D_1$. We have to estimate $r(\phi(z_1)) - r(\phi(z_2))$ in terms of some power of $|z_1 - z_2|$. Two cases will be distinguished according to whether $|z_1 - z_2|^{1/2} < \max(\text{dist}(z_1, \partial D_1), \text{dist}(z_2, \partial D_1))$ or not.

In the first case we also have

$$|z_1 - z_2|^{1/2} < 2 \,\text{dist}(z, \partial D_1)$$

for any z on the segment $z_1 z_2$, at least if z_1, z_2 are sufficiently near to ∂D_1. Hence

$$\begin{aligned} |r(\phi(z_1)) - r(\phi(z_2))| &\le c|\phi(z_1) - \phi(z_2)| \le \\ &\le c|z_1 - z_2| \sup\{\|\phi'(z)\| : z \in z_1 z_2\} \le \\ &\le c|z_1 - z_2| \, |z_1 - z_2|^{-1/2} = c|z_1 - z_2|^{1/2} . \end{aligned}$$

In the second case we use Lemma 2.1:

$$\begin{aligned} |r(\phi(z_1)) - r(\phi(z_2))| &\le c\{\text{dist}(\phi(z_1), \partial D_2) + \text{dist}(\phi(z_2), \partial D_2)\} \le \\ &\le c\{\text{dist}(z_1, \partial D_1)^{1/\eta} + \text{dist}(z_2, \partial D_1)^{1/\eta}\} \le \\ &\le c|z_1 - z_2|^{1/(2\eta)} . \end{aligned}$$

This proves Lemma 2.3.

3. The relative position of a real analytic hypersurface and a holomorphic curve (Review of results of J. D'Angelo and K. Diederich - J.E. Fornaess)

The result of K. Diederich - J.E. Fornaess that we are going to need is the following (see [D-F1], Theorem 4).

Theorem 3.1: Suppose S is an arbitrary compact real analytic variety in $\mathbb{C}^n$. Then S does not contain any nontrivial germs of complex varieties.

Using the tools developed by J. D'Angelo in [D1] and [D2], this theorem can be given a quantitative form. We are now going to review some of the concepts he introduced and theorems he proved.

Let $\mathcal{C}^*$ stand for the set of nonconstant germs of holomorphic mappings $(\mathbb{C},0) \longrightarrow (\mathbb{C}^n,p)$, p being a fixed point in $\mathbb{C}^n$. If g is an (eventually vector valued) smooth function near $0 \in \mathbb{C}$, $\nu(g) \leq \infty$ is its order in 0 , i.e., the largest ν such that $g(\xi) - g(0) = O(\xi^\nu)$ $(\xi \longrightarrow 0)$.

Let now r be a real analytic function defined in a neighborhood of $p \in \mathbb{C}^n$, $dr(p) \neq 0$. We put

$$\Delta(r,p) = \sup_{f \in \mathcal{C}^*} \nu(r \circ f)/\nu(f) \leq \infty$$

and call it the maximal order of contact of the hypersurface $\{z : r(z) = r(p)\}$ with holomorphic curves.

The order of contact of a proper ideal I contained in $\mathcal{O}_p = \mathcal{O}$, the ring of germs of holomorphic functions at p , is an analogous concept, defined by

$$\tau^*(I) = \sup_{f \in \mathcal{C}^*} \inf_{h \in I} \nu(h \circ f)/\nu(f) \leq \infty .$$

By the Nullstellensatz $\tau^*(I)$ is finite if and only if the variety of I consists of the single point p , and in this case I contains some power of the maximal ideal $\mathcal{O}$. One also has the estimate

$$(3.1) \qquad \tau^*(I) \le \dim_{\mathbb{C}} \mathcal{O}/_I \le \tau^*(I)^n$$

(see [D1], Theorem 2.7).

In [D2] D'Angelo shows how to reduce the computation of $\Delta(r,p)$ to the computation of $\tau^*(I)$ for certain ideals $I \subset \mathcal{O}_p$. The way he achieves that is the following. For the sake of simplicity we shall assume $r(p) = 0$.

First D'Angelo shows that there exists a separable Hilbert-space $\mathcal{H}$ - which we are going to identify with ℓ_2 - and three germs of holomorphic mappings

$$H : (\mathbb{C}^n,p) \longrightarrow (\mathbb{C},0) \ ,$$

$$F = (F_1,F_2,\ldots) : (\mathbb{C}^n,p) \longrightarrow (\mathcal{H},0) \ ,$$

$$G = (G_1,G_2,\ldots) : (\mathbb{C}^n,p) \longrightarrow (\mathcal{H},0)$$

such that

$$r(z) = 2 \operatorname{Re} H(z) + \|F(z)\|^2 - \|G(z)\|^2 =$$

$$= 2 \operatorname{Re} H(z) + \Sigma(|F_j(z)|^2 - |G_j(z)|^2) \ .$$

Next for any unitary transformation $U : \mathcal{H} \longrightarrow \mathcal{H}$ he introduces the ideals $I(U,p) \subset \mathcal{O}_p$ generated by H and the components of $F - UG$, i.e., $F_j - \sum_i u_{ji}G_i$, u_{ji} being the entries of (the matrix of) U . Then he proves

$$(3.2) \qquad \sup_U \tau^*(I(U,p)) \le \Delta(r,p) \le 2 \sup_U \tau^*(I(U,p)) \ ,$$

where U runs over all unitary mappings $\mathcal{H} \longrightarrow \mathcal{H}$ (see [D2], Theorem 10).

He also proves the following important theorem:

Theorem 3.2: Suppose $\Delta(r,p_0)$ is finite for some p_0. Then $\Delta(r,p)$ is bounded for p near p_0. (See [D1], Theorem 4.11.)

Using these tools, it will be easy to prove the following theorem, which essentially says that analytically bounded domains are of finite type (in the sense of D'Angelo). This theorem has also been found by D'Angelo and possibly by others, too, but it does not seem to have ever been published.

Theorem 3.3: Let $D \subset \mathbb{C}^n$ be an analytically bounded domain, $r : \mathbb{C}^n \longrightarrow \mathbb{R}$ a real analytic defining function of D. Then there is a neighborhood O of ∂D and a positive integer k such that for $p \in O$ we have $\Delta(r,p) \leq k$.

Proof: By Theorem 3.2 and the compactness of ∂D it suffices to prove $\Delta(r,p) < \infty$ for $p \in \partial D$. Suppose that $\Delta(r,p) = \infty$. Then by (3.2) there is a sequence $U^\nu = (u^\nu_{ji})$ of unitary matrices such that

$$\tau^*(I(U^\nu,p)) \longrightarrow \infty \qquad (\nu \longrightarrow \infty) \quad . \tag{3.3}$$

It can be assumed that for each i,j, the sequence u^ν_{ji} has a limit u_{ji}. Then the operator norm of $U = (u_{ji})$ will be ≤ 1; however, U need not be unitary.

Let us now define the ideal

$$J = (H, F_j - \sum_i u_{ji}G_i \ , \ G_j - \sum_i \overline{u_{ij}}\, F_i, \ j = 1,2,\ldots) \ .$$

We claim that the variety of J consists of the single point p. Indeed, let z be a point in this variety. Then

$$H(z) = 0 \ , \quad F(z) = UG(z) \ , \quad G(z) = U^*F(z) \ ,$$

where $U^* = (\overline{u_{ij}})$ is the adjoint of U. Since both U and U^* are of norm at most 1, this implies $\|F(z)\| = \|G(z)\|$, whence $z \in \partial D$. Thus the variety of J is contained in ∂D, so that by Theorem 3.1 it reduces to the point p. By the Nullstellensatz we have then

$$\dim_{\mathbb{C}} \mathcal{O}/J = d < \infty .$$

Now by the Noether property of $\mathcal{O}_p$ there is an N such that

$$J = (H, F_j - \sum_i u_{ji} G_i , G_j - \sum_i \overline{u_{ij}} F_i , j = 1,\dots,N) .$$

Let

$$J^\nu = (H, F_j - \sum_i u^\nu_{ji} G_i , G_j - \sum_i \overline{u^\nu_{ij}} F_i , j = 1,\dots,N) .$$

By one half of the Banach-Steinhaus theorem (the trivial half), there is a neighborhood of p on which

$$F_j - \sum_i u^\nu_{ji} G_i \longrightarrow F_j - \sum_i u_{ji} G_i ,$$

$$G_j - \sum_i \overline{u^\nu_{ij}} F_i \longrightarrow G_j - \sum_i \overline{u_{ij}} F_i$$

uniformly, as $\nu \longrightarrow \infty$. Applying the theorem about the upper semi-continuity of $\dim_{\mathbb{C}} \mathcal{O}/I$ as a function of I ([To], Chap. 11, Proposition 5.3), we deduce that for ν large

$$\dim \mathcal{O}/J^\nu \leq d .$$

On the other hand

$$J^\nu \subset (H, F_j - \sum_i u^\nu_{ji} G_i . G_j - \sum_i \overline{u^\nu_{ij}} F_i , j = 1,2,\dots) = I(U^\nu, p) ,$$

since $(U^\nu)^* = (U^\nu)^{-1}$. Therefore $\dim \mathcal{O}/I(U^\nu,p) \leq d$, contradicting (3.3). This contradiction proves Theorem 3.3.

4. More on the relative position of a real analytic hypersurface and a holomorphic curve

Theorem (3.3) implies in particular that if $f : (\mathbb{C},0) \longrightarrow (\mathbb{C}^n, p)$ is a holomorphic germ with $f'(0) \neq 0$ (i.e., $\nu(f) = 1$) then the k'th Taylor polynomial of $r \circ f$ (about 0) will contain at least one non-vanishing nonconstant term. Some general properties of real polynomials and analytic functions permit us to estimate this nonconstant term from below.

<u>Lemma 4.1</u>: Let $G \subset \mathbb{R}^N$ be open, let K be a compact subset of G, and let $W : G \times \mathbb{R}^M \longrightarrow \mathbb{R}$ be a positive real analytic function such that for all $x \in G$ the function $y \longmapsto W(x,y)$ is a polynomial of degree d at most. Then there are positive numbers a, b such that

$$W(x,y) > a(1 + |y|^2)^{-b} \quad . \tag{4.1}$$

<u>Proof</u>: Let $y_1,\ldots,y_M$ denote the coordinates of y. It will be enough to prove (4.1) for y's such that $1 \leq |y_1| = \max_i |y_i|$. In the sequel we shall assume that this is the case.

Introduce the new variables $\eta_1 = 1/y_1$, $\eta_i = y_i/y_1$ $(i=2,\ldots M)$.

Then

$$y_1^{-d} W(x,y) = V(x,\eta)$$

will be a polynomial in η, analytic in x, η, and positive when $\eta_1 \neq 0$. Hence the Łojasiewicz inequality (see [Lo] yields

$$y_1^{-d} W(x,y) = V(x,\eta) \geq a|\eta_1|^{2b} = a|y_1|^{-2b}$$

with some positive a, b, for $x \in K$, $|\eta_1|,\ldots,|\eta_M| \leq 1$. This gives (4.1) in the case considered.

We are now ready to estimate a nonvanishing Taylor coefficient of $r \circ f$ from below. We shall use the notation of Theorem 3.3.

<u>Lemma 4.2</u>: There are a neighborhood L of ∂D and positive numbers a, b such that whenever $f : (\mathbb{C},0) \longrightarrow (\mathbb{C}^n,p)$ is a holomorphic germ with $p \in L$ and $|f'(0)| = 1$ then

$$(4.2) \qquad \sum_{0<\mu+\nu\leq r} \left| \left(\frac{\partial^{\mu+\nu} r(f(\xi))}{\partial\xi^{\mu}\,\partial\bar{\xi}^{\nu}} \right)_{\xi=0} \right|^2 \geq a \left(\sum_{0<\mu\leq k} \left| f^{(\mu)}(0) \right| \right)^{-b}$$

<u>Proof</u>: Let L be any compact neighborhood of ∂D contained in O of Theorem 3.3. For $x = (x'x'') \in \mathbb{C}^n \times \mathbb{C}^n$ in a neighborhood of $L \times S^{2n-1} \subset \mathbb{C}^n \times \mathbb{C}^n$, $y = (y_2,y_3,\ldots,y_k) \in \mathbb{C}^{(k-1)n}$ $(y_\mu \in \mathbb{C}^n, \mu=2,\ldots,k)$ define

$$W(x,y) = \sum_{0<\mu+\nu\leq k} \left| \left(\frac{\partial^{\mu+\nu}}{\partial\xi^{\mu}\,\partial\bar{\xi}^{\nu}} \right)_{\xi=0} r(x' + x''\xi + \sum_{2}^{k} y_\mu \xi^\mu/\mu!) \right|^2 .$$

W is then a real analytic function, polynomial in y, and Theorem 3.3 implies that it vanishes nowhere. Hence Lemma 4.1 applies and we obtain (4.2).

5. Proof of Theorem 0.1

Lemma 5.1: Given a positive integer k, there is a $c > 0$ with the following property. If $q(t,\bar{t}) = \sum_{0 \le i+j \le k^2} q_{ij} t^i \bar{t}^j / i!j!$ is a polynomial of degree at most k^2 such that

$$\sum_{0 < i+j \le k^2} |q_{ij}|^2 \ge 1 \tag{5.1}$$

then

$$\max_{|t| \le 1} |q(t)| - \min_{|t| \le 1} |q(t)| > c \tag{5.2}$$

Proof: It will be enough to prove (5.2) for polynomials q satisfying $q(0) = 0$ and $\Sigma |q_{ij}|^2 = 1$. The set of such polynomials is, however, a compact set, and (5.2) follows by Weierstrass' theorem.

Proof of Theorem 0.1: Let r be a real analytic defining function of D_2. Choose $a < 1$, b, k so that (4.2) of Lemma 4.2 be satisfied with D_2 substituted for D. We claim that there is a $\delta > 0$ such that for $z \in D_1$ sufficiently near ∂D_1

$$\| \phi'(z) \| < \operatorname{dist}(z, \partial D_1)^{\delta - 1}, \tag{5.1}$$

where $\| \ \|$ stands for operator norm.

Suppose that (5.1) does not hold for some z near ∂D_1 and some $\delta > 0$. Then there is a unit vector $v \in \mathbb{C}^n$ such that

$$|\phi'(z)v| \ge \operatorname{dist}(z, \partial D_1)^{\delta - 1}. \tag{5.2}$$

We are going to show that this cannot be the case if δ is small enough. For brevity, we introduce positive numbers λ, Δ so that the left hand side of (5.2) is $1/\lambda$ and the right hand side is $\Delta^{\delta-1}$. Then (5.2), together with the Cauchy estimate, yields

$$(5.3) \qquad c\Delta < \lambda \le \Delta^{1-\delta} .$$

For $\xi \in \mathbb{C}$, $|\xi| < \Delta/\lambda$ $(\ge \Delta^{\delta})$ put $f(\xi) = \phi(z+\lambda v \xi)$. We have then

$$(5.4) \qquad |f'(0)| = 1 , \quad |f^{(j)}(0)| < c_j \Delta^{-j}\lambda^j \le c_j \Delta^{-j\delta}$$

for $j = 2,3,\ldots$, by the Cauchy estimates and the boundedness of ϕ .

Let furthermore F denote the k'th Taylor polynomial of f about 0 , R denote the k'th Taylor polynomial of r about $z = f(0)$, and let

$$Q(\xi) = R(F(\xi)) = \sum_{0 \le i+j \le k^2} Q_{ij}\, \xi^i\, \bar{\xi}^j /i!j!$$

Then $r \circ f$ and Q agree at 0 to order k . Hence by Lemma 4.2 and (5.4) we have for z near ∂D_1

$$\sum_{0 < i+j \le k} |Q_{ij}|^2 \ge a\, \Delta^{2bk\delta}$$

Put $\alpha = 2\delta k(b+1)$ (any small α greater than that would also do) and

$$q(t) = \Sigma\, q_{ij} t^i\, \bar{t}^j /i!j! = a^{-1}\Delta^{-bk\delta}(\lambda/\Delta^{1+\alpha})^k\, Q(t\Delta^{1+\alpha}/\lambda)$$

Then

$$\sum_{0 < i+j \le k^2} |q_{ij}|^2 \ge a^{-2}\Delta^{-2bk\delta}(\lambda/\Delta^{1+\alpha})^{2k} \sum_{0 < i+j \le k} |Q_{ij}|^2 (\Delta^{(1+\alpha)}/\lambda)^{2(i+j)} \ge 1 ,$$

so that Lemma 5.1 can be applied and we obtain

$$\max_{|t| \le 1} q(t) - \min_{|t| \le 1} q(t) > c ,$$

which means

$$(5.5)\qquad \max_{|\xi|\le \Delta^{1+\alpha}/\lambda} Q(\xi) - \min_{|\xi|\le \Delta^{1+\alpha}/\lambda} Q(\xi) \ge c\,\Delta^{k(1+\alpha+b\delta)}\lambda^{-k} \ge$$

$$> c\,\Delta^{k(b\delta+\delta+\alpha)} = c\,\Delta^{\alpha(k+1/2)}$$

We now wish to show that a similar estimate holds with $Q = R\circ F$ replaced by $r\circ f$. To see this, for $|\xi|\le \Delta^{1+\alpha}/\lambda$ we estimate

$$|f(\xi) - F(\xi)| \le \max_{|\xi'|\le \Delta^{1+\alpha}/\lambda} \frac{|f^{(k+1)}(\xi')|}{(k+1)!}(\Delta^{1+\alpha}/\lambda)^{k+1} .$$

Since f is bounded on the disc $|\xi| < \Delta/\lambda$, $|f^{(k+1)}(\xi')| \le c(\Delta/\lambda)^{-(k+1)}$, so that $|f(\xi)-F(\xi)| \le c\,\Delta^{\alpha(k+1)}$, whence

$$(5.6)\qquad |r(f(\xi))-r(F(\xi))| < c\,\Delta^{\alpha(k+1)} \qquad (|\xi| \le \Delta^{1+\alpha}/\lambda) .$$

Furthermore, since for $|\xi| \le \Delta^{1+\alpha}/\lambda$

$$|F(\xi)-F(0)| \le \sum_{j=1}^{k} c_j\Delta^{-j}\lambda^{j}(\Delta^{(1+\alpha)}/\lambda)^{j}/j! < c\Delta^{\alpha} ,$$

we also have

$$(5.7)\qquad |r(F(\xi))-R(F(\xi))| \le c|F(\xi)-F(0)|^{k+1} < c\,\Delta^{\alpha(k+1)} .$$

(5.5), (5.6) and (5.7) together imply

$$\max_{|\xi|\le \Delta^{1+\alpha}/\lambda} r(f(\xi)) - \min_{|\xi|\le \Delta^{1+\alpha}/\lambda} r(f(\xi)) > c\,\Delta^{\alpha(k+1/2)} .$$

Let ξ_1 resp. ξ_2 denote the two points where the maximum resp. minimum is attained. Then for the points $z_1 = z + \lambda v\xi_i$ $(i = 1,2)$ we have

(5.8) $$r(\phi(z_1)) - r(\phi(z_2)) > c\ \Delta^{\alpha(k+1/2)}$$

On the other hand,

$$|z_1 - z_2| \le 2\lambda\Delta^{1+\alpha}/\lambda = 2\Delta^{1+\alpha} \quad ,$$

so that if ε denotes the exponent for which $r \circ \phi$ is ε-Hölder-continuous (see Lemma 2.3), then

$$c'\Delta^{\varepsilon(1+\alpha)} > r(\phi(z_1)) - r(\phi(z_2)) \ .$$

This, compared with (5.8), implies $\alpha > \varepsilon/(k+1)$. But then we proved that if (5.2) holds then $\delta > \varepsilon/\{2(k+1)k(b+1)\}$; in other words (5.1) is proved to hold for $z \in D_1$ in a neighborhood of ∂D_1 , provided δ is sufficiently small.

This finishes the proof, since by a theorem of Hardy and Littlewood (5.1) is equivalent to the δ-Hölder-continuity of ϕ , see [G]. (Although Hardy and Littlewood proved their theorem for holomorphic functions defined in the unit disc, their proof with obvious modifications applies to holomorphic mappings defined in smoothly bounded domains in $\mathbb{C}^n$.) Exchanging the roles of D_1 and D_2 , we obtain that also ϕ^{-1} is Hölder continuous.

6. Concluding Remarks

Our proof actually proved slightly stronger theorems than Theorem 0.1. For instance, we have

Theorem 6.1: Let D_1 be strictly star-shaped and C^2-bounded, D_2 analytically bounded. Then any biholomorphic map $\phi : D_1 \longrightarrow D_2$ extends to a Hölder-continuous map $\bar{D}_1 \longrightarrow \bar{D}_2$.

Strict star-shapedness can also be replaced by an analytic condition.

Theorem 6.2: Let D_1 be such that its Bergman projection maps $C_0^\infty(D_1)$ into $L^{2+\varepsilon}(D_1)$ with some $\varepsilon > 0$, and ∂D_1 is C^2 . If D_2 and ϕ are as in Theorem 6.1, then the conclusion of Theorem 6.1 still holds.

In [Ba2] D. Barrett constructed a C^∞-bounded domain D_1 for which the assumption of Theorem 6.2 fails for every $\varepsilon > 0$, and also for every ε he constructed analytically bounded domains for which the assumption of Theorem 6.2 fails for this particular ε . It is, however, feasible that for an arbitrary domain D_1 with analytic boundary there is an $\varepsilon > 0$ for which the assumption of Theorem 6.2 holds.

In yet another version the essential condition is imposed on the mapping ϕ rather than on D_1 :

Theorem 6.3: Let D_1 be C^2 bounded, D_2 and ϕ as in Theorem 6.1, and assume in addition that with some $c > 0$, $\varepsilon > 0$

$$\operatorname{dist}(\phi(z),\partial D_2) < c \operatorname{dist}(z,\partial D_1)^\varepsilon , \quad z \in D_1 .$$

Then, again, ϕ is Hölder continuous.

Indeed, this is what we prove in Chapters 3, 4, 5.

The condition of ∂D_2 being analytic is very heavily used. In the proof presented above, we relied on it on three independent occasions.

In Lemma 2.2 the function $\varphi \in C_0^\infty(D_2)$ is found by using the Cauchy-Kovalevskaya theorem, which is false in the C^∞ category. Also Theorem 3.3 fails to hold for just smooth domains. Finally, and somewhat surprisingly, Lemma 4.1 also needs that the function W be real analytic. Otherwise it may fail, even if $N = M = 1$. Indeed, let $e(x) = e^{-x^{-2}}$ if $x \neq 0$, $e(0) = 0$. Then $W(x,y) = (1-xy)^2 + e(x)$ does not satisfy (4.1) if K contains a neighborhood of 0. Indeed

$$\min_{x \in K} W(x,y) \leq W(y^{-1},y) = e^{-y^2} .$$

References

[Ba1] D. Barrett: Regularity of the Bergman projection on domains with transverse symmetries. Math. Ann. 258 (1982) 441-446.

[Ba2] D. Barrett: Irregularity of the Bergman projection on a smooth bounded domain in $\mathbb{C}^2$. Ann. of Math. 119 (1984) 431-436.

[Ba3] D. Barrett: Regularity of the Bergman projection and local geometry of domains. To appear

[Be] S. Bell: Analytic hypoellipticity of the $\bar{\partial}$-Neumann problem and the extendability of holomorphic mappings. Acta Math., Vol. 147 (1981) 109-116.

[B-B] S. Bell, H. Boas: Regularity of the Bergman projection in weakly pseudoconvex domains. Math.Ann. 257 (1981) 23-30.

[B-L] S. Bell, E. Ligocka: A simplification and extension of Fefferman's theorem on biholomorphic mappings. Invent. Math. 57 (1980) 283-289.

[C] D. Catlin: Global regularity of the $\bar{\partial}$-Neumann problem. Proc. Symp. Pure Math. Vol. 41, Amer. Math. Soc. R.1, 1984.

[D1] J. D'Angelo: Real hypersurfaces, orders of contact and applications. Ann. of Math. 115 (1982) 615-637.

[D2] J. D'Angelo: Intersection theory and the $\bar{\partial}$-Neumann problem. Proc. Symp. Pure Math. Vol. 41, Amer. Math. Soc. R.1, 1984.

[D-F1] K. Diederich, J.E. Fornaess: Pseudoconvex domains with real analytic boundary. Ann. of Math. 107 (1978) 371-384.

[D-F2] K. Diederich, J.E. Fornaess: Biholomorphic mappings between certain real analytic domains in $\mathbb{C}^2$. Math. Ann. 245 (1979) 255-272.

[D-F3] K. Diederich, J.E. Fornaess: Biholomorphic mappings between two-dimensional Hartogs domains with real-analytic boundaries. Recent Developments in Several Complex Variables, Princeton University Press, 1981.

[D-T] M. Derridj, D. Tartakoff: On the global real analyticity of solutions to the $\bar{\partial}$-Neumann problem. Comm. Partial Diff. Equ. 1 (1976) 435-601.

[G] G.M. Golusin: Geometric theory of functions of a complex variable. Moscow-Leningrad, GITTL, 1952 (in Russian).

[Ka] W. Kaup: Über das Randverhalten von holomorphen Automorphismen beschränkter Gebiete. Manuscripta Math. 3 (1970) 257-270.

[Ke] N. Kerzmann: A Monge-Ampère equation in complex analysis. Proc. Symp. Pure Math., 30 (1977) 161-167.

[Ko] G. Komatsu: Global analytic-hypoellipticity of the $\bar{\partial}$-Neumann problem. Tohoku Math. J., Ser. 2, 28 (1976) 145-156.

[Le] H. Lewy: On the boundary behavior of holomorphic mappings. Att. Acad. Naz. deiLincei 35 (1977).

[Lo] S. Łojasiewicz: Sur le problème de la division. Studia Math. 18 (1959) 87-136.

[P] S.I. Pinchuk: On the analytic continuation of holomorphic mappings. Math. Sb. Vol. 27 (1975) 375-392.

[Ta] D. Tartakoff: On the global real analyticity of solutions to $\Box_b$. Comm. Partial Diff. Equ. 1 (1976) 283-311.

[To] J.C. Tougeron: Idéaux de fonctions différentiables. Springer 1972.

Integral Geometry of the Monge-Ampère Operator

R. Molzon
Department of Mathematics
University of Kentucky
Lexington, Kentucky 40506

1. INTRODUCTION

In 1971 W. Stoll proved an average Bezout estimate for complete intersections of analytic varieties. In simplified terms, the results may be described as follows. Suppose $X \subset \mathbb{C}^n$ is an analytic variety of pure dimension p and $q \geq n-p$. Let $G(q,n)$ denote the Grassmannian of q-dimensional linear subspaces of $\mathbb{C}^n$. We measure the "growth" of a variety Y of dimension p by computing $\mathrm{vol}_{2p}(Y \cap B^n(r))$ where vol_{2p} denotes the 2p-Hausdorff measure. Stoll's result roughly says that one can compute the growth of X from the average of the growth of $X \cdot \ell$, $\ell \in G(q,n)$ if we average over all of $G(q,n)$.

Stoll's theorem is an example of a result in integral geometry. Note that instead of averaging over $G(q,n)$ we could fix $\ell_o \in G(q,n)$, and let Γ be the group of complex affine transformations on $\mathbb{C}^n$, and then average the growth of $X \cdot \gamma \ell_o$ as γ ranges over Γ. This is the manner in which integral geometry results are often stated.

The intersection of two analytic varieties, X and Y, can be reduced to studying intersection with a linear space by considering $X \times Y$ and intersecting with the diagonal.

One can ask if averaging is really necessary to determining growth. For example, could one compute the growth of X from the growth of $X \cdot \ell$ for some fixed $\ell \in G(q,n)$? This question is answered in the negative by an example of Cornalba and Shiffman. They produced two curves in $\mathbb{C}^2$ with order zero growth whose common zero locus is zero dimensional but of infinite order.

In 1976 J. Carlson considered the problem of intersection in a different way. In $\mathbb{C}^2$ suppose we have the families of varieties $X_\alpha = \{f(z) = \alpha\}$ and $Y_\beta = \{g(z) = \beta\}$. For a particular

α and β nothing can be said about the growth of the intersection $X_\alpha \bullet Y_\beta$ in terms of the growth of X_α and Y_β. Carlson considered the growth of the intersection $X_\alpha \circ Y_\beta$ averaged over a set of $(\alpha,\beta) \in E \subset \mathbb{C}^2$ which has positive Newtonian $2+\varepsilon$ capacity. This capacity is defined in terms of a kernel function $K(z,w)$. This concept of positive capacity is closely related to the Hausdorff measure of a set.

Carlson's result is an integral geometry result in which the averaging is done over a much smaller set of linear spaces than $G(q,n)$ and is even "smaller" than any open subset of $G(q,n)$. This method works since we already have quite a bit of structure on the X_α and Y_β; that is, they are analytic varieties.

More recently L. Gruman has obtained results on average growth in which the integration is done over sets even smaller than those considered by Carlson. These sets are pluripolar; a property which is detected by a capacity associated with the Monge-Ampère operator. We will first look at some properties of this capacity and then the methods for applying it to integral geometry.

It is clear that the entire subject of integral geometry in several complex variables owes a great deal to the work of Professor Stoll; my work in this area is no exception. In addition, his support and encouragement have been extremely valuable to me and I am grateful for the opportunity to participate in this conference in his honor.

2. CAPACITY OF THE MONGE-AMPÈRE OPERATOR

If $\Omega \subset \mathbb{C}^n$ is open, a set $E \subset \Omega$ is called pluripolar if it is the $-\infty$ set of a plurisubharmonic function. We let $PSH(\Omega)$ denote the plurisubharmonic functions on Ω. There is a capacity function which detects pluripolar sets; it is defined via the nonlinear complex Monge-Ampère operator.

If $E \subset \Omega$ is compact, one associates to E an L^∞_{loc} plurisubharmonic function u_E as follows. Let

Let

$$\tilde{\mu}_E(\zeta) = \sup\{v(\zeta):\ v\in PSH(\Omega),\ v\le 0 \text{ on } E \text{ and } v<1 \text{ on } \Omega\}\ .$$

Then

$$u_E(z) = \limsup_{\zeta\to z} \tilde{u}_E(\zeta)\ .$$

The Monge-Ampère operator acts on μ_E to define a nonnegative measure which is supported on E,

$$d\mu_E = (dd^c u_E)^n\ .$$

The following result is basic [1].

Theorem (Bedford and Taylor).

$$\int_{\mathbb{C}^n} d\mu_E > 0 \quad \text{if and only if E is not pluripolar.}$$

If $E\subset\Omega\subset\mathbb{C}^n$ is not pluripolar, one may normalize the measure μ_E so it is a probability measure. The method may be extended to define the property of positive capacity on a complex manifold.

In contrast to polar sets and Newtonian capacity, there is no nontrivial relationship between positive Monge-Ampère capacity and Hausdorff dimension. This fact is demonstrated by the following result which extends an example of Diederich and Fornaess. The idea is the same as in their paper [4].

Theorem. There is a C^∞, real, closed curve γ in $\mathbb{C}^{N+1}$ which is not pluripolar.

Proof. The construction of the curve proceeds as follows. First construct N uncountable families of disjoint Cantor sets in the unit interval, each having positive logarithmic capacity. The Cantor sets will be indexed as $C_\alpha = C_{\alpha_1,\dots,\alpha_N}$ where $a_i\in J_i$ with J_i an uncountable set. Next construct N smooth functions $f_1,\dots,f_N$ on $\mathbb{R}$ such that the following two conditions hold.

(i) $f_i|C_{\alpha_1,\dots,\alpha_N} = c_{\alpha_i}$ is a real constant depending only on the index α_i.

(ii) The set $\{c_{\alpha_i}:\ \alpha_i\in J_i\}$ is a Cantor set of positive capacity.

The curve γ may then be given in parametric form as:

$\gamma = \{z \in \mathbb{C}^{N+1}: \text{Im } z_1 = 0, \text{ Re } z_1 = t, \ z_2 = f_1(t), \ldots, z_{n+1} = f_N(t)\}$.

We first show that γ is not a pluripolar set in $\mathbb{C}^{N+1}$ and then construct the Cantor sets and the functions f_i.

Suppose ϕ is a plurisubharmonic function, thus not identically $-\infty$ by definition, and $\phi|\gamma \equiv -\infty$. Consider $\phi(\zeta, c_{\alpha_1}, \ldots, c_{\alpha_N})$ for fixed $\alpha = (a_1, \ldots, a_N)$. Then $\phi(t, c_{\alpha_1}, \ldots, c_{\alpha_N}) \equiv -\infty$ for $t \in C_{\alpha_1, \ldots, \alpha_N}$ which is a set of positive capacity. It follows that $\phi(\zeta, c_{\alpha_1}, \ldots, c_{\alpha_N}) \equiv -\infty$ as a function of ζ. Next consider $\phi(z_1, \zeta, c_{\alpha_2}, \ldots, c_{\alpha_N})$ for fixed $\alpha_2, \ldots, \alpha_N$. As a function of ζ, $\phi(z_1, \zeta, c_{\alpha_2}, \ldots, c_{\alpha_N}) \equiv -\infty$ if $\zeta \in \{c_{\alpha_1}: \alpha_1 \in J_1\}$ which is a set of positive capacity and thus $\phi(z_1, \zeta, c_{\alpha_2}, \ldots, c_{\alpha_N}) \equiv -\infty$. Continuing in this manner gives $\phi(z_1, z_2, \ldots, z_{N+1}) \equiv -\infty$ which is a contradiction and thus γ must be non-pluripolar.

Construction of the Cantor Sets. The construction of the Cantor sets uses a recursive process in the usual manner. If I denotes a closed interval in $\mathbb{R}$, denote by (*) the following process.

(*) Remove the open middle 1/3 of I, the open middle 1/3 of the remaining closed intervals, and continue removing middle 1/3's until 2^{N+1} closed intervals remain, each having length $(1/3^{N+1})|I|$ where $|I|$ denotes the length of I.

We may label these intervals with the lexicographic ordering on N-tuples of elements of $\mathbb{Z}_2$. The first 2^N intervals are labeled left to right as $(0,0,\ldots,0)$, $(0,0,\ldots,0,1)$, $\ldots$, $(0,1,1,\ldots,1)$, $(1,1,1,\ldots,1)$. The second 2^N intervals are labeled in the same way so there are pairs of intervals with the same label. To distinguish these pairs, label them $I_{b,1}$ and $I_{b,2}$ where $b \in \mathbb{Z}_2^N = \mathbb{Z}_2 \times \ldots \times \mathbb{Z}_2$ (N times). Now define intervals $I_{b^1, b^2, \ldots, b^n, k}$ with $b^i \in \mathbb{Z}_2^N$ and $k = 1, \ldots, 2^N$ recursively by applying (*) n times. The index k is used to distinguish intervals with identical lexicographic label. If the $I_{b^1, \ldots, b^n, k}$ are defined, define the $I_{b_1^1, \ldots, b^{n+1}, \tilde{k}}$ by applying (*) to each of the intervals $I_{b_1^1, \ldots, b^n, k}$. There are thus $(2^{N+1})^n$ intervals $I_{b^1, \ldots, b^n, k}$, each of length $(1/3^{N+1})^n$.

We now define the Cantor sets. The index set will be $J = J_1 \times \ldots \times J_N$ where J_i consists of all possible infinite sequences in $\mathbb{Z}_2$. If $\alpha=(\alpha_1,\ldots,\alpha_N)\in J$ write $\alpha_i=\{a_i^j\}_{j=1,2,\ldots}$ where $a_i^j\in\mathbb{Z}_2$ and $a^j=(a_1^j,\ldots,a_N^j)$. Now suppose α is fixed. Define $C_{\alpha,1}$ to be those intervals $I_{b^1,k}$ with $b^1=a^1$, $k=1,2$. $C_{\alpha,n}$ is defined to be the intervals $I_{b^1,\ldots,b^n,k}$ with $b^j=a^j$, $j=1,\ldots,n$ and $k=1,\ldots,2^n$. Therefore $C_{\alpha,n}$ consists of 2^n intervals each of length $(1/3^{N+1})^n$. Furthermore, the minimum distance between the intervals is $(2\cdot 3^N-1)\cdot(1/3^{N+1})^n$. Let $C_\alpha=\bigcap_n C_{\alpha,n}$. Then C_α is a generalized Cantor set in the sense of Tsuji [3] or Carleson [1] and has positive logarithmic capacity. (See for example Tsuji, Theorem III.6.3 or Carleson, Theorem IV.3.)

Construction of the Parameter Functions. First define $\gamma_{b^1,\ldots,b^n k}$ to be the initial points of the intervals $I_{b^1,\ldots,b^n,k}$, $k=1,\ldots,2^n$. Define smooth functions χ_i on $\mathbb{R}$ as follows.

$$\chi_i(t) = \begin{cases} 1 & \text{if } t\in I_{b^1} = I_{(b_1^1,\ldots,b_N^1)} \quad \text{where } b_i^1=1 \\ 0 & \text{if } t\in I_{b^1} \text{ with } b_i^1=0 . \end{cases}$$

Further, we require $0\le\chi_i(t)\le 1$ everywhere and $\chi_i(t)\equiv 0$ outside an ε-neighborhood of $[0,1]$. Note that there are 2^N intervals on which $\chi_i(t)\equiv 1$. Now define the functions f_i as:

$$f_i(t) = \chi_i(t) + \sum_{n=1}^{\infty} \sum_{b^1,\ldots,b^n,k} (1/10^{n^2})\chi_i((3^{N+1})^n(t-\gamma_{b^1,\ldots,b^n,k})) .$$

First note that $f_i(t)$ is C^∞ since for any fixed n, the derivatives up to order ℓ of the nth term of the series are bounded by

$$\frac{(2^{N+1})^n(3^{N+1})^{\ell\cdot N}}{10^{n^2}} \cdot \sup_{\substack{t\in\mathbb{R}\\ 0\le j\le\ell}} \left\{ \frac{d^j\chi_i}{dt^j}(t) \right\} .$$

Now fix $\alpha=(\alpha_1,\ldots,\alpha_N)$ and suppose $t\in C_\alpha$. Then $t\in I_{a^1,\ldots,a^n,k}$ for all n and some $k\in\{1,2,\ldots,2^n\}$ where k of course depends on n. Consequently, if $t\in I_{a^1,\ldots,a^{n+1},\tilde{k}}$ with $a_i^{n+1}=1(0)$, then $\chi_i((3^{N+1})^n(t-\gamma_{a^1,\ldots,a^n,k})) = 1(0)$. For all $(b^1,\ldots,b^n)\neq(a^1,\ldots,a^n)$, $\chi_i((3^{N+1})^n(t-\gamma_{a^1,\ldots,a^n,k}))=0$. Therefore

$$f_i(t) = \sum_{n=0}^{\infty} \frac{1}{10^{n^2}} \cdot a_i^{n+1} .$$

Define $c_{\alpha_i} = \sum_{n=0}^{\infty}(1/10^{n^2})\cdot a_i^{n+1}$. It remains to show that the set $\{c_{\alpha_i}: \alpha_i \in J_i\}$ has positive capacity. To do this we exhibit the set as a generalized Cantor set. Let $C_o=[0,2]$. Suppose C_n has been defined as a union of 2^n disjoint closed intervals. Apply to each of these closed intervals, I, the following procedure. Divide I into $2\cdot 10^{2n+1}$ intervals of equal length and keep the first two intervals and the $10^{2n+1}+1^{st}$ and $10^{2n+1}+2^{nd}$ intervals (closed). Let C_{n+1} be the set of all subintervals so obtained from the intervals of C_n. Then C_n has the following properties:

(i) $C_{n+1} \subset C_n$;

(ii) C_n consists of 2^n closed subintervals of length $2/10^n$;

(iii) The minimum distance between any two of the intervals is $(1/10^{(n-1)^2})\cdot(1-(2/10^{2n-1}))$;

(iv) If $C = \bigcap_n C_n$ then by Theorem IV.3. of Carleson [1], C has positive logarithmic capacity.

Now suppose $\alpha \in J$ is fixed and $c_\alpha = \sum_{n=0}^{\infty}(1/10^{n^2})\cdot a_{n+1}$. (We drop the subscript i for simplicity.) Let $c_{\alpha,n} = \sum_{k=0}^{n}(1/10^{k^2})\cdot a_{k+1}$. Then $c_{\alpha,n}$ lies on the initial point of some closed interval in C_n. Furthermore, $c_\alpha = \lim_{n\to\infty} c_{\alpha,n}$ and $c_{\alpha,n} \in C_\ell$ for all n,ℓ. Since C is a compact set $c_\alpha \in C$. Conversely suppose $d \in C$. Then $d \in C_n$ for all n and it belongs to a unique maximal subinterval of C_n. Call this interval I_n. Since $d \in C_{n+1}$, $d \in I_{n+1}$ and $I_{n+1} \subset I_n$. Let d_n denote the initial point of I_n. Then $d = \lim_{n\to\infty} d_n$ where $d_n = \sum_{k=0}^{n-1}(1/10^{k^2})\cdot a_{k+1}$ with $a_k \in \mathbb{Z}_2$. Therefore $d \in \{c_\alpha: \alpha \in J\}$ and $C=\{c_\alpha: \alpha \in J\}$ which concludes the construction.

3. INTEGRAL GEOMETRY

The integral geometry problems I want to consider involve averaging over sets of linear spaces which are not pluripolar. The sets of linear spaces are sets in G(k,n), k-dimensional planes in $\mathbb{C}^n$. If $X \subset \mathbb{C}^n$ is a pure p-dimensional variety, let $q = n-p$ and consider intersection with the q-dimensional linear subspaces of $\mathbb{C}^n$. The idea is to make an assumption on the growth properties of $X \bullet \ell$ for all $\ell \in E \subset G(q,n)$ where E is not locally pluripolar. One then tries to recover information about the growth of X from the growth on the average of $X \circ \ell$.

The growth of X is measured by

$$(3.1) \qquad \sigma_X(r) = \mathrm{vol}_{2p}(X \cap B^n(r))$$

where $B^n(r)$ is the ball of radius r centered at the origin in $\mathbb{C}^n$. We say that the function $\chi(r)$ has finite order growth ρ if

$$(3.2) \qquad \limsup_{r\to\infty} \frac{\log \chi(r)}{\log r} = \rho < \infty .$$

X is said to be of finite order ρ if $r^{-2p}\sigma_X(r)$ is of finite order ρ.

Following Gruman [5], we say that $\rho(r)$ is a *precise order* if:

$$(3.3) \qquad \begin{array}{ll} \text{(i)} & \lim_{r\to\infty} \rho(r) = \rho \qquad \text{and} \\ \text{(ii)} & \lim_{r\to\infty} r \log r \cdot \rho'(r) = 0 . \end{array}$$

We say that $\chi(r)$ is of normal type λ with respect to the precise order $\rho(r)$ provided

$$(3.4) \qquad \lambda = \limsup_{r\to\infty} \frac{\chi(r)}{r^{\rho(r)}} .$$

If $\ell \in G(q,n)$ let

$$(3.5) \qquad \sigma_X(\ell,r) = \mathrm{card}(X \bullet \ell \cap B^n(r)) .$$

Note that except for a pluripolar set of $\ell \in G(q,n)$, X intersects ℓ in a discrete set of points. The idea now is to obtain estimates above and below for $\sigma_X(\ell,r)$ by $\sigma_X(r)$ *on the average*.

L. Gruman has obtained very interesting and strong results in this direction [5] and so I will describe some of his work. My aim is to indicate how the Monge-Ampère operator comes into the picture. The upper estimate referred to above is easier, so I will describe it first.

Theorem (Gruman). Given $\varepsilon>0$ and $\beta>1$, the set

$$(3.6)\qquad E=\left\{\ell\in G(q,n):\ \limsup_{r\to\infty}\frac{\sigma_X(\ell,r)}{(\log r)^{\beta}\, r^{-2p}\,\sigma_X((1+\varepsilon)r)}\neq 0\right\}$$

is locally pluripolar in $G(q,n)$.

The two following results are basic starting points for proving the above result and for the integral geometry of the Monge-Ampère operator in general.

Basic Integration by Parts Result. Let θ be a closed positive current of bidegree (p,p) with C^{∞} and let $V\in PSH(\mathbb{C}^n)$. Let $\beta=\sqrt{-1}\,\partial\bar{\partial}|z|^2$. Then

$$(3.7)\qquad \theta\wedge\beta^{n-p-1}\wedge\sqrt{-1}\,\bar{\partial}|z|^2 \text{ is a positive measure on } \partial B^n(r)\ .$$

$$(3.8)\qquad \int_{\partial B^n(r)} V\cdot\theta\wedge\beta^{n-p-1}\wedge\bar{\partial}|z|^2$$

$$= \int_{B^n(r)} V\cdot\theta\wedge\beta^{n-p} + \int_{B^n(r)} (r^2-|z|^2)\sqrt{-1}\,\partial\bar{\partial}V\theta\wedge\beta^{n-p-1}\ .$$

Basic Inequality for the Monge-Ampere Operator. Let $M_V(r)=\sup_{|z|\le r}V(z)$ if $V\ge 0$. Let $V\in PSH(\mathbb{C}^n)$ and θ_X be the current of integration over X. Then, if $\gamma<1$ is given, there exists a constant $C(\gamma)$ such that

$$(3.9)\qquad \int_{B^n(\gamma^q r)}\theta_X\wedge\beta^{p-q}\wedge(\sqrt{-1}\,\partial\bar{\partial}V)^q\le C(\gamma)\,[M_V(r)]^q\cdot\sigma_X(r)\cdot r^{-2q}\ .$$

The upper estimate, (3.6), is obtained by the following argument. Let $E\subset G(q,n)$ be a non-locally pluripolar set. Let u_E be the associated potential function. One then has an estimate for $|u_E|$. Let $\tilde{\theta}$ be a smoothing of θ_X, the current of integration over X. Extend $\tilde{\theta}$ to a current on $U\times\mathbb{C}^n$ where $U\subset G(q,n)$. Then apply the basic inequality for the Monge-Ampère operator where one takes $V=u_E$. This in turn gives an estimate above on

$$\int \sigma_X(\ell,r)\ d\mu_E(\ell)$$

with $d\mu_E = (\sqrt{-1}\ \partial\bar{\partial} u_E)^k$.

Remark. The upper estimate holds without any assumption on the growth of X.

We now consider the lower estimate for the average growth of $\sigma_X(\ell,r)$. Let $E\subset G(q,n)$ be non-locally pluripolar. The lower estimate is given by the following inequality. There exist constants c_1 and c_2 and a probability measure μ supported on E such that if $t>0$, then

$$(3.10)\quad \int_E \sigma_X(\ell,r(t+1))d\mu(\ell)$$

$$\geq r^{-2p}\left\{c_1\sigma_X(r)-[\sigma_X(r(t+1))-\sigma_X(tr)]-c_2(1/r)\sigma_X(3r)\right\} .$$

Again this estimate holds without any assumption on the growth of X.

If the analytic variety X has finite order growth, then it is possible to obtain a nicer form for the right side of the above inequality. If the function $r^{-2p}\sigma_X(r)$ is of finite order ρ and of normal type λ, and if $\{r_m\}$ is a sequence increasing to ∞, then there exist constants M, T, and C such that if $m\geq M$ then

$$\int \sigma_X(\ell,Tr_m)d\mu(\ell) \geq \lambda\cdot C\cdot r_m^{\rho(r_m)} .$$

Corollary. Suppose $X\subset\mathbb{C}^n$ is of pure dimension p, of finite order ρ and normal type λ. Then if there exists a set $E\subset G(q,n)$ which is non-locally pluripolar such that $X\circ\ell$ is finite for all $\ell\in E$, then X is algebraic.

During the conference in honor of Professor Stoll, Professor Lelong informed me that L. Gruman has obtained an estimate which would imply the above corollary without the assumptions on growth of X; one would need only that $X\bullet\ell$ be finite for a non-pluripolar set of $\ell\in G(q,n)$.

I would like to end with the following question. Does a converse to the above result hold? Given $E \subset G(q,n)$, which is locally pluripolar, can one construct $X \subset \mathbb{C}^n$ of dimension $p = n-q$ such that $X \cdot \ell$ is finite for all $\ell \in E$ but X is transcendental?

REFERENCES

1. E. Bedford and B.A. Taylor, *Potential theoretic properties of plurisubharmonic functions*, Acta Math. **149** (1982), pp. 1-40.

2. J. Carlson, *A moving lemma for the transcendental Bezout problem*, Annals of Math. **130** (1976), pp. 305-330.

3. M. Cornalba and B. Shiffman, *A counterexample to the transcendental Bezout problem*, Annals of Math. **96** (1972), pp. 402-406.

4. K. Diederich and J.E. Fornaess, *A smooth curve in $\mathbb{C}^2$ which is not a pluripolar set*, Duke Math. J. **49** (1982), pp. 931-936.

5. L. Gruman, *Ensembles exceptionnels pour les applications holomorphes dans $\mathbb{C}^n$*, Lecture Notes in Math. **1028**, Séminaire d'Analyse, P. Lelong - P. Dolbeault - H. Skoda, pp. 126-162.

6. W. Stoll, *A Bezout estimate for complete intersections*, Annals of Math. **96** (1972), pp. 361-401.

Logarithmic Jet Spaces and Extensions of de Franchis' Theorem

Dedicated to Professor Wilhelm Stoll on the occasion
of his sixtieth birthday

Junjiro Noguchi

University of Notre Dame, Department of Mathematics
Notre Dame, Indiana 46556, U.S.A.

Tokyo Institute of Technology, Department of
Mathematics, Oh-okayama, Meguro, Tokyo 152, JAPAN.

Jet bundles and logarithmic 1-forms play an important role in the study of the value distribution of meromorphic mappings into algebraic varieties (cf. [O1] [G-G1] and [N1∼4]). On the other hand, logarithmic vector fields as well as logarithmic 1-forms have been used in the study of Gauss-Manin connection and singularities (cf. [S1]). In this paper we define the sheaf $\mathcal{J}_k(X;\log D)$ of germs of logarithmic jet fields along a hypersurface D over a complex space X, which induces the logarithmic jet space $J_k(X; \log D)$ along D, if D has only normal crossings (see section 1 for details). Then we investigate their functorial properties.

In section 2 we apply the logarithmic jet space to prove the following theorem which generalizes de Franchis' Theorem to the higher dimensional case:

(2.5) Theorem. Let A be a quasi-Abelian variety and X a subvariety of general type of A. Let Y be a Zariski open subset of a compact complex space $\overline{Y}$. Then there are only finitely many holomorphic mappings $f: Y \longrightarrow X$ which are non-degenerate with respect to $\Lambda(X)$ (cf. (2.3)).

Here $\Lambda(X)$ is, roughly speaking, the family of the supports of logarithmic cannonical divisors on X and the above non-degeneracy of f means that $f(Y) \not\subset Z$ for all $Z \in \Lambda(X)$. It is noted that the non-degeneracy condition for f is necessary (see Example (2.14)). Since a compact smooth algebraic curve C with genus ≥ 2 is imbedded into the Jacobian variety, Theorem

(2.5) implies that there are only finitely many non-constant rational mappings from an algebraic variety into C (de Franchis' Theorem).

H. Fujimoto [F1] obtained a finiteness theorem for a family of linearly non-degenerate meromorphic mappings from the m-dimensional complex vector space $\mathbb{C}^m$ into the n-dimensional projective space $\mathbb{P}^n(\mathbb{C})$. We will give a remark on the relationship between Fujimoto's result and Theorem (2.5) (see Remark (2.15)).

In section 3, we will prove another extension of de Franchis' Theorem, which generalizes the result of [N-S1 , Main Theorem (1.2)] in the case of the complex number field $\mathbb{C}$. Let $\bar{V}$ be a complete smooth algebraic variety, D a hypersurface of $\bar{V}$ with normal crossings and $V = \bar{V}-D$. Let $T(\bar{V};\log D)$ be the vector bundle of logarithmic vector fields along D and $T^q(\bar{V};\log D)$ denote the q-th exterior power $\wedge^q T(\bar{V};\log D)$. We say that the vector bundle $T^q(V;\log D) \longrightarrow \bar{V}$ is <u>quasi-negative over</u> V if there is a proper morphism $\Phi: T^q(\bar{V};\log D) \to \mathbb{C}^N$ into the complex affine N-space $\mathbb{C}^N$ such that the restriction of Φ over $T^q(\bar{V};\log D)|V$ minus the zero section is an isomorphism onto its image. Let W be another algebraic variety and $F_q(W,V)$ the set of proper rational mappings $f: W \longrightarrow V$ with rank $f \geq q$.

(3.1) <u>Theorem</u>. If $T^q(\bar{V};\log D)$ is quasi-negative over V, then $F_q(W,V)$ is finite.

<u>Acknowledgement</u>. The main part of this paper was written during the author's visit to the University of Notre Dame, 1984/85. He expresses his sincere gratitude to the Department of Mathematics of the University of Notre Dame for the hospitality, and especially to Professor W. Stoll for numerous discussions on the subjects of this paper and related topics.

§1 LOGARITHMIC JET SPACES

a) <u>Jet bundle</u>. Let $\mathbb{C}$ be the Gaussian plane with the standard coordinate z. Let M be a complex manifold and $f,g: (\mathbb{C},0) \to (M;x)$ germs of holomorphic mappings from neighborhoods of the origin $0 \in \mathbb{C}$ into M with $f(0) = g(0) = x \in M$. For a positive integer $k \in \mathbb{Z}$ we write $f \underset{k}{\sim} g$ if f and g have the same Taylor expansions in z up to order k for some holomorphic local coordinate system around x. Then it is easily checked that the relation, "$\underset{k}{\sim}$" is independent of the choice of the holomorphic local coordinate system around x and defines an equivalence relation on the set $\{f:(\mathbb{C};0) \to (M;x)\}$. Let $j_k(f)$ denote the equivalence class of f and set

$$(1.1)\qquad \begin{aligned} J_k(M)_x &= \{j_k(f);\ f:\ (\mathbb{C};0) \to (M;x)\}, \\ J_k(M) &= \bigcup_{x\in M} J_k(M)_x . \end{aligned}$$

Then $J_k(M)$ naturally carries the structure of a holomorphic fibre bundle over M with the canonical projection $\pi: J_k(M) \longrightarrow M$ (cf. [O1] and [G-G1]). The bundle $(J_k(M), \pi, M)$ is called the <u>jet bundle</u> of order k and $j_k(f) \in J_k(M)$ is called a <u>jet</u> of order k. It is noted that $J_1(M)$ is isomorphic to the holomorphic tangent bundle T(M) over M, and that $J_k(M)$ has a structure of flag with the natural projection $J_k(M) \longrightarrow J_{k-1}(M)$

$$(1.2)\qquad J_k(M) \longrightarrow J_{k-1}(M) \longrightarrow \cdots \longrightarrow J_1(M) \cong T(M) \longrightarrow M$$

such that for a holomorphic section $s \in \Gamma(U, J_{k-1}(M))$ over an open subset $U \subset M$, the restriction $J_k(M)|s(U)$ of $J_k(M) \longrightarrow J_{k-1}(M)$ over s(U) is isomorphic to $J_1(M)|U \cong T(M)|U$. If M is a complex algebraic manifold, then $(J_k(M), \pi, M)$ is also a complex algebraic fibre bundle over M. Let W be an open subset of $\mathbb{C}$ and $G: W \longrightarrow M$ a holomorphic mapping. Then G naturally induces the lifting

$$(1.3)\qquad J_k(G): W \longrightarrow J_k(M)$$

such that $\pi \circ J_k(G) = G$ (cf. [O1 , p. 86]).

Let Ω^1_M denote the sheaf of germs of holomorphic 1-forms over M. Take a holomorphic section $\omega \in \Gamma(U, \Omega^1_M)$ over an open subset U of M. For $j_k(f) \in J_k(M)|U$, put

$$f^*\omega = A(z)dz .$$

Then the derivatives $d^jA/dz^j(0)$, $0 \le j \le k-1$, are well-defined, independently of the representative f for $j_k(f)$. Hence we have a mapping

$$(1.4)\qquad \tilde{\omega}: J_k(M)|U \ni j_k(f) \longmapsto \left(\frac{d^jA}{dz^j}(0)\right)_{0\le j\le k-1} \in \mathbb{C}^k$$

which is holomorphic. Let $\omega^1,\dots,\omega^m$ with $m = \dim M$ be holomorphic 1-forms on U such that $\omega^1 \wedge \dots \wedge \omega^m$ does not vanish anywhere. Then we have a biholomorphic mapping

(1.5) $\quad \pi \times (\tilde{\omega}^1,\dots,\tilde{\omega}^m) : J_k(M)|U \longrightarrow U \times (\mathbb{C}^k)^m$

which we call the trivialization associated with $\{\omega^1,\dots,\omega^m\}$. Let Ξ_M^1 denote the sheaf of germs of meromorphic 1-forms and take $\xi \in \Gamma(U, \Xi_M^1)$. Then as in (1.4), ξ induces a meromorphic vector function

(1.6) $\quad \tilde{\xi} : J_k(M)|U \longrightarrow \mathbb{C}^k$.

b) Jet spaces. Let X be a complex space with structure sheaf $\mathcal{O}_X$, which is, in this paper, always assumed to be irreducible and reduced unless otherwise mentioned. We assume for a while that X is biholomorphically imbedded into a complex manifold M. Let $\mathcal{J}(X)$ denote the ideal sheaf of X. We say that a jet $j_k(f) \in J_k(M)_x$ with $x \in X$ is tangent to X if $P \circ f$ have zero of order $\geq k$ at 0 for all $P \in \mathcal{J}(X)_x$. Let $J_k(X)_x$ be the set of all jets $j_k(f) \in J_k(M)_x$ which are tangent to X and set

$$J_k(X) = \bigcup_{x \in X} J_k(X)_x,$$

$$\pi : J_k(X) \longrightarrow X ,$$

where π denotes the natural projection. It follows from the coherence of the ideal sheaf $\mathcal{J}(X)$ ([Cl]) that $J_k(X)$ is a complex subspace of $J_k(M)$ and $\pi : J_k(X) \longrightarrow X$ is a holomorphic fibre space. Let $X \longrightarrow M'$ be another imbedding of X into a complex manifold M'. Then, in the same way as above, we have another holomorphic fibre space $\pi' : J_k(X)' \longrightarrow X$, which is however isomorphic to $\pi : J_k(X) \longrightarrow X$ as fibre space. Hence for general X, we define the holomorphic fibre space $\pi : J_k(X) \longrightarrow X$ by making use of local imbeddings of X into open subsets of complex vector spaces. We call $(J_k(X),\pi,X)$ the jet space of order k over X. The jet space $J_1(X)$ of order 1 is isomorphic to the Zariski tangent space $\Theta(X)$. As in (1.2), $J_k(X)$ carries a sequence of fibrations.

(1.7) $\quad J_k(X) \longrightarrow J_{k-1}(X) \longrightarrow \dots \longrightarrow J_1(X) \cong \Theta(X) \longrightarrow X$

such that for a holomorphic sections $s \in \Gamma(U, J_{k-1}(X))$ on an open subset U of X

$$J_k(X)|s(U) \cong H\ (X)|U\ .$$

Let $G: W \longrightarrow X$ be a holomorphic mapping from an open subset W of $\mathbb{C}$ into X. Then as in (1.3), we have the lifting of G

(1.8) $$J_k(G): W \longrightarrow J_k(X)$$

such that $\pi \circ J_k(G) = F$.

c) Sheaf and space of logarithmic jet fields over manifolds. Let D be a hypersurface of a complex manifold M and $\Omega^1_M(\log D)$ the sheaf of germs of logarithmic 1-forms along D (cf. [D1], [I1] and [I2]). For convenience, we recall the definition of $\Omega^1_M(\log D)$. For $x \in M-D$, the stalk $\Omega^1_M(\log D)_x$ is identical to the stalk $\Omega^1_{M,x}$ of Ω^1_M at x. For $x \in D$, take a neighborhood U of x and irreducible holomorphic functions $\sigma_1,\ldots,\sigma_\ell$ on U such that

$$U \cap D = \{\sigma_1 \ldots \sigma_s = 0\}\ .$$

Then we define

(1.9) $$\Omega^1_M(\log D)_x = \sum_{j=1}^{\ell} \mathcal{O}_{M,x} \frac{d\sigma_j}{\sigma_j} + \Omega^1_{M,x}\ .$$

If D has only normal crossings, then $\Omega^1_M(\log D)$ is locally free. If M is bimeromorphic to a compact Kähler manifold, the global sections of $\Omega^1_M(\log D)$ are d-closed (cf. [D1]). Let N be another complex manifold, E a hypersurface of N and $\phi: N \longrightarrow M$ a holomorphic mapping such that $\phi^{-1}(D) \subset E$. Then ϕ naturally induces a sheaf morphism

(1.10) $$\phi^*: \Omega^1_M(\log D) \longrightarrow \Omega^1_N(\log E).$$

If D and E have only normal crossings, then for a meromorphic mapping $\psi: N \longrightarrow M$ with $\psi^{-1}(D) \subset E$ we have

(1.11) $$\psi^*: \Gamma(M, \Omega^1_M(\log D)) \longrightarrow \Gamma(N, \Omega^1_N(\log E)).$$

Let $s \in \Gamma(U, J_k(M))$ be a holomorphic section on an open subset U of M.

(1.12) Definition. We say that s is a logarithmic jet field along D if $\tilde{\omega} \circ s|U : U \longrightarrow \mathbb{C}^k$ are holomorphic for all $\omega \in \Gamma(U', \Omega^1_M(\log D))$, where U' is an arbitrary open subset of U (see (1.6) for the definition of $\tilde{\omega}$).

The sets of logarithmic jet fields along D over open subsets of M form a complete presheaf which defines a sheaf $\mathcal{J}_k(M; \log D)$ over M. We call $\mathcal{J}_k(M; \log D)$ the sheaf of germs of logarithmic jet fields along D over M.

Assume that D has only normal crossings. Take a point $x_o \in D$ and a holomorphic local coordinate neighborhood $U(x^1,\dots,x^m)$ around x_o so that

$$x_o = (0,\dots,0),$$
$$D \cap U = \{x^1 \cdots x^\ell = 0\} \quad (1 \leq \ell \leq m).$$

Then any logarithmic 1-form ω along D on an open subset U' of U is written as

$$\omega = a_1\frac{dx^1}{x^1} + \dots + a_\ell\frac{dx^\ell}{x^\ell} + a_{\ell+1}dx^{\ell+1} + \dots + a_m dx^m,$$

where a_j are holomorphic functions on U'. Let $J_k(M)|U \cong U \times (\mathbb{C}^k)^m$ be the trivialization associated with $\{dx^1,\dots,dx^m\}$. Then a section $s \in \Gamma(U, J_k(M))$ is given by $s(x) = (x, Z(x)): U \longrightarrow U \times (\mathbb{C}^k)^m$ with

$$Z(x) = \begin{pmatrix} Z_1^1(x) & \dots & Z_1^m(x) \\ \cdot & \cdots & \cdot \\ Z_k^1(x) & \dots & Z_k^m(x) \end{pmatrix},$$

where $Z_j^i(x)$ are holomorphic functions on U and the indices j correspond to the orders of derivatives. Put

$$f(z) = x + \frac{1}{1!}Z_1(x)z + \dots + \frac{1}{k!}Z_k(x)z^k$$

for $x \in U$, which is a holomorphic mapping from a neighborhood of $0 \in \mathbb{C}$ into U. Then $j_k(f) = s(x)$. Let $x \in U'$ and put $f^*\omega = A(z)dz$. We have

$$A(z) = \sum_{i=1}^{\ell} a_i Z_1^i \frac{1}{x^i} + \sum_{i=\ell+1}^{m} a_i Z_1^i + \dots$$
$$+ \frac{1}{(k-1)!}\Big(\sum_{i=1}^{\ell} a_i Z_k^i \frac{1}{x^i} + \sum_{i=\ell+1}^{m} a_i Z_k^i\Big) z^{k-1}.$$

It follows that

$$(1.13) \qquad \tilde{\omega}\circ s = \begin{pmatrix} \sum_{i=1}^{\ell} a_i Z_1^i \frac{1}{x^i} + \sum_{i=\ell+1}^{m} a_i Z_1^i \\ \cdot \quad \cdot \quad \cdot \quad \cdot \\ \sum_{i=1}^{\ell} a_i Z_k^i \frac{1}{x^i} + \sum_{i=\ell+1}^{m} a_i Z_k^i \end{pmatrix}$$

Therefore one sees that

(1.14) $s \in \Gamma(U, \mathcal{J}_k(M;\ \log D))$ if and only if
$z^i_j = x^i z'^i_j$ for $1 \le i \le \ell$ and $1 \le j \le k$,

where z'^i_j are holomorphic functions on U. Hence we have the following:

(1.15) Proposition. i) If D has only normal crossings, then there is a subbundle

$$\pi: J_k(M: \log D) \longrightarrow M$$

of $J_k(M)$ such that the sheaf of germs of holomorphic sections of $J_k(M;\ \log D)$ is isomorphic to $\mathcal{J}_k(M;\ \log D)$.

ii) Let N be another complex manifold and E a hypersurface of N with normal crossings. Let $\phi : N \longrightarrow M$ be a holomorphic mapping such that $\phi^{-1}(D) \subset E$. Then ϕ induces a bundle morphism

$$\phi^*: J_k(N;\ \log E) \longrightarrow J_k(M;\ \log D).$$

The assertion ii) follows from (1.10).

We call the above $J_k(M;\ \log D)$ the logarithmic jet bundle of order k along D. In the case of $k = 1$, we also write

$$J_1(M;\ \log D) = T(M;\ \log D),$$

which is called the vector bundle of logarithmic vector fields along D (cf. [S1]). As in (1.2), the fibre bundle $J_k(M{:}\log D)$ has a structure of flag:

(1.16) $J_k(M;\ \log D) \longrightarrow \dots \longrightarrow J_1(M;\ \log D) \cong T(M;\log D) \longrightarrow M$
such that for a holomorphic section $s \in \Gamma(U, J_{k-1}(M))$ over an open subset $U \subset M$

$$J_k(M;\ \log D)|s(U) \cong T(M{:}\ \log D)|U.$$

We continue to assume that D has only normal crossings. By definition, every $\omega \in \Gamma(U, \Omega^1_M(\log D))$ defines a holomorphic vector function

(1.17) $\tilde{\omega} : J_k(M;\ \log D)|U \longrightarrow \mathbb{C}^k$

and if $\omega^1, \dots, \omega^m \in \Gamma(U, \Omega^1_M(\log D))$ are logarithmic 1-forms along D such that $\omega^1 \wedge \dots \wedge \omega^m$ does not vanish anywhere, then

(1.18) $\pi \times (\tilde{\omega}^1, \ldots, \tilde{\omega}^m) : J_k(M; \log D)|U \longrightarrow U \times (\mathbb{C}^k)^m$

is a biholomorphism, which we call the trivialization of $J_k(M; \log D)$ over U associated with $\{\omega^1, \ldots, \omega^m\}$.

d) Logarithmic 1-forms and logarithmic jet fields over complex spaces: Imbedded case. Let X be a complex subspace of M. Then $J_k(X)$ is a complex subspace of $J_k(X)$. Let D be a hypersurface of M such that $D \not\subset X$. Let $\Delta: X \longrightarrow X \times X$ be the diagonal imbedding and $J(\Delta(X))$ the ideal sheaf of the diagonal $\Delta(X)$. The sheaf Ω_X^1 of germs of holomorphic 1-forms is defined by

$$\Omega_X^1 = \Delta^{-1}(J(\Delta(X))/J^2(\Delta(X))).$$

Let M_X denote the sheaf of germs of meromorphic functions over X and set

$$\Xi_X^1 = M_X \otimes_{\mathcal{O}_X} \Omega_X^1 .$$

which is called the sheaf of germs of meromorphic 1-forms over X. By definitions, Ω_X^1 and Ξ_X^1 are defined for general complex spaces which are not necessarily imbedded into complex manifolds. The inclusion mapping $i_X: X \longrightarrow M$ induces

$$i_X^* : \Omega_M^1 \longrightarrow \Omega_X^1 \longrightarrow 0,$$

$$i_X^* : \Omega_M^1(\log D) \longrightarrow \Xi_\xi .$$

We set

(1.19) $$\Omega_X^1(\log D) = i_X^*(\Omega_M^1(\log D)),$$

which is a coherent sheaf over X.

Let U be an open subset of X and $\omega \in \Gamma(U, \Omega_X^1(\log D))$. Then as in (1.6), ω induces a meromorphic vector function

(1.20) $$\tilde{\omega}: J_k(X)|U \longrightarrow \mathbb{C}^k.$$

Let $s \in \Gamma(U', J_k(X))$ be a holomorphic section on an open subset $U' \subset X$. We say that s is a logarithmic jet field along D if $\tilde{\omega} \circ s|U: U \longrightarrow \mathbb{C}^k$ are holomorphic for all $\omega \in \Gamma(U, \Omega_X^1(\log D))$ with $U \subset U'$. We denote by $\mathcal{J}_k(X; \log D)$ the sheaf of germs of logarithmic jet fields along D over X. Assume that D has only normal crossings. Let

(1.21) $$J_k(X;\ \log D) = J_k(X) \cap J_k(M;\ \log D),$$

which we call the logarithmic jet space of order k along D over X with imbedding $X \subset M$.

(1.22) Proposition. i) The sheaf of germs of holomorphic sections of $J_k(X;\ \log D)$ is isomorphic to $\mathcal{J}_k(X;\ \log D)$.

ii) Let Y be a complex subspace of a complex manifold N and E a hypersurface with normal crossings such that $E \not\supset Y$. Let $f\colon N \longrightarrow M$ be a holomorphic mapping such that $f(Y) \subset X$ and $f^{-1}D \subset E$. Then f induces a fibre morphism

$$f_*\colon J_k(Y;\ \log E) \longrightarrow J_k(X;\ \log D).$$

Proof. Take an arbitrary section $s \in \Gamma(U, J_k(X;\ \log D))$ over an open subset U of X. For any $x \in U$ and $\omega \in \Gamma(U', \Omega^1_X(\log D))$ with a small open neighborhood U' of x in U, we may choose a neighborhood V of x in M and $\eta \in \Gamma(V, \Omega^1_M(\log D))$ such that

$$V \cap X = U' \text{ and } i_X^* \eta = \omega .$$

It follows from (1.17) and (1.21) that $\tilde{\eta} \circ s$ is holomorphic on U'. Therefore $s \in \Gamma(U, J_k(X;\ \log D))$. On the other hand, take any $s \in \Gamma(U, J_k(X;\ \log D))$. Then, of course, $s \in \Gamma(U, J_k(X)) \subset \Gamma(U, J_k(M))$. By the definition of $J_k(X;\ \log D)$ we have

$$s \in \Gamma(U, J_k(M;\ \log D)).$$

Hence, $s \in \Gamma(U, J_k(X;\ \log D))$ by (1.21).

The second part ii) follows from Proposition (1.15).

Q.E.D.

e) Logarithmic forms and logarithmic jet fields over complex spaces. For the purpose here, it suffices to consider the case of a complex space X imbedded into a complex manifold M. However, it is worthwhile to give the intrinsic definitions of logarithmic forms and logarithmic jet fields over general complex spaces.

Let X be a complex space and A a proper analytic subset of X. Let S(X) denote the set of singular points of X and put $R(X) = X - S(X)$. Let $\lambda\colon \tilde{X} \longrightarrow X$ be a blowing-up of X such that $\tilde{A} = \lambda^{-1}A$ is of codimension 1 and has only normal crossings and that

(1.23) $\lambda | (\tilde{X} - \lambda^{-1}(A \cup S(X))) : X - \lambda^{-1}(A \cup S(X)) \longrightarrow X - (A \cup S(X))$

is biholomorphic (Hironaka's theorem). Then we set

$$(1.24) \qquad \Omega^1_X(\log A) = \lambda_* \, \Omega^1_{\tilde{X}}(\log \tilde{A}),$$

$$\Omega^\ell_X(\log A) = \overset{\ell}{\wedge} \, \Omega^1_X(\log A) \quad (\ell \in \mathbb{Z}, > 0)$$

which are coherent analytic sheaves over X. We see by (1.11) that $\Omega^1_X(\log A)$ is independent of the choice of $\tilde{X}$ and so is $\Omega^\ell_X(\log A)$. Let Y be another complex space, B a proper analytic subset of Y and $f: Y \longrightarrow X$ a holomorphic mapping such that $f^{-1}A \subset B$. Then it follows from (1.11) that f induces a sheaf morphism

$$(1.25) \qquad f^*: \Omega^1_X(\log A) \longrightarrow \Omega^1_Y(\log B).$$

Take a section $\omega \in \Gamma(U, \Omega^1_X(\log A))$ over an open subset U of X. Then by (1.23) ω induces a holomorphic vector function

$$(1.26) \qquad \tilde{\omega} : J_k(X) | (U - (A \cup S(X))) \longrightarrow \mathbb{C}^k .$$

Let $s \in \Gamma(U', J_k(X))$ be a section over an open subset U' of X. We call s a logarithmic jet field along A over U' if $\tilde{\omega} \circ s | U - (A \cup S(X))$ have holomorphic extensions over U for all $\omega \in \Gamma(U, \Omega^1_X(\log A))$ with $U \subset U'$. We denote by $_k(X; \log A)$ the sheaf of germs of logarithmic jet fields along A over X.

We consider the case where X is imbedded into a complex manifold M so that A is the intersection of a hypersurface D of M with X. It immediately follows that

$$\Omega^\ell_X(\log D) \subset \Omega^\ell_X(\log A),$$

$$\mathcal{J}_k(X; \log A) \subset \mathcal{J}_k(X; \log D).$$

(1.27) Proposition. If X is normal, then

$$\mathcal{J}_k(X; \log A) = \mathcal{J}_k(X; \log D).$$

Proof. Let $s \in \Gamma(U, \mathcal{J}_k(X; \log D))$ be a section over an open subset of U and $\omega \in \Gamma(U', \Omega^1_X(\log A))$. Then $\tilde{\omega} \circ s | U' \cap R(X) : U' \cap R(X) \longrightarrow \mathbb{C}^k$ is holomorphic. Since X is normal, $\omega \circ s | U' \cap R(X)$ extends holomorphically over U'. Hence it follows that $s \in \Gamma(U, \mathcal{J}_k(X; \log A))$. Q.E.D.

Assume that X is normal and A is a hypersurface such that for every point $x \in A$, there are neighborhood U of x, an imbedding of U into a domain M of some $\mathbb{C}^N$ and a hypersurface D of M with normal crossings satisfying $D \cap X = A \cap U$. Then Propositions (1.22) and (1.27) imply the following:

(1.28) Corollary. Let the notation be as above. Then

i) there is a complex fibre subspace $J_k(X; \log A)$ of $J_k(X)$ such that $\mathcal{J}_k(X; \log A)$ is isomorphic to the sheaf of germs of holomorphic sections of $J_k(X; \log A)$ over X:

ii) Let Y be another normal complex space and B a hypersurface of Y satisfying the same condition as A. Let $f: Y \longrightarrow X$ be a holomorphic mapping such that $f^{-1}A \subset B$. Then we have

$$f_*: J_k(Y; \log B) \longrightarrow J_k(X; \log A).$$

We call the above $J_k(X; \log A)$ the logarithmic jet space of order k along A over X.

§2 HOLOMORPHIC MAPPINGS INTO SUBVARIETIES OF GENERAL TYPE OF QUASI-ABELIAN VARIETIES

Let A be a quasi-Abelian variety; i.e., A is a complex abelian algebraic group carrying the exact sequence:

$$0 \longrightarrow (\mathbb{C}^*)^{\ell} \longrightarrow A \longrightarrow A_o \longrightarrow 0 \quad (\ell \in \mathbb{Z}, \geq 0),$$

where $\mathbb{C}^*$ denotes the multiplicative group of non-zero complex numbers and A_o is an Abelian variety (cf. [I2]). Let X be an irreducible algebraic subvariety of A. We note that

(2.1) X is of general type if and only if the group $\{a \in A;\ a + X = X\}$ is finite (cf. [I1] and [I2]).

Taking the compactification $(\mathbb{P}^1)^{\ell} \supset (\mathbb{C}^*)^{\ell}$, we have

$$\bar{A} = A \times_{(\mathbb{C}^*)^{\ell}} (\mathbb{P}^1)^{\ell}$$

which gives a smooth compactification of A such that $D = \bar{A} - A$ is a hypersurface with normal crossings. Let $\bar{X}$ be the Zariski closure of X in $\bar{A}$. Then we have logarithmic jet spaces along D

$$J_k(X; \log D) \subset J_k(A; \log D).$$

Let $\omega^1, \ldots, \omega^N$ (N = dim A) be the linearly independent invariant 1-forms on A. Then $\{\omega^j\}_{j=1}^N$ is a basis of $H^0(\bar{A}, \Omega^1_{\bar{A}}(\log D))$.

All ω^j are d-closed and $\omega^1 \wedge \ldots \wedge \omega^N$ does not vanish anywhere on $\bar{A}$(cf. [I1] and [I2]). Therefore we have the global triviality of $J_k(\bar{A};\ \log D)$:

$$\pi \times \tilde{\omega} : J_k(\bar{A};\ \log D) \longrightarrow \bar{A} \times (\mathbb{C}^k)^N,$$

where $\pi : J_k(\bar{A};\ \log D) \longrightarrow \bar{A}$ is the projection and $\tilde{\omega} = (\tilde{\omega}^1, \ldots, \tilde{\omega}^N)$ (cf. (1.18)). It follows that $\tilde{\omega} : J_k(\bar{A};\ \log D) \longrightarrow (\mathbb{C}^k)^N$ is a proper morphism and so is the restriction

$$(2.2) \qquad I_k = \tilde{\omega}|J_k(\bar{X};\ \log D): J_k(\bar{X};\ \log D) \longrightarrow (\mathbb{C}^k)^N.$$

Set $n = \dim X$. Let $S(X)$ be the set of singular points of X and $R(X) = X - S(X)$. Let $i_{R(X)} : R(X) \longrightarrow A$ be the inclusion mapping. Let H_o be the linear subspace of $\Gamma(R(X), K_{R(X)})$ spanned by $i^*_{R(X)}(\omega^{j_1} \wedge \ldots \wedge \omega^{j_n})$, $1 \leq j_1 < \ldots < j_n \leq N$ over $\mathbb{C}$, where $K_{R(X)}$ is the canonical bundle over $R(X)$. We denote by $Z(\eta, X)$ the closure in X of the zero locus of $\eta \in H_o$. If $\eta \neq 0$, then $Z(\eta, X)$ is a proper algebraic subset of X. Set

$$\Lambda(X) = \{S(X)\} \cup \{Z(\eta, X);\ \eta \in H_o - \{0\}\}.$$

(2.3) <u>Definition</u>. A holomorphic mapping $f: Y \longrightarrow X$ from a complex space Y into X is said to be non-degenerate with respect to $\Lambda(X)$ if $f(Y) \not\subset Z$ for any $Z \in \Lambda(X)$.

By [N4], Lemma(2.4) we have the following:

(2.4) <u>Lemma</u>. Assume that X is of general type. Then the differential dI_n of

$$I_n\colon\ J_n(\bar{X};\ \log D) \longrightarrow (\mathbb{C}^n)^N$$

has maximal rank at some points. Moreover, if a holomorphic mapping $F : W \longrightarrow X$ from a domain W of $\mathbb{C}$ into X is non-degenerate with respect to $\Lambda(X)$, then the lifting $J_n(F): W \longrightarrow J_n(X) \subset J_n(X;\ \log D)$ generically takes values at which dI_n has maximal rank.

(2.5) <u>Theorem</u>. Assume that X is of general type. Let Y be a Zariski open subset of a compact complex space $\bar{Y}$. Then there are only finitely many holomorphic mappings $f: Y \longrightarrow X$ which are non-generate with respect to $\Lambda(X)$.

<u>Proof</u>. Taking a desingularization of $\bar{Y}$, we may assume

that $\bar{Y}$ is smooth. It follows from [N3, Corollary (4.7)] that any f: $Y \longrightarrow X$, non-degenerate with respect to $\Lambda(X)$ can be meromorphically extended from $\bar{Y}$ into $\bar{X}$. Then by [N-S1, Lemma (2.3)] we may assume that $\bar{Y}$ is projective, so that f: $Y \longrightarrow X$ is a (rational) morphism from Y into X.

Assume that there are infinitely many such f's; to say, $f_1, f_2, \ldots$. By Lemma (2.12) proved later, we take an algebraic irreducible curve C in Y such that the restrictions $f_j|C : C \longrightarrow X$, $j=1, 2, \ldots$, are distinct and non-degenerate with respect to $\Lambda(X)$. Therefore we may assume that $\bar{Y}$ is a smooth curve. Put $E = \bar{Y}-Y$. Taking out several points of Y if necessary, we may assume that there is a rational function t on $\bar{Y}$ such that t is regular on Y and the differential dt does not vanish anywhere on Y. Let

(2.6) $$J_n(Y) \cong Y \times \mathbb{C}^n$$

be the trivialization associated with dt (cf. (1.5)), where $n = \dim X$. By Proposition (1.22), any morphism f: $\bar{Y} \longrightarrow \bar{X}$ induces

$$f^*: J_n(\bar{Y}; \log f^{-1}D) \longrightarrow J_n(\bar{X}; \log D).$$

By using (2.6) we have

$$J_n(\bar{Y}; \log f^{-1}D)|(Y-f^{-1}D) \cong (Y-f^{-1}D) \times \mathbb{C}^n .$$

Put

(2.7) $$J_n(f|Y-f^{-1}D): Y-f^{-1}D \ni y \longmapsto f^*\left(y, \begin{pmatrix} 1 \\ 0 \\ \vdots \\ 0 \end{pmatrix}\right) \in J_n(\bar{X}; \log D).$$

We use t as holomorphic local coordinates on $Y-f^{-1}D$ which yields the lifting of $f|Y-f^{-1}D$ by (1.3); this lifting coincides with $J_n(f|Y-f^{-1}D)$. If $f|Y-f^{-1}D$ is non-degenerate with respect to $\Lambda(X)$, it follows from Lemma (2.4) that $J_n(f|Y-f^{-1}D)$ generically takes values at which dI_n has maximal rank. Since $I_n: J_n(\bar{X}; \log D) \longrightarrow (\mathbb{C}^n)^N$ is proper, we apply the Stein factorization to I_n:

$$\begin{array}{ccc} J_n(\bar{X}, \log D) & & \\ \mu \downarrow & \searrow^{I_n} & \\ Z & \xrightarrow[I'_n]{} & (\mathbb{C}^n)^N, \end{array}$$

where Z is a normal algebraic variety, μ is a surjective morphism with connected fibres, I'_n is a finite morphism and $I_n = I'_n \circ \mu$. Hence Z is affine algebraic. Adding several rational functions on Z to I'_n, we have a proper imbedding

$$\phi : Z \longrightarrow \mathbb{C}^{N'}.$$

Put

$$\psi = \phi \circ \mu : J_n(\overline{X};\ \log D) \longrightarrow \mathbb{C}^{N'}. \tag{2.9}$$

We note that the fibres of ψ are connected. By (2.7) and (2.9) we have a rational vector function $\eta(f) = \psi \circ J_n(f|Y - f^{-1}D)$ on $\overline{Y}$ for any $f: \overline{Y} \longrightarrow \overline{X}$, such that $\eta(f)$ may have poles on $E \cup f^{-1}D$ and the orders of poles of $\eta(f)$ are bounded from above by an integer $\ell_o > 0$, independent of f.

Let $L(\ell_o E)$ be the finite dimensional vector space of rational functions on $\overline{Y}$ with polar divisors $\leq \ell_o E$, and put

$$H = \prod_{1}^{N'} L(\ell_o E).$$

Let F be the family of all morphism $f: Y \longrightarrow X$, non-degenerate with respect to $\Lambda(X)$. Then we see that the mapping

$$F \ni f \longmapsto \eta(f) \in H \tag{2.10}$$

is injective. Put

$$G = \{(y, w, \alpha) \in Y \times J_n(X;\ \log D) \times H;\ \alpha(y) = \psi(w)\}$$

which is an algebraic subset of $Y \times J_n(\overline{X};\ \log D) \times H$. Since ψ is proper, the projection

$$p : G \ni (y, \omega, \alpha) \longmapsto (y, \alpha) \in Y \times H$$

is proper. Put $G' = p(G)$, which is an algebraic subset of $Y \times H$. Let S be the algebraic subset of points of $J_n(X;\ \log D)$ at which $d\psi$ does not have maximal rank, and put $Z = \psi(S)$. Put

$$H' = \{\alpha \in H \quad Y \times \{\alpha\} \subset G'\}.$$

$$H'' = \{\alpha \in H';\ \alpha(Y) \not\subset Z\}.$$

Then H' is an algebraic subset of H and H" is a Zariski open subset of H'. For $\alpha \in H''$ we put

$$\alpha^{\#} = \psi^{-1} \circ \alpha : Y \longrightarrow J_n(\overline{X};\ \log D),$$

$$\alpha_b = \pi \circ \alpha^{\#} : Y \longrightarrow \overline{X}$$

which are rational mappings. Consider the equation

(2.11) $$\eta(\alpha_b) = \alpha .$$

Denote by H''' the algebraic subset of H'' defined by (2.11). Then $\eta(f) \in H'''$ for $f \in F$ and $F \ni f \longmapsto \eta(f) \in H'''$ is injective by (2.10). We are going to show that dim H''' = 0. Assume that dim H''' > 0. Take an irreducible curve $R \subset H'''$ and the normalization $\mu: R' \longrightarrow R$ of R. Then we have a rational mapping

$$\Psi : R' \times Y \ni (\alpha,y) \longmapsto \alpha_b(y) \in \overline{X}.$$

Taking a smooth compactification $\overline{R}'$ of R', we have the rational extension $\overline{\Psi}$ of Ψ:

$$\overline{\Psi}: R' \times Y \longrightarrow \overline{X}.$$

Since the indeterminancy locus $I(\overline{\Psi})$ of $\overline{\Psi}$ is a finite set of points, $I(\Psi) \not\supset \{\alpha\} \times \overline{Y}$ for all $\alpha \in \overline{R}'$, so that $\overline{\Psi}|\{\alpha\} \times \overline{Y}$ define morphisms from $\overline{Y}$ to $\overline{X}$. For generic $\alpha \in \overline{R}'$ we have

$$\eta(\overline{\Psi}|\{\alpha\} \times Y) \in H$$

and hence a morphism

$$\gamma: \overline{R}' \ni \alpha \longmapsto \eta(\overline{\psi}|\{\alpha\} \times \overline{Y}) \in H.$$

Since $\overline{R}'$ is compact and H is an affine space, γ must be constant. This contradicts (2.11). Q.E.D.

Now we show the following lemma used in the above proof:

(2.12) Lemma. Let Y be an algebraic variety and $f_i: Y \longrightarrow X$, $i = 1,2 \ldots,$ distinct holomorphic mappings which are non-degenerate with respect to $\Lambda(X)$. Then there is an irreducible algebraic curve C in Y such that $f_i|C: C \longrightarrow X$ are all distinct and non-degenerate with respect to $\Lambda(X)$.

Proof. Put

$$Z_{ij} = \{y \in Y;\ f_i(y) = f_j(y)\} \qquad \text{for } i < j.$$

Then Z_{ij} are proper algebraic subsets of Y. Put

$$Z = \bigcup_{i<j} Z_{ij} \cup \bigcup_i f_i^{-1}S(X) \cup S(Y).$$

Then the subset Z of Y does not contain a non-empty open set of Y. Take a point $y_o \in Y-Z$ and a local coordinate neighborhood $U \subset R(Y)$ such that U is biholomorphic to the polydisc

$$\Delta^m = \{(y^1,\ldots,y^m);\ |y^j| < 1\} \subset \mathbb{C}^m,$$
$$y_o = (0,\ldots,0) = 0,$$

where $m = \dim Y$. By [N4, Remark (2.3)] we have an analytically non-degenerate holomorphic mapping $\phi: \Delta \longrightarrow \Delta^m$ with $\phi(0) = 0$. Here ϕ is said to be analytically non-degenerate if the image $\phi(\Delta)$ is not contained in any proper analytic subset of Δ^m. Let $\{\sigma_1,\ldots,\sigma_q\}$ be the maximal linearly independent system of $\{i^*_{R(X)}(\omega^{j_n} \wedge\ldots\wedge \omega^{j_n});\ 1 \leq j_1 < \ldots < j_n \leq N\}$. Then f_i are non-degenerate with respect to $\Lambda(X)$ if and only if $f_i^{-1}\sigma_1,\ldots,f_i^{-1}\sigma_q \in \Gamma(U,\ f_i^{-1}K_{R(X)})$ are linearly independent. Therefore f_i are non-degenerate with respect to $\Lambda(X)$ if and only if $(f_i\circ\phi)^{-1}\sigma_1,\ldots,(f_i\circ\phi)^{-1}\sigma_q \in \Gamma(\Delta,\ (f_i\circ\phi)^{-1}K_{R(X)})$ are linearly independent. Fix a trivialization $f_i^{-1}K_{R(X)} \cong \Delta \times \mathbb{C}$ for every i. Then $f_i^{-1}\sigma_j$ are given by holomorphic functions τ_{ij} on U. Put $\tau^*_{ij} = \tau_{ij}(\phi(z))$. Then it follows that $\tau_{i1},\ldots,\tau_{iq}$ are linearly independent if and only if the Wronskians

$$d_j(z) = \begin{vmatrix} \tau^*_{i1}(z) & \cdots & \tau^*_{iq}(z) \\ \tau^{*\prime}_{i1}(z) & \cdots & \tau^{*\prime}_{iq}(z) \\ & \cdots & \\ \tau^{*(q-1)}_{i1}(z) & \cdots & \tau^{*(q-1)}_{iq}(z) \end{vmatrix} \not\equiv 0;$$

that is, the sets $\{d_j(z) = 0\}$ are discrete. Note that $\phi^{-1}(\ )$ is the countable union of discrete subsets. We take a point $z_o \in \Delta - (\cup_j\ \{d_j(z) = 0\} \cup \phi^{-1}(Z))$ and put $y'_o = \phi(z_o)$.

Set

$$\phi(z) = (\phi^1(z),\ldots,\phi^m(z)),$$

$$\phi^\nu(z) = \sum_{k=0}^{\infty} a^\nu_k\,(z - z_o)^k,$$

$$y'_o = (a^1_o,\ldots,\ a^m_o).$$

Now we may assume that $y^1,\dots,y^m$ are the restrictions of affine coordinate functions over an affine open subset W of R(Y) with $U \subset W \subset \mathbb{C}^\ell$. Let $y^1,\dots,y^\ell$ be the affine coordinate functions of $\mathbb{C}^\ell$. Consider the affine equations:

$$(2.13)\qquad \begin{aligned} y^1 &= a_o^1 + a_1^1 z + \dots + a_{q-1}^1 z^{q-1},\\ &\ \ \vdots\\ y^m &= a_o^m + a_1^m z + \dots + a_{q-1}^m z^{q-1}, \end{aligned}$$

where z is considered as the affine coordinate function over $\mathbb{C}$. Let V be the intersection of $W \times \mathbb{C}$ and the algebraic subset of $\mathbb{C}^\ell \times \mathbb{C}$ defined by (2.13), and V' the irreducible component of V containing the set of points defined by (2.13) with $z \in \Delta$ and $(y^1,\dots,y^m) \in \Delta^m \cong U$. Let $p_1\colon V' \ni (y,z) \longmapsto y \in W$ and $p_2\colon V' \ni (y,z) \longmapsto z \in \mathbb{C}$ be the projections. Then p_1 and p_2 are locally biholomorphic around $(y_o',0)$. Let C be the Zariski closure of $p_1(V')$ in Y. Then C is smooth at y_o', around which z gives the local coordinate. The Wronskians of $\tau_{i1}|C \cap U, \dots, \tau_{iq}|C \cap U$ at $y_o' \in C$ with respect to z are $d_i(z_o)$ by the construction. Since $d_i(z_o) \neq 0$, $t_{i1}|C \cap U, \dots, \tau_{iq}|C \cap U$ are linearly independent. Therefore $f_i|C\colon C \longrightarrow X$ are distinct and non-degenerate with respect to $\Lambda(X)$. Q.E.D.

(2.14) <u>Example</u>. Let C be a compact smooth curve with genus 3 and $\alpha\colon C \longrightarrow A$ the Albanese mapping. Put

$$X = \alpha(C) + \alpha(C) \subset A.$$

Then X is an algebraic surface of general type. For $y \in C$, we have morphisms:

$$\alpha_y\colon \ C \ni x \longmapsto \alpha(x) + \alpha(y) \in X.$$

Moving $y \in C$, we have infinitely many morphisms $\alpha_y\colon C \longrightarrow X$ which are degenerate with respect to $\Lambda(X)$.

(2.15) <u>Remark</u>. Let $H_1,\dots,H_{n+2}$ be hyperplanes in general position of the n-dimensional complex projective space $\mathbb{P}^n(\mathbb{C})$ and $Z_1,\dots,Z_{n+2}$ divisors on $\mathbb{C}^m$ which may be transcendental. Then Fujimoto [F1] proved that there are only finitely many linearly non-degenerate meromorphic mappings $f\colon \mathbb{C}^m \longrightarrow \mathbb{P}^n(\mathbb{C})$ such that $f^*H_j = Z_j$ for $j = 1,2,\dots,n+2$. Suppose that Z_j are algebraic. Then the second main theorem ([St1] and [V1]) implies that f are necessarily rational. Put

$$X = \mathbb{P}^n(\mathbb{C}) - \bigcup_{j=1}^{n+2} H_j$$

Then X is of general type. Take a homogeneous coordinate system $(u^1, \ldots, u^{n+1})$ of $\mathbb{P}^n(\mathbb{C})$ so that

$$H_j = \{u^j = 0\},\ 1 \le j \le n+1,\ H_{n+2} = \{\textstyle\sum_{j=1}^{n+1} u^j = 0\}.$$

Set $z^j = u^j/u^{n+1}$ for $j = 1, 2, \ldots, n$, and

$$\omega^j = \frac{dz^j}{z^j},\ 1 \le j \le n,$$

$$\omega^{n+1} = \frac{dz^1 + \ldots + dz^n}{1 + z^1 + \ldots + z^n}.$$

Then $\{\omega^1, \ldots, \omega^{n+1}\}$ forms a basis of $\Gamma(\mathbb{P}^n(\mathbb{C}), \Omega^1_{\mathbb{P}^n(C)}(\log \bigcup_{j=1}^{n+2} H_j))$

and the quasi-Albanese mapping

$$\alpha:\ X \ni x \longrightarrow \left(\int^x \omega^1, \ldots, \int^x \omega^{n+1}\right) \in (C^*)^{n+1}$$

is an imbedding. We identify X with $\alpha(X) \subset (\mathbb{C}^*)^{n+1}$. By [N1, Proposition 4.1] a holomorphic mapping f into X is non-degenerate with respect to $\Lambda(X)$ if and only if f is a linearly non-degenerate holomorphic mappings into $X \subset \mathbb{P}^n(C)$. Therefore Theorem (2.5) implies that there are only finitely many linearly non-degenerate holomorphic mappings $f\colon \mathbb{C}^m - \bigcup_{j=1}^{n+2} \mathrm{Supp}\, Z_j \longrightarrow X \subset \mathbb{P}^n(\mathbb{C})$. This is the same as the above theorem of Fujimoto in the case where Z_j are algebraic.

§3 RATIONAL MAPPINGS INTO ALGEBRAIC VARIETIES WITH QUASI-NEGATIVE $T^q(\bar{V};\ \log D)$

In this section, we prove a finiteness theorem of another type, which generalizes the Main Theorem (1.2) of [N-S1] in the case of the complex number field. Let $\bar{V}$ be a compact complex algebraic variety and V a Zariski open subset of $\bar{V}$. Put $D = \bar{V} - V$. Let $F \longrightarrow \bar{V}$ be a vector bundle over $\bar{V}$.

(3.1) Definition. We say that F is quasi-negative over V if there is a proper morphism $\Phi\colon F \longrightarrow C^N$ such that $\Phi|((F|V)-O)\colon (F|V)-O \longrightarrow \Phi(F)-\Phi(F|D)$ is an isomorphism, where O denotes the zero section of F.

Remark. If $D = \phi$, then the above quasi-negativity is equivalent to the weak negativity of Grauert [G1].

Assume that $\overline{V}$ is smooth and D is a hypersurface with normal crossings. Set

$$T^q(V;\ \log D) = \bigwedge_1^q T(\overline{V};\ \log D)$$

which is the dual of the locally free sheaf $\Omega^q_{\overline{V}}(\log D)$. Let W be another complex algebraic variety and $f: W \longrightarrow V$ a rational mapping. Then f is said to be proper if the first projection of the graph $G(f) \subset W \times V$ onto W is proper. Set

$$\operatorname{rank} f = \sup \{\dim W - \dim_x f^{-1}f(x);\ x \in W\}.$$

We denote by $F_q(W,V)$ the set of all proper rational mappings $f: W \longrightarrow V$ with rank $f \geq q$.

(3.2) Theorem. Assume that $T^q(\overline{V};\ \log D)$ is quasi-negative over V. Then $F_q(W,V)$ is finite.

Proof. We may assume that W is a Zariski open subset of a compact algebraic variety $\overline{W}$ such that $E = \overline{W} - W$ is a hypersurface with normal crossings; moreover we may assume that there are rational functions $x^1,\dots,x^m$ (m=dim W) over $\overline{W}$ such that x^j are regular on W and $dx^1 \wedge \dots \wedge dx^m$ does not vanish anywhere on W. Every $f \in F_q(W,V)$ defines a rational mapping from $\overline{W}$ into $\overline{V}$ which is denoted by the same f. Let I(f) denote the indeterminacy locus of $f: \overline{W} \longrightarrow \overline{V}$. Then $f \in F_q(W,V)$ induces the non-trivial bundle morphism.

(3.3) $\quad (f|W-I(f))_*: T^q(W;\log E)\,|\,(W-I(f)) \longrightarrow T^q(\overline{V};\ \log D).$

If $f,g \in F_q(W,V)$ are distinct, then $(f|(W-(I(f) \cup I(g)))_* \neq (g|W-(I(g) \cup I(f)))_*$. Since $T^q(V;\ \log D)$ is quasi-negative over V, we have a proper morphism $\Phi: T^q(V;\ \log D) \longrightarrow \mathbb{C}^N$ as in Definition (3.1). As in the proof of Theorem (2.5), we may assume that $\Phi^{-1}\Phi(y)$ are connected for $y \in T^q(V;\ \log D)$. By making use of $x^1,\dots,x^m$, we have a trivialization

(3.4) $$T^q(\overline{W};\ \log E)\,|W \cong W \times \mathbb{C}^{\binom{m}{q}}.$$

By using (3.3), (3.4) and $\Phi: T^q(\overline{V};\ \log D) \longrightarrow \mathbb{C}^N$, we apply the same arguments as in the proofs of Theorem (2.5) in section 2 and the Main Theorem (1.2) of [N-S1] to complete the proof.

Q.E.D.

(3.5) Example. Let $H_0, \dots, H_4$ be 5 hyperplanes in general position of $\mathbb{P}^2(\mathbb{C})$ and $A = \cup_{j=0}^4 H_j$. Taking a suitable homogeneous coordinate system (u^0, u^1, u^2), we write

$$H_j = \{u^j = 0\},\ 0 \le j \le 2,$$

$$H_3 = \{a_0 u^0 + a_1 u^1 + a_2 u^2 = 0\},$$

$$H_4 = \{b_0 u^0 + b_1 u^1 + b_2 u^2 = 0\},$$

where $a_i, b_i \in \mathbb{C}^*$. Set $z^i = u^i/u^0$. Then a basis of $\Gamma(\mathbb{P}^2(\mathbb{C}), \Omega^1_{\mathbb{P}^2(\mathbb{C})}(\log A))$ is given by

$$\phi^1 = \frac{dz^1}{z^1},\quad \phi^2 = \frac{dz^2}{z^2},$$

$$\phi^3 = \frac{a_1 dz^1 + a_2 dz^2}{a_0 + a_1 z^1 + a_2 z^2},$$

$$\phi^4 = \frac{b_1 dz^1 + b_2 dz^2}{b_0 + b_1 z^1 + b_2 z^2}.$$

The vector fields $z^1\partial/\partial z^1$ and $z^2\partial/\partial z^2$ form a local frame of $T(\mathbb{P}^2(\mathbb{C}): \log A)$ around $(0,0)$. Put

$$\Phi(v) = (\phi^1(v), \dots, \phi^4(v)) = \Big(v^1, v^2, \frac{a_1 v^1 z^1 + a_2 v^2 z^2}{a_0 + a_1 z^1 + a_2 z_2}, \frac{b_1 v^1 z^1 + b_2 v^2 z^2}{b_0 + b_1 z^1 + b_2 z^2}\Big)$$

for $v = v^1 z^1 \frac{\partial}{\partial z^1} + v^2 z^2 \frac{\partial}{\partial z^2} \in T(\mathbb{P}^2(\mathbb{C}); \log A)$. Then the morphism

(3.6) $\Phi\colon T(\mathbb{P}^2(\mathbb{C}); \log A) \longrightarrow \mathbb{C}^4$ is proper.

We calculate the Jacobian of Φ with respect to the coordinate system (z^1, z^2, v^1, v^2) of $T(\mathbb{P}^2(\mathbb{C}); \log A)$:

$$(3.7) \qquad \frac{\partial \Phi}{\partial (z^1, z^2, v^1, v^2)} = (v^2 - v^1)(b_1 v^1 z^1 + b_2 v^2 z^2) a_1 a_2$$

$$- \{b_2 v^2 + (v^2 - v^1) b_1 b_2 z^1\} \, v^1 a_1$$

$$+ \{b_1 v^1 + (v^1 - v^2) b_1 b_2 z^2\} \, v^2 a_2 \, .$$

Now take 6 hyperplanes $H_o, \ldots, H_5$ in general position of $\mathbb{P}^2(\mathbb{C})$ so that

$$H_j = \{u^2 = 0\}, \quad 0 \le j \le 2 \, ,$$

$$H_3 = \{u^0 + u^1 + u^2 = 0\},$$

$$H_4 = \{u^0 + a_1^1 u^1 + a_2^1 u^2 = 0\},$$

$$H_5 = \{u^0 + a_1^2 u^1 + a_2^2 u^2 = 0\}.$$

Put

$$\psi^1 = \frac{dz^1}{z^1} \, , \quad \psi^2 = \frac{dz^2}{z^2} \, ,$$

$$\psi^3 = \frac{dz^1 + dz^2}{1 + z^1 + z^2} \, ,$$

$$\psi^4 = \frac{a_1^1 \, dz^1 + a_2^1 \, dz^2}{1 + a_1^1 z^1 + a_2^1 z^2} \, ,$$

$$\psi^5 = \frac{a_1^2 \, dz^1 + a_2^2 \, dz^2}{1 + a_1^2 \, z^1 + a_2^2 \, z^2}$$

and

$$\Psi : T(\mathbb{P}^2(\mathbb{C}); \log D) \ni v \longmapsto (\psi^1(v), \ldots, \psi^5(v)) \in \mathbb{C}^5,$$

where $D = \cup_{j=0}^{5} H_j$. Then, as in (3.6), Ψ is a proper morphism.

Consider two equations:

$$\frac{\partial(\psi^1, \psi^2, \psi^3, \psi^4)}{\partial(z^1, z^2, v^1, v^2)} = 0,$$

$$\frac{\partial(\psi^1, \psi^2, \psi^3, \psi^5)}{\partial(z^1, z^2, v^1, v^2)} = 0 \, .$$

By (3.7) we have

$$(3.8)\quad (v^2 - v^1)(v^1 z^1 + v^2 z^2) a_1^i a_2^i - \{v^2 + (v^2 - v^1) z^1\} v^1 a_1^i$$

$$+ \{v^1 + (v^1 - v^2) z^2\} v^2 a_2^i = 0, \quad i = 1,2 .$$

We want to choose a_k^i so that $\Psi | (T(\mathbb{P}^2(\mathbb{C}); \log D) | ((\mathbb{P}^2(\mathbb{C}) - D) - O)$ is an isomorphism onto its image and the degenerating locus of Ψ is contained in the union of the zero section and $T(\mathbb{P}^2(\mathbb{C}); \log D) | D$. We may assume that

$$v^1 = 1 \quad \text{and} \quad v^2 \neq 1 .$$

Put

$$Z^1 = (v^2 - 1) z^1 \quad (\neq 0 \text{ on } \mathbb{P}^2(\mathbb{C}) - D) ,$$

$$Z^1 = (v^2 - 1) v^2 z^2 .$$

Substituting these to (3.8), we have

$$(3.9)\quad (a_1^i a_2^i - a_1^i) Z^1 + (a_1^i a_2^i - a_2^i) Z^2 + (a_2^i - a_1^i) v^2 = 0 ,$$

$$i = 1,2 .$$

If $v^2 = 0$, then $Z^2 = 0$, so that

$$a_1^i (a_2^i - 1) = 0 .$$

This contradicts the assumption that $H_j, j=0,\dots,5$, are in general position. If the determinant δ of coefficients of (3.9) in the variables Z^1, Z^2, v^2, is not zero, then Ψ satisfies our requirements. By calculation we have

$$\delta = - \begin{vmatrix} a_1^1 a_2^1 - a_1^1 & a_1^1 a_2^1 - a_2^1 \\ a_1^2 a_2^2 - a_1^2 & a_1^2 a_2^2 - a_2^2 \end{vmatrix}^2$$

Therefore if

$$\begin{vmatrix} a_1^1 a_2^1 - a_1^1 & a_1^1 a_2^1 - a_2^1 \\ a_1^2 a_2^2 - a_1^2 & a_1^2 a_2^2 - a_2^2 \end{vmatrix} \neq 0 ,$$

then $T(\mathbb{P}^2(\mathbb{C}); \log D)$ is quasi-negative over $\mathbb{P}^2(\mathbb{C}) - D$.

REFERENCES

[C1] H. Cartan, Idéaux de fonctions analytiques de n variables, Ann. École Normale, (3) 61(1944), 149-197.

[D1] P. Deligne, Theorie de Hodge II, Inst. Hautes Etudes Sci. Publ. 40(1971), 5-57.

[F1] H. Fujimoto, Remarks to the uniqueness problem of meromorphic maps into $\mathbb{P}^N(\mathbb{C})$, IV, Nagoya Math. J. 83(1981), 153-181.

[G1] H. Grauert, Über Modifikationen und exzeptionelle analytische Mengen, Math. Ann.146 (1962), 331-368.

[G-G1] M. Green and P. Griffiths, Two applications of algebraic geometry to entire holomorphic mappings, The Chern Symposium 1979, pp. 41-74, Springer-Verlag, New York-Heidelberg-Berlin, 1980.

[I1] S. Iitaka, Logarithmic forms of algebraic varieties, J. Fac. Sci. Univ. Tokyo Sect. IA 23(1976), 525-544.

[I2] S. Iitaka, On logarithmic Kodaira dimension of algebraic varieties, Complex Analysis and Algebraic Geometry, pp. 175-189, Iwanami, Tokyo, 1977.

[N1] J. Noguchi, Holomorphic curves in algebraic varieties, Hiroshima Math. J. 7(1977), 833-853.

[N2] J. Noguchi, Supplement to "Holomorphic curves in algebraic varieties", Hiroshima Math. J. 10(1980), 229-231.

[N3] J. Noguchi, Lemma on logarithmic derivatives and holomorphic curves in algebraic varieties, Nagoya Math. J. 83(1981), 213-233.

[N4] J. Noguchi, On the value distribution of meromorphic mappings of covering spaces over $\mathbb{C}^m$ into algebraic varieties, to appear in J. Math. Soc. Japan.

[N-S1] J. Noguchi and T. Sunada, Finiteness of the family of rational and meromorphic mappings into algebraic varieties, Amer. J. Math 104(1982), 887-900.

[O1] T. Ochiai, On holomorphic curves in algebraic varieties with ample irregularity, Invet. Math. 43(1977), 83-96.

[S1] K. Saito, Theory of logarithmic differential forms and logarithmic vector fields, J. Fac. Sci. Univ. Tokyo, Sec. IA 27(1980), 265-291.

[St1] W. Stoll, Die beiden Hauptsätze der Wertverteilungstheorie bei Funktionen mehrerer komplexer Veränderlichen (I), Acta Math. 90 (1953), 1-115; ibid (II), Acta Math. 92(1954), 55-169.

[V1] A.L. Vitter, The lemma of logarithmic derivatives in several complex variables, Duke Math. J. 44(1977), 89-104.

Remarks on the Nakano Vanishing Theorem

Bernard Shiffman
Johns Hopkins University
Baltimore, Maryland 21218, U.S.A.

In this note we give a numerical version of k-ampleness for line bundles (Definition 1) and prove a vanishing theorem (Theorem 2) of Nakano type for these bundles. This vanishing theorem yields a Lefschetz-type theorem (Theorem 3). We begin by reviewing the Nakai-Moishezon-Kleiman criterion for ampleness on which our numerical condition is based.

Let L be a holomorphic line bundle on a compact n-dimensional complex manifold X, and let D be the divisor of a meromorphic section of L. We can write $D = \sum m_i D_i$, where the $m_i \in \mathbb{Z}$ and the D_i are complex hypersurfaces. We can regard D as a current (form with distribution coefficients) of type $(1,1)$ on X by the formula

$$(D,\alpha) = \sum_i m_i \int_{D_i} \alpha \,, \tag{1}$$

for C^∞ $(n-1,n-1)$-forms α on X. It is well known that D represents the Chern class (with real coefficients) of L, $c_1(L)$. One way to see this is as follows: Consider a hermitian metric $\| \ \|$ on L, and let s be a meromorphic section of L with divisor D. We note that

$$\frac{\sqrt{-1}}{\pi} \partial\bar\partial \log \|s\| = -\frac{\sqrt{-1}}{2\pi} \Theta + D \,, \tag{2}$$

where Θ is the curvature form of L. To verify (2), we let $\{h_\alpha\}$ represent the metric and $\{s_\alpha\}$ represent the section s with respect to a local covering $\{U_\alpha\}$, so that $\|s\|^2 = h_\alpha |s_\alpha|^2$. The identity (2) then follows from the curvature formula

$$\Theta = -\partial\bar\partial \log h_\alpha$$

and the Poincaré-Lelong formula

$$\frac{\sqrt{-1}}{\pi}\,\partial\bar{\partial}\log|s_\alpha| = \operatorname{Div} s_\alpha .$$

Since $\frac{\sqrt{-1}}{2\pi}\Theta$ is a representative of $c_1(L)$ and the left side of (2) is exact, it follows that D is also a representative of $c_1(L)$ as claimed.

It is the observation (2) which led to the Carlson-Griffiths [2] generalization of the First and Second Main Theorems of Stoll [13] to maps into algebraic varieties.

Suppose L is an ample line bundle on a compact complex (and hence projective) manifold X, and let Y be a d-dimensional algebraic subvariety of X $(1 \leq d \leq n)$. Since L is ample, there is a hermitian metric on L with $\sqrt{-1}\,\Theta > 0$ at all points of X, and hence

(3) $$\int_Y (\sqrt{-1}\,\Theta)^d > 0 .$$

Since $\frac{\sqrt{-1}}{2\pi}\Theta$ is a representative of $c_1(L)$, (3) is valid for any hermitian metric on L and is equivalent to

(4) $$(c_1(L)^k, Y) > 0 ,$$

where Y is regarded in (4) as a real (homology) 2p-cycle. We shall restate (4) in algebraic terms. Suppose D is a divisor on X. We let $[D]$ denote the line bundle associated with the divisor D; $[D]$ has a global meromorphic section whose divisor is D. Hence D is a representative of $c_1([D])$, and the intersection number

$$D^d \cdot Y = (c_1([D])^d, Y) .$$

(See [7] or [11, pp. 135-139].) Thus (4) is equivalent to the inequality

(4′) $$D^d \cdot Y > 0$$

where D is the divisor of a meromorphic section of L. The Nakai-Moishezon-Kleiman criterion [7], [9], [10] (see also [4, pp. 29-33]) gives the converse: If L is a line bundle on a projective manifold such that (4) or (4′) holds for all d-dimensional subvarieties Y of X for $1 \leq d \leq n$, then L is ample.

The Nakano Vanishing Theorem [1] states that for an ample

line bundle L on a projective manifold X,

(5) $$H^q(X,\Omega^p(L)) = 0 \quad \text{for } p+q>n,$$

where $\Omega^p(L)$ denotes the sheaf of germs of L-valued holomorphic p-forms. We shall also write

(6) $$H^{p,q}(X,L) = H^q(X,\Omega^p(L)) \ ,$$

which by Hodge theory can be identified with the space of harmonic (p,q)-forms on X (for X compact Kähler). Recall that by Serre duality, (5) is equivalent to

(7) $$H^{p,q}(X,L^{-1}) = 0 \quad \text{for } p+q<n \ .$$

Hodge theory also provides a proof of the First Lefschetz Theorem (with rational coefficients). This observation, originally due to Kodaira and Spencer [8] is embodied in the following lemma:

<u>Lemma 1</u>. Let D be a smooth hypersurface in a compact Kähler manifold X, and let $k\geq 0$ be fixed. If

$$H^{p,q}(X,[-D]) = H^{p,q-1}(D,[-D]|_D) = 0 \quad \text{for } p+q=k \ ,$$

then

$$H^k(X,D;\mathbb{Q}) = 0 \ .$$

For a proof of Lemma 1, see [11, Th. 3.44]. (Note that $[-D]|_D$ is the dual to the normal bundle of D.) Lemma 1 and the Nakano Vanishing Theorem (7) immediately imply the following theorem of Lefschetz:

<u>First Lefschetz Theorem</u>. Let X be an n-dimensional projective manifold and let D be a smooth hyperplane section of L. Then the restriction maps

$$H^j(X;\mathbb{Q}) \to H^j(D;\mathbb{Q})$$

are bijective for $j\leq n-2$ and injective for $j=n-1$.

We recall next the concept of k-ampleness due to Sommese [12]: Let L be a holomorphic line bundle on X. We say that L is k-ample if there exists a positive integer N and global holomorphic sections $s_1,\dots,s_m$ of L^N such that at least one of the s_j does not vanish at each point of X and the map

(8) $$\Phi_L = (s_1:\ldots:s_m): X \to \mathbb{C}P^{m-1}$$

has at most k-dimensional fibres.

Note that 0-ample corresponds to ample. A differential-geometric analogue of k-ampleness is k-positivity. A holomorphic line bundle L is defined to be k-positive if L carries a hermitian metric whose curvature form Θ is semi-positive and has at least $n-k$ positive eigenvalues at each point. The following summarizes two generalizations of the Nakano Vanishing Theorem:

Theorem 1. If either

a) L is a k-ample line bundle on a projective manifold X,

or

b) L is a k-positive line bundle on a compact Kähler manifold X,

then

$$H^{p,q}(X,L) = 0 \quad \text{for} \quad p+q>n+k,$$
$$H^{p,q}(X,L^{-1}) = 0 \quad \text{for} \quad p+q<n-k.$$

Part (a) of Theorem 1 is due to Sommese [12] and part (b) is due to Girbau [3]. (See also [11, Ch. III].) We shall generalize Theorem 1(a) to *numerically k-ample* line bundles, which we define below.

First, we recall the following definition: Suppose X is an algebraic variety. We say that $\{A_s\}_{s\in S}$ is an algebraic family of subvarieties of X if S is an algebraic variety and there is an algebraic subvariety $Y\subset X\times S$ such that $Y\cap\pi^{-1}(s)=A_s\times\{s\}$ for all $s\in S$, where $\pi: X\times S\to X$ is the projection. (Here we may consider only reduced varieties.)

Definition 1. Let D be a divisor on a projective manifold X. We say that D is numerically k-ample if there exists an algebraic family $\{A_s\}_{s\in S}$ of subvarieties of dimension $\leq k$ in X such that for every irreducible algebraic subvariety

$Y \subset X$ such that $\dim Y = d \geq 1$ and $Y \cap A_s$ is empty or 0-dimensional for all $s \in S$, we have $D^d \cdot Y > 0$. We say that a holomorphic line bundle L is numerically k-ample if any divisor of a meromorphic section of L is numerically k-ample, or equivalently if $(c_1(L)^d, Y) > 0$ for all Y as above.

We have the following observation:

<u>Proposition 1</u>. Let L be a holomorphic line bundle on a projective manifold. If L is k-ample, then L is numerically k-ample.

<u>Proof:</u> Suppose L is a k-ample line bundle on a projective manifold X, and let $\Phi_L : X \to \mathbb{CP}^{m-1}$ be given as in (8). Then $L = \Phi^* H$, where H is the hyperplane section bundle on $\mathbb{CP}^{m-1}$. We consider the algebraic family of fibres $A_s = \Phi^{-1}(s)$ for $s \in \Phi(X)$. Let $Y \subset X$ be given as in Definition 1. Choose $s \in \Phi(Y)$; since

$$\dim Y \cap \Phi^{-1}(s) = \dim Y \cap A_s = 0,$$

it follows that

$$\dim \Phi(Y) = \dim Y = d.$$

Thus

$$(c_1(L)^d, [Y]) = (\Phi^*\{c_1(H)^d\}, [Y]) = (c_1(H)^d, \Phi_*[Y]) > 0 ,$$

and hence L is numerically k-ample.

The following result generalizes Theorem 1(a):

<u>Theorem 2</u>: Let L be a numerically k-ample line bundle on a projective manifold X. Then

$$H^{p,q}(X,L) = 0 \quad \text{for} \quad p+q>n+k,$$
$$H^{p,q}(X,L^{-1}) = 0 \quad \text{for} \quad p+q<n-k.$$

Theorem 2 yields the following generalization of the First Lefschetz Theorem:

<u>Theorem 3</u>: If D is a numerically k-ample smooth hypersurface in a projective manifold X, then

$$H^j(X,D;\mathbb{Q}) = 0 \quad \text{for} \quad j<n-k.$$

We first prove that Theorem 2 implies Theorem 3: Let D, X be as in Theorem 3 and let $j<n-k$. Let $L=[D]$. By Theorem 2,

$$H^{p,q}(X,L^{-1}) = 0 \quad \text{for} \quad p+q=j.$$

One easily checks that $L|_D$ is a numerically k-ample line bundle on D. Thus

$$H^{p,q-1}(D,L^{-1}|_D) = 0 \quad \text{for} \quad p+q-1<n-1-k ,$$

and thus for $p+q=j<n-k$. The conclusion of Theorem 3 then follows from Lemma 1.

Note that the conclusion of Theorem 3 is also valid for k-positive line bundles on compact Kähler manifolds. This follows from the Girbau vanishing theorem 1(b) and the above proof. In order to prove Theorem 2, we need the following slicing technique of Sommese (see [12], [11, (3.24)]):

Lemma 2. Let D be a smooth hypersurface in a compact Kähler manifold X and let E be a holomorphic vector bundle on X. Let $p,q\geq 0$. If

a) $H^{p,q}(X, [D]\otimes E) = 0$,

b) $H^{p-1,q-1}(D, E|_D) = 0$,

c) $H^{p,q-1}(D, ([D]\otimes E)|_D) = 0$,

then

$$H^{p,q}(X,E) = 0 .$$

Proof: Consider the short exact sequences of sheaves

(9) $$0 \to \Omega_X^p\otimes E \xrightarrow{\cdot s} \Omega_X^p\otimes[D]\otimes E \to (\Omega_X^p\otimes[D]\otimes E)|_D \to 0 ,$$

(10) $$0 \to \Omega_D^{p-1}\otimes E|_D \xrightarrow{\wedge ds} (\Omega_X^p\otimes[D]\otimes E)|_D \to \Omega_D^p\otimes([D]\otimes E)|_D \to 0 ,$$

where $s \in \Gamma(X,[D])$ with $\mathrm{Div}\ s = D$. By (b), (c), and the long exact sequence associated with (10), it follows that

(11) $$H^{q-1}(D, (\Omega_X^p\otimes[D]\otimes E)|_D) = 0 .$$

The desired vanishing then follows from (a), (11), and the long exact cohomology sequence associated with (9).

In order to prove Theorem 2 by induction on k, we need the following result of Hironaka:

Lemma 3 (Hironaka [5]): Let L be an ample line bundle on a projective variety X, and let $\{A_s\}_{s\varepsilon S}$ be an algebraic family of subvarieties of X. Then there exists a positive integer N such that the generic element of $\Gamma(X,L^N)$ does not vanish identically on any positive-dimensional irreducible component of A_s for all $s\varepsilon S$.

For a proof of Lemma 3, see [5, Sec. 2] or [11, pp.62-63].

We now prove Theorem 2: Let L, X be as in Theorem 2, and let $\{A_s\}_{s\varepsilon S}$ be as in the definition of numerically k-ample. We use induction on k. If $k=0$, then L is ample by the Nakai-Moishezon-Kleiman criterion, and the conclusion follows by the Nakano Vanishing Theorem. Thus we can assume that $k\geq 1$ and the theorem is valid for numerically $(k-1)$-ample line bundles. Choose an ample line bundle E on X such that $E\otimes L$ is ample. By Lemma 3 and Bertini's Theorem, we can find a positive integer N and a global holomorphic section of E^N whose divisor is a smooth hypersurface D such that

$$\dim D\cap A_s \leq k-1$$

for all $s\varepsilon S$. Then $L|_D$ is a numerically $(k-1)$-ample line bundle on D. Let $p+q>n+k$. Then

$$H^{p,q}(X,E^N\otimes L) = 0 \ , \tag{12}$$

$$H^{p,q-1}(D,(E^N\otimes L)|_D) = 0 \ , \tag{13}$$

by the Nakano Vanishing Theorem. Furthermore, by our inductive hypothesis,

$$H^{p-1,q-1}(D,L|_D) = 0 \ , \tag{14}$$

since $p-1+q-1>n-1+k-1$. The first conclusion of Theorem 2 follows from (12),(13),(14), and Lemma 3. The second conclusion follows by Serre duality.

The following special case of Theorem 2 is a Nakano-type

version of the Kawamata-Viehweg vanishing theorem.

Corollary 1. Let D be a divisor on a projective manifold X, such that

$$D^d \cdot Y > 0$$

for all pure d-dimensional subvarieties Y of X, for $1 \leq d \leq n-k$. Then

$$H^{p,q}(X,[D]) = 0 \quad \text{for} \quad p+q>n+k,$$
$$H^{p,q}(X,[-D]) = 0 \quad \text{for} \quad p+q<n-k.$$

Proof: By Theorem 2, it suffices to show that D is numerically k-ample. Let $H_1,\ldots,H_{n-k}$ be hyperplane sections in X such that the intersection $A=H_1\cap\ldots\cap H_{n-k}$ has dimension k. We consider the algebraic family consisting of the single subvariety A. It suffices to show that the set of subvarieties Y such that $\dim Y > n-k$ and $\dim A\cap Y \leq 0$ is empty. Suppose

$$\dim Y = d > n - k.$$

Choose $H_{n-k+1},\ldots,H_d$ such that $H_1\cap\ldots\cap H_d$ has dimension $n-d$. Then the intersection number

$$Y\cdot H_1\cdot\ldots\cdot H_d = (c_1([H])^d, Y) > 0$$

(where $[H]$ is the hyperplane section bundle). Therefore

$$Y\cap A \supset Y\cap H_1\cap\ldots\cap H_d \neq \emptyset$$

which implies that

$$\dim Y\cap A \geq d+k-n > 0,$$

which completes the proof.

Corollary 1 is an analogue of the following vanishing theorem of Kawamata [6] and Viehweg [14]:

Theorem 4 (Kawamata-Viehweg). Let D be a numerically effective divisor on a projective manifold X such that

$$D^{n-k}\cdot H^k > 0 ,$$

where H is a hyperplane section of X. Then

$$H^q(X,[-D]) = 0 \quad \text{for} \quad q < n - k .$$

Recall that D is numerically effective if $D\cdot C\geq 0$ for all algebraic curves $C\subset X$. Note that the hypotheses of Theorem 4 are weaker than the hypotheses of Corollary 1.

We end this report by giving a variation of Theorem 2:

<u>Theorem 5</u>. Let L be a holomorphic line bundle on a projective manifold X. Suppose there exists an algebraic family $\{A_s\}_{s\in S}$ of subvarieties of dimension $\leq k$ in X such that $(c_1(L),C) > 0$ for all irreducible algebraic curves $C\subset X$ such that $C\not\subset A_s$ for all $s\in S$. Suppose further that there exists a positive integer N such that the dimension of the base point set of L^N is $\leq k$. Then the conclusion of Theorem 2 holds.

<u>Proof</u>: We note that if $k=0$ then L is ample. (See [11, (3.38), (7.5)].) The theorem then follows by induction exactly as in the proof of Theorem 2.

Theorem 5 is the numerical analogue of Theorem (3.37) in [11].

REFERENCES

1. Akizuki, Y., and Nakano, S., Note on Kodaira-Spencer's proof of Lefschetz theorems, *Proc. Jap. Acad.* 30 (1954), 266-272.
2. Carlson, J., and Griffiths, P.A., A defect relation for equidimensional holomorphic mappings between algebraic varieties, *Ann. of Math.* 95 (1972), 557-584.
3. Girbau, J., Sur le théorème de Le Potier d'annulation de la cohomologie, *C. R. Acad. Sci. Paris* 283 (1976), Serie A., 355-358.
4. Hartshorne, R., *Ample subvarieties of algebraic varieties*, Lecture Notes in Math. 156, Springer-Verlag, New York, 1970.
5. Hironaka, H., Smoothing of algebraic cycles of small dimensions, *Amer. J. Math.* 90 (1968), 1-54.
6. Kawamata, Y., A generalization of Kodaira-Ramanujam's vanishing theorem, *Math. Ann.* 261 (1982), 43-46.
7. Kleiman, S.L., Toward a numerical theory of ampleness, *Ann. of Math.* 84 (1966), 293-344.

8. Kodaira, K., and Spencer, D., On a theorem of Lefschetz and the lemma of Enriques-Severi-Zariski, *Proc. Nat. Acad. Sci. U.S.A.* 39 (1953), 1273-1278.

9. Moishezon, B.G., A criterion for projectivity of complete algebraic abstract varieties, *Amer. Math. Soc. Translations* 63 (1967), 1-50.

10. Nakai, Y., A criterion of an ample sheaf on a projective scheme, *Amer. J. Math.* 85 (1963), 14-26.

11. Shiffman, B., and Sommese, A.J., *Vanishing theorems on complex manifolds*, Progress in Math. 56, Birkhäuser, Boston, 1985.

12. Sommese, A.J., Submanifolds of abelian varieties, *Math. Ann.* 233 (1978), 229-250.

13. Stoll, W., Die beiden Hauptsätze der Wertverteilungstheorie bei Funktionen merherer komplexen Veränderlichen (I), (II), *Acta Math.* 90 (1953), 1-115, and 92 (1954), 55-169.

14. Viehweg, E., Vanishing theorems, *J. Reine Angew. Math.* 335 (1982), 1-8.

Curvature of the Weil-Petersson Metric in the Moduli Space of Compact Kähler-Einstein Manifolds of Negative First Chern Class

Yum-Tong Siu

Department of Mathematics, Harvard University,

Cambridge, Massachusetts 02138 USA

For compact Riemann surfaces of genus at least two, using Petersson's Hermitian pairing for automorphic forms, Weil introduced a Hermitian metric for the Teichmüller space, now known as the Weil-Petersson metric. Ahlfors [1,2] showed that the Weil-Petersson metric is Kähler and that its Ricci and holomorphic section curvatures are negative. By using a different method of curvature computation, Royden [8] later showed that the holomorphic sectional curvature of the Weil-Petersson metric is bounded away from zero and conjectured the best bound to be $-\frac{1}{2\pi(g-1)}$, where g is the genus. Recently Wolpert [12] and also Royden proved Royden's conjecture on the bound of the holomorphic sectional curvature and obtained in addition the negativity of the Riemannian sectional curvature. Wolpert's method used some SL(2,R) invariant first-order differential operators obtained by Maass [7]. Royden's computation is based on the fact that the Poincaré metric on a compact Riemann surface of genus at least two is Einstein.

Since Yau's proof [13] of the existence of a Kähler-Einstein metric on a Kähler manifold of negative first Chern class, the question naturally arose about the negativity of the curvature of the Weil- Petersson metric induced by the Kähler-Einstein metric on the moduli space of such a manifold. In this case the square of the length of a tangent vector in the moduli space with the Weil-Petersson metric is simply the L^2 norm, computed from the Kähler-Einstein metric, of the harmonic

Research partially supported by a grant from the National Science Foundation

(0,1)-form with coefficients in the tangent bundle representing it. Koiso [6] showed that this Weil-Petersson metric is always Kahler.

In this paper we introduce a method of computing the curvature of the Weil-Petersson metric in the higher-dimensional case. We compute the curvature of the Weil-Petersson metric for any local complex suomanifold in the moduli space even when the moduli space itself is singular. The method involves canonically lifting a vector field on the moduli space to a vector field on the total space of the family of deformations. This method of canonical lifting greatly simplifies the computation of the full curvature tensor and makes it possible to express the full curvature tensor in simple explicit terms. In the one-dimensional case the final explicit form of the full curvature tensor agrees with that obtained by Wolpert [12] which so far is the most manageable form of the full curvature tensor. To a certain extent this seems to indicate that our final explicit form of the full curvature tensor in the general higher-dimensional case may be the most manageable form. From this explicit expression of the full curvature tensor we conclude that, for a compact Kähler-Einstein manifold M_0 with negative first Chern class and with tangent bundle T_{M_0}, the holomorphic bisectional curvature of the Weil-Petersson metric in the direction of $\xi,\eta \in H^1(M_0,T_{M_0})$ is negative if the elment in $H^2(M_0,\Lambda^2 T_{M_0})$ defined by $\xi \wedge \eta$ in the natural way vanishes, where $\Lambda^2 T_{M_0}$ is the exterior product of two copies of T_{M_0}. In particular, if $H^2(M_0,\Lambda^2 T_{M_0})$ vanishes, then the holomorphic bisectional curvature of the Weil-Petersson metric for any local complex submanifold of the moduli space of M_0 is negative. It is not clear whether the sufficient condition of the vanishing of $H^2(M_0,\Lambda^2 M_0)$ (or the vanishing of the exterior product of two copies of $H^1(M_0,T_{M_0})$) can be substantially weakened, because the formula for the curvature explicitly contains an obstruction term from the exterior product of two copies of $H^1(M_0,T_{M_0})$. Though we have no counter-example, from the explicit expression for the curvature it seems rather impossible to absorb the obstruction term into the other terms. The vanishing of

$H^2(M_0, \Lambda^2 T_{M_0})$ is very restrictive. For example no two-dimensional compact Kähler surface with negative first Chern class can satisfy it, because for such a surface the nonvanishing of the second plurigenus follows from the theorem of Riemann-Roch and the Kodaira vanishing theorem.

The canonical lifting of a vector field on the moduli space to a vector field on the total space of the family of deformations is defined by the property that the tangent-bundle-valued (0,1)-form computed at each fiber from the lifting is harmonic with respect to the Kähler-Einstein metric on that fiber. The use of the canonical lifting makes it unnecessary to use the computationally complicated process of going to the harmonic projection. Moreover, by using the canonical lifting, one gets the vanishing of all variations of the volume form and a very simple expression for the variations of the Kähler-Einstein metric.

Having a canonical lifting is the same as having a canonical smooth trivialization of the total space of the family of deformations with one real parameter, or equivalently a canonical way of regarding the family of deformations as obtained by deforming the complex structure on a fixed underlying smooth manifold. Our concept of canonical lifting is motivated by Hitchin's method of deforming the complex structure on a K3 surface with a Kähler-Einstein metric [3]. The smooth trivialization given by his deformation of the complex structure corresponds precisely to the canonical lifting though in his case the Ricci curvature tensor is zero instead of negative. In the case of K3 surfaces there is a possibility of explicitly constructing a Kähler-Einstein metric and obtaining an alternative direct proof of the Kähler property of every K3 surface by trying to find a canonical smooth trivialization in a given holomorphic deformation. We will, however, not pursue this possibility in this paper.

Royden [9] told me that he could prove the negativity of the holomorphic sectional curvature and the nonpositivity of the bisectional curvature under the assumption of the vanishing of the Lie bracket on $H^1(M_0,T_{M_0})$, by using his proof for the case of a compact hyperbolic Riemann surface and adding the following two new ingredients. The first ingredient is that on a compact Kähler-Einstein manifold of negative Ricci curvature the tensor of covariant rank two obtained by lowering the contravariant index of a harmonic (0,1)-form is always symmetric. (This ingredient was also used by Koiso in his proof that the Weil-Petersson metric is Kähler.) The second ingredient is to use, in the place of the Bers coordinates in the one-dimensional case, the holomorphic local coordinates of the moduli space constructed by Kodaira-Nirenberg-Spencer [5] in their proof of the existence of deformations of complex structures. The assumption of the vanishing of the Lie bracket on $H^1(M_0,T_{M_0})$ is needed for the second ingredient.

Recently Schumacher [10] considered the Weil-Petersson metric for polarized compact Kähler manifolds with zero first Chern class and showed explicitly that in the case of polarized tori and polarized symplectic manifolds respectively it agrees with the Maass metric on the Siegel upper half space and the Bergman metric on a symmetric space of the third type and is therefore Kähler-Einstein with negative curvature.

I would like to thank Koji Cho, Alan Fekete, and Antonella Nannicini for having very carefully checked my lengthy curvature computations and spotted some misplaced signs, missing complex conjugations, and missing curvature terms in an earlier version.

Table of Contents

§1. Canonical Lifting of Vector Fields.

(1.1) Let I be the unit interval (-1,1) in $\mathbb{R}$ and π: $M \to I$ be a smooth family of compact complex manifolds. For $t \in I$ let $M_t = \pi^{-1}(t)$. We cover M by coordinate charts $(\tilde{U}_i, w_i^\alpha, t)$. Let n be the complex dimension of M_0. We have coordinate transformations

$$w_i^\alpha = f_{ij}^{\ \alpha}(w_j^1, \ldots, w_j^n, t).$$

Let $U_i = M_0 \cap \tilde{U}_i$ and $z_i^\alpha = w_i^\alpha | U_i$. Take a diffeomorphism Φ: $M_0 \times I \to M$ so that $\Phi | M_0 \times \{0\}$ is the identity on M_0. In local coordinates Φ is given by

$$w_i^\alpha = F_i^\alpha(z_i^1, \ldots z_i^n, t),$$

where F_i^α is C^∞ in z_i and t. The diffeomorphism Φ corresponds to a lifting of the vector field $\frac{\partial}{\partial t}$ on I to the vector field

$$v = \frac{\partial}{\partial t} + \sum_\alpha \frac{\partial F_i^\alpha}{\partial t} \frac{\partial}{\partial w_i^\alpha} + \sum_\alpha \frac{\partial \overline{F_i^\alpha}}{\partial t} \frac{\partial}{\partial \overline{w_i^\alpha}}$$

on M. The T_{M_0}-valued (0,1)-form representing the infinitesimal deformation is

$\phi = \sum_\alpha \phi_i^\alpha \frac{\partial}{\partial z_i^\alpha}$ on U_i with $\phi_i^\alpha = \bar{\partial}\left(\frac{\partial F_i^\alpha}{\partial t}\right)$. In other words, ϕ is equal to the $\bar{\partial}$ of the (1,0)-component of v at M_0. Note that, while v is globally defined, the (1,0)-component of v is not globally defined because its projection onto a vector tangential to M_0 depends on the local coordinate chart. So ϕ is not $\bar{\partial}$-exact in general. However, if we have chosen another lifting $\tilde{v}$, then the (1,0)-component of v-v' is globally defined on M_0 and the $\tilde{\phi}$ obtained by using $\tilde{v}$ differs from ϕ by the $\bar{\partial}$ of the global vector field v-v'.

The infinitesimal deformation can also be described by the Cech cohomology class

defined by $\{\xi_{ij}\}$ with $\xi_{ij} = \sum_\alpha \xi_{ij}^\alpha \frac{\partial}{\partial z_i^\alpha}$ and $\xi_{ij}^\alpha = -\frac{\partial}{\partial t} f_{ij}^\alpha$ on $U_i \cap U_j$. Geometrically ξ_{ij}

is the limit as $t \to 0$ of $\frac{1}{t}$ times the discrepancy of going from M_0 to M_t along w_i^α = constant in $\tilde{U}_i$ and going along w_j^α = constant in $\tilde{U}_j$. In this paper we will not use the description of the infinitesimal deformation by Čech cohomology.

(1.3) Now assume that M_0 carries a Hermitian metric. In the cohomology class represented by ϕ there is a unique harmonic representative ψ. The difference $\phi-\psi$ is given by the $\bar\partial$ of a global (1,0) vector field u on M_0. Write $u = \Sigma u_i^\alpha \frac{\partial}{\partial z_i^\alpha}$ on U_i. Define a new vector field $\tilde{v}$ on M_0 by

$$\tilde{v} = v - \sum_\alpha u_i^\alpha \frac{\partial}{dz_i^\alpha} - \sum_\alpha \overline{u_i^\alpha} \frac{\partial}{\partial \overline{z_i^\alpha}}$$

We call $\tilde{v}$ on M_0 the <u>canonical lifting</u> of $\frac{\partial}{\partial t}$. <u>It is characterized by the fact that the $\bar\partial$ of its (1,0)-component</u> (with respect to the charts w_i^α, t, $\tilde{U}_i$) <u>is the harmonic representative of the infinitesimal deformation.</u>

When the first Chern class of M_0 is negative, the canonical lifting is unique, because on M_0 there exists no nonzero holomorphic vector field. One way to see the nonexistence of nonzero holomorphic vector field ξ^α is to use Yau's result on the existence of a Kähler-Einstein metric $g_{\alpha\bar\beta}$ on M_0, because

$$g^{\mu\bar\nu}\partial_\mu\partial_{\bar\nu}(g_{\alpha\bar\beta}\xi^\alpha\overline{\xi^\beta}) = R_{\alpha\bar\beta}\xi^\alpha\overline{\xi^\beta} + g^{\mu\bar\nu}g_{\alpha\bar\beta}(\nabla_\mu\xi^\alpha)(\overline{\nabla_\nu\xi^\beta}) > 0$$

and $g_{\alpha\bar\beta}\xi^\alpha\overline{\xi^\beta}$ is a strictly subharmonic function on the compact manifold M_0. Here, as in the rest of the paper the summation convention is used and the following commonly used notations in Kähler geometry especially those in [4] are used without special explanations.

$$\partial_\mu = \frac{\partial}{\partial z^\mu}, \qquad \partial_{\overline{\nu}} = \frac{\partial}{\partial z^{\overline{\nu}}}$$

$R_{\alpha\overline{\beta}} = \partial_\alpha \partial_{\overline{\beta}}$ of the determiniant of $(g_{\alpha\overline{\beta}})$

∇_μ = covariant differentiation in the direction of $\frac{\partial}{\partial z^\mu}$

$\nabla_{\overline{\mu}}$ = covariant differentiation in the direction of $\frac{\partial}{\partial z^{\overline{\mu}}}$

$(g^{\mu\overline{\nu}})$ = the inverse matrix of $(g_{\alpha\overline{\beta}})$

The coordinates $z^1,\dots,z^n$ mean some coordinate chart $z^1_i,\dots,z^n_i$ with the subscript i suppressed.

§2. Lie Derivatives.

(2.1) We continue with the notations of §1 and assume that each M_t has negative first Chern class. We endow each M_t with a Kähler-Einstein metric $g_{\alpha\overline{\beta}}(t)$ such that $R_{\alpha\overline{\beta}}(t) = e\, g_{\alpha\overline{\beta}}(t)$ where e is a positive number independent of t. Let v be the unique vector field on M which at every M_t is the canonical lifting of $\frac{\partial}{\partial t}$ with respect to the Kähler-Einstein metric of M_t.

Let $T_a^b(t)$ be a tensor of rank two on M_t, where a,b and other lower case Latin letters in this paper denote indices with the range $1,2,\dots,n,\overline{1},\overline{2},\dots,\overline{n}$ (in contrast to lower case Greek letters α,β, etc. with the range $1,2,\dots,n$). We now compute the Lie derivative of T_a^b with respect to the vector field v and evaluate it at $t = 0$. Choose a smooth trivialization Φ: $M_0 \times I \to M$ given by $w_i^\alpha = F_i^\alpha(z_i^1,\dots,z_i^n,t)$ so that

$$v = \frac{\partial}{\partial t} + \sum_\alpha \frac{\partial F_i^\alpha}{\partial t}\frac{\partial}{\partial w_i^\alpha} + \sum_\alpha \frac{\partial \overline{F_i^\alpha}}{\partial t}\frac{\partial}{\partial w_i^{\overline{\alpha}}}.$$

That is, the smooth trivialization Φ corresponds to the vector field v. We use here and in the rest of the paper the following notation. The dot • in the superscript position denotes the differentiation of a scalar function or a component of a tensor with respect to t when $z^1,\dots,z^n$ are kept constant. The operator $\mathbf{L}_v$ denotes the Lie derivative with respect to the vector field v. For example, when t is set to be 0,

$$
\begin{aligned}
(\mathbf{L}_v T)^d_c &= \frac{\partial}{\partial t}\Big(T^b_a \frac{\partial w^a}{\partial z^c}\frac{\partial z^d}{\partial w^b}\Big) \\
&= (T^d_c)^\bullet + T^d_a\Big(\frac{\partial w^a}{\partial z^c}\Big)^\bullet + T^b_c\Big(\frac{\partial z^d}{\partial w^b}\Big)^\bullet \\
&= (T^d_c)^\bullet + T^d_a \partial_c v^a - T^b_c \partial_b v^d
\end{aligned}
$$

Here $\mathbf{L}_v T$ means the derivative with respect to t of the pullback of $T^b_a(t)$ by the diffeomorphism $\Phi_t\colon M_0 \to M_t$ induced by Φ. The term $(T^d_c)^\bullet$ means the derivative with respect to t with fixed $z^1,\dots,z^n$ of the component T^d_c of T for $dw^d \otimes \frac{\partial}{\partial w^c}$. The factor $\Big(\frac{\partial w^a}{\partial z^c}\Big)^\bullet$ means the derivative with respect to t with fixed $z^1,\dots,z^n$ of the function $\frac{\partial w^a}{\partial z^c}$.

Let J(t) denote the almost complex structure of M_t. Using $J^\beta_\alpha = \sqrt{-1}\,\delta^\beta_\alpha$, $J^{\bar\beta}_{\bar\alpha} = -\sqrt{-1}\,\delta^\beta_\alpha$ (where δ^β_α is the Kronecker delta), $J^\beta_{\bar\alpha} = 0$, and $J^{\bar\beta}_\alpha = 0$, we obtain at t = 0 from the above formula for $\mathbf{L}_v T$

$$
\begin{aligned}
(\mathbf{L}_v J)^\beta_\alpha &= 0 \\
(\mathbf{L}_v J)^\beta_{\bar\alpha} &= 2\sqrt{-1}\,\partial_{\bar\alpha} v^\beta = 2\sqrt{-1}\,\phi^\beta_{\bar\alpha} \\
(\mathbf{L}_v J)^{\bar\beta}_\alpha &= -2\sqrt{-1}\,\overline{\phi^\beta_{\bar\alpha}} \\
(\mathbf{L}_v J)^{\bar\beta}_{\bar\alpha} &= 0
\end{aligned}
$$

Here $\phi^\beta_{\bar\alpha} = \partial_{\bar\alpha} v^\beta$ denotes the harmonic $T^{1,0}_{M_0}$-valued (0,1)-form representing the infinitesimal deformation at t = 0.

(2.2) We apply the same method to compute the Lie derivative of the Kähler-

Einstein metrics g_{ab} and get at $t = 0$

$$(L_v g)_{\alpha\beta} = g_{\bar\gamma\beta}\phi^{\bar\gamma}_{\alpha} + g_{\alpha\bar\gamma}\phi^{\bar\gamma}_{\beta}$$

$$(L_v g)_{\alpha\bar\beta} = (g_{\alpha\bar\beta})^{\bullet} + g_{\gamma\bar\beta}\partial_{\alpha}v^{\gamma} + g_{\alpha\bar\gamma}\partial_{\bar\beta}\overline{v^{\gamma}}.$$

The other components of $L_v g$ can be obtained by taking conjugates, because both g and v are real. Here we have used $\phi^{\bar\gamma}_{\alpha}$ to denote $\overline{\phi^{\gamma}_{\bar\alpha}}$. The expression for $(L_v g)_{\alpha\bar\beta}$ is unsatisfactory. We are going to use the Einstein condition $R_{\alpha\bar\beta} = e\, g_{\alpha\bar\beta}$ to prove that $(L_v g)_{\alpha\bar\beta} = 0$. We have at $t = 0$

$$(L_v R)_{\alpha\bar\beta} = (R_{\alpha\bar\beta})^{\bullet} + R_{\gamma\bar\beta}\partial_{\alpha}v^{\gamma} + R_{\alpha\bar\gamma}\partial_{\bar\beta}\overline{v^{\gamma}}.$$

$$(R_{\alpha\bar\beta})^{\bullet} = \frac{\partial}{\partial t}\left(\frac{\partial^2}{\partial w^{\alpha}\partial \bar w^{\beta}} \log\det(g_{\gamma\bar\delta})\right)$$

$$= \frac{\partial}{\partial t}\left(\left(\frac{\partial z^{\lambda}}{\partial w^{\alpha}}\partial_{\lambda} + \frac{\partial \overline{z^{\lambda}}}{\partial w^{\alpha}}\partial_{\bar\lambda}\right)\left(\frac{\partial z^{\mu}}{\partial \bar w^{\beta}}\partial_{\mu} + \frac{\partial \overline{z^{\mu}}}{\partial w^{\beta}}\partial_{\bar\mu}\right)\log\det(g_{\gamma\bar\delta})\right)$$

$$= \left(-(\partial_{\alpha}v^{\lambda})\,\partial_{\lambda} - (\partial_{\alpha}\overline{v^{\lambda}})\,\partial_{\bar\lambda}\right)\partial_{\bar\beta}\log\det(g_{\gamma\bar\delta})$$

$$+ \partial_{\alpha}\left(-(\partial_{\bar\beta}v^{\mu})\,\partial_{\mu} - \partial_{\bar\beta}\overline{v^{\mu}}\,\partial_{\bar\mu}\right)\log\det(g_{\gamma\bar\delta}) + \partial_{\alpha}\partial_{\bar\beta}(\log\det(g_{\gamma\bar\delta}))^{\bullet}$$

At the point under consideration we use normal coordinates with $dg_{\alpha\bar\beta} = 0$ and $\partial_{\gamma}\partial_{\bar\delta}g_{\alpha\bar\beta} = 0$. Then

$$(R_{\alpha\bar\beta})^{\bullet} = -(\partial_{\alpha}v^{\lambda})R_{\lambda\bar\beta} - \partial_{\bar\beta}\overline{v^{\mu}}\,R_{\alpha\bar\mu} + \partial_{\alpha}\partial_{\bar\beta}(\log\det(g_{\gamma\bar\delta}))^{\bullet}.$$

Hence

$$(L_v R)_{\alpha\bar\beta} = \partial_{\alpha}\partial_{\bar\beta}(\log\det g_{\gamma\bar\delta})^{\bullet}$$

$$= \partial_{\alpha}\partial_{\bar\beta}(g^{\gamma\bar\delta}(g_{\gamma\bar\delta})^{\bullet})$$

$$= \partial_{\alpha}\partial_{\bar\beta}\left(g^{\gamma\bar\delta}\left((L_v g)_{\gamma\bar\delta} - g_{\lambda\bar\delta}\partial_{\gamma}v^{\lambda} - g_{\gamma\bar\mu}\partial_{\bar\delta}\overline{v^{\mu}}\right)\right)$$

$$=\partial_\alpha\partial_{\bar\beta}(g^{\gamma\bar\delta}(\mathbf{L}_V g)_{\gamma\bar\delta}) - \nabla_\alpha(g^{\gamma\bar\delta}\nabla_\gamma\phi_{\bar\beta\bar\delta}) - \nabla_{\bar\beta}(g^{\gamma\bar\delta}\nabla_{\bar\delta}\phi_{\alpha\gamma}),$$

where $\phi_{\bar\beta\bar\lambda} = g_{\mu\bar\lambda}\phi^\lambda_{\bar\beta}$ and $\phi_{\alpha\gamma} = \overline{\phi_{\bar\alpha\bar\gamma}}$. Using $\mathbf{L}_V R = e\,\mathbf{L}_V g$, we get

$$\partial_\alpha\partial_{\bar\beta}(g^{\gamma\bar\delta}(\mathbf{L}_V g)_{\gamma\bar\delta}) = e(\mathbf{L}_V g)_{\alpha\bar\beta} + \nabla_\alpha(g^{\gamma\bar\delta}\nabla_\gamma\phi_{\bar\beta\bar\delta}) + \nabla_{\bar\beta}(g^{\gamma\bar\delta}\nabla_{\bar\delta}\phi_{\alpha\gamma}).$$

At this point we need the fact that $\phi_{\bar\alpha\bar\beta}$ is symmetric in α and β. One way to see it is to apply the Bochner-Kodaira formula to the harmonic T_{M_0}-valued (0,1)-form $\phi^\alpha_{\bar\beta}$ and get

$$0 = (\Box\phi)^\alpha_{\bar\beta} = -g^{\lambda\bar\mu}\nabla_{\bar\mu}\nabla_\lambda\phi^\alpha_{\bar\beta} - R_\sigma{}^{\alpha\bar\tau}{}_{\bar\beta}\phi^\sigma_{\bar\tau} + R_\sigma{}^\alpha\phi^\sigma_{\bar\beta}$$

(see e.g. [11, (1.3.4)]). Here $\Box = \bar\partial\bar\partial^* + \bar\partial^*\bar\partial$ and $R_{\alpha\bar\beta\gamma\bar\delta} = \partial_\alpha\partial_{\bar\beta}g_{\gamma\bar\delta}$ in normal coordinates. Lowering the index α and using the Einstein condition, we get

$$0 = -g^{\lambda\bar\mu}\nabla_{\bar\mu}\nabla_\lambda\phi_{\bar\alpha\bar\beta} - R_{\sigma\bar\alpha}{}^{\bar\tau}{}_{\bar\beta}\phi^\sigma_{\bar\tau} + e\,\phi_{\bar\alpha\bar\beta}.$$

Since $R_{\sigma\bar\alpha}{}^{\bar\tau}{}_{\bar\beta}$ is symmetric in α and β, it follows that

$$0 = -g^{\lambda\bar\mu}\nabla_{\bar\mu}\nabla_\lambda(\phi_{\bar\alpha\bar\beta} - \phi_{\bar\beta\bar\alpha}) + e(\phi_{\bar\alpha\bar\beta} - \phi_{\bar\beta\bar\alpha}).$$

Multiplying both sides by $g^{\bar\alpha\sigma}g^{\bar\beta\tau}(\phi_{\sigma\tau}-\phi_{\tau\sigma})$ and integrating over M_0 and using integration by parts we conclude from $e > 0$ that $\phi_{\bar\alpha\bar\beta} - \phi_{\bar\beta\bar\alpha}$ must vanish identically on M_0.

From the symmetry of $\phi_{\bar\alpha\bar\beta}$ in α and β and from the harmonicity of $\phi^\alpha_{\bar\beta}$ it follows that $g^{\gamma\bar\delta}\nabla_\gamma\phi_{\bar\beta\bar\delta}$ and $g^{\gamma\bar\delta}\nabla_{\bar\delta}\phi_{\alpha\gamma}$ both vanish identically on M_0. Hence

$$\partial_\alpha\partial_{\bar\beta}(g^{\gamma\bar\delta}(\mathbf{L}_V g)_{\gamma\bar\delta}) = e\,(\mathbf{L}_V g)_{\alpha\bar\beta}$$

and contracting it with $g^{\alpha\bar\beta}$ we conclude that $g^{\gamma\bar\delta}(\mathbf{L}_V g)_{\gamma\bar\delta}$ must vanish identically otherwise it is a nonzero subharmonic function on M_0. The vanishing of $g^{\gamma\bar\delta}(\mathbf{L}_V g)_{\gamma\bar\delta}$ in turn implies that $(\mathbf{L}_V g)_{\alpha\bar\beta}$ vanishes identically. Our final result is that

$$(L_v g)_{\alpha\beta} = 2\phi_{\alpha\beta}$$

$$(L_v g)_{\alpha\bar{\beta}} = 0.$$

From $g_{ab}g^{bc} = \delta^c_a$ (where δ^c_a is the Kronecker delta) and these two equations direct computation yields

$$(L_v g)^{\alpha\bar{\beta}} = 0$$

$$(L_v g)^{\alpha\beta} = -2\phi^{\alpha\beta}$$

(where $\phi^{\alpha\beta}$ is obtained by raising the indices in $\phi_{\bar{\alpha}\bar{\beta}}$).

(2.3) We now compute the Lie derivative of the volume form of the Kähler-Einstein metric. Let

$$\omega = \frac{1}{2}\omega_{ab}dw^a \wedge dw^b = g_{\alpha\bar{\beta}}dw^\alpha \wedge \overline{dw^\beta}$$

with ω_{ab} skew-symmetric in a and b. Then

$$\omega_{\alpha\bar{\beta}} = g_{\alpha\bar{\beta}}$$

$$\omega_{\bar{\alpha}\beta} = -g_{\beta\bar{\alpha}}$$

$$\omega_{\alpha\beta} = 0$$

$$\omega_{\bar{\alpha}\bar{\beta}} = 0.$$

Using the suymmetry of $\phi_{\alpha\beta}$ in α and β and using $(L_v g)_{\alpha\bar{\beta}} = 0$, we conclude that $L_v\omega$ vanishes identically on M_0. The volume form is, up to a constant, the exterior product of n copies of ω. Hence the Lie derivative with respect to v of the volume form of the Kähler-Einstein metric is zero.

§3. Kähler Condition.

(3.1) Though the Kähler property of the Weil-Petersson metric was proved already by Kosio [6], to compute the curvature of the Weil-Petersson metric we have to first differentiate the Weil-Petersson metric once as an intermediate step. This differentiation immediately yields the Kähler property. We are going to do this intermediate step of first-order differentiation of the Weil-Petersson metric here in this section.

Now assume that Ω is an open neighborhood of O in $\mathbb{C}^N$ with holomorphic coordinates $t = (t_1,\ldots,t_N)$. We now modify the situation in §1 as follows. We replace the unit open interval I by Ω so that $\pi\colon M \to \Omega$ is a holomorphic family of compact Kähler-Einstein manifolds with negative Ricci curvature. We keep, with obvious modifications, the notations of §1. Let t'_k and t''_k be respectively the real part and imaginary part of t_k. Let v'_k and v''_k denote respectively the canonical liftings of $\frac{\partial}{\partial t'_k}, \frac{\partial}{\partial t''_k}$. Let $\phi(\frac{\partial}{\partial t'_k}), \phi(\frac{\partial}{\partial t''_k})$ denote respectively the tangent-bundle-valued hamonic (0,1)-forms representing the infinitesimal deformation in the direction of $\frac{\partial}{\partial t'_k}, \frac{\partial}{\partial t''_k}$. Define

$$L_{v_k} = \frac{1}{2}(L_{v'_k} - \sqrt{-1}L_{v''_k})$$

$$L_{\overline{v_k}} = \frac{1}{2}(L_{v'_k} + \sqrt{-1}L_{v''_k}).$$

Note that in general the Lie bracket $[v_k', v_k'']$ is not zero so that corresponding to v_k', v_k'' there is no smooth trivialization of the family M over a two-real-dimensional plane in Ω with t'_k, t''_k as variables. Let

$$\phi(\frac{\partial}{\partial t_k}) = \frac{1}{2}(\phi(\frac{\partial}{\partial t'_k}) - \sqrt{-1}\phi(\frac{\partial}{\partial t''_k}))$$

$$\phi(\frac{\partial}{\partial \overline{t_k}}) = \frac{1}{2}(\overline{\phi(\frac{\partial}{\partial t'_k})} + \sqrt{-1}\,\overline{\phi(\frac{\partial}{\partial t''_k})}).$$

Since

$$\phi(\frac{\partial}{\partial t''_k}) = \sqrt{-1}\phi(\frac{\partial}{\partial t'_k}),$$

it follows that

$$(L_{v_k} J)^{\beta}_{\overline{\alpha}} = 2\sqrt{-1}\phi(\frac{\partial}{\partial t_k})^{\beta}_{\overline{\alpha}}$$

$$(L_{v_k}J)^{\overline{\beta}}_{\alpha} = (L_{v_k}J)^{\beta}_{\alpha} = (L_{v_k}J)^{\overline{\beta}}_{\overline{\alpha}} = 0.$$

The tensor $L_{\overline{v}_k}J$ is the complex conjugate of $L_{v_k}J$. The tensor $\phi(\frac{\partial}{\partial \overline{t}_k})$ is the complex conjugate of $\phi(\frac{\partial}{\partial t_k})$.

(3.2) Let dV denote the volume form of the Kähler-Einstein metric $g_{\alpha\overline{\beta}}$. We define the <u>Weil-Petersson metric</u> $\sum_{i,j=1}^{N} h_{i\overline{j}}\, dt_i \otimes \overline{dt_j}$ on Ω by

$$h_{i\overline{j}}(t) = \int_{M_t} (L_{v_i}J)^p_a \overline{(L_{v_j}J)^q_b}\, g_{p\overline{q}} g^{a\overline{b}} dV$$

$$= 4\int_{M_t} \phi(\frac{\partial}{\partial t_i})^p_a \overline{\phi(\frac{\partial}{\partial t_j})^q_b}\, g_{p\overline{q}} g^{a\overline{b}} dV$$

In order for the metric to be positive definite we assume from now on the element of $H^1(M_t, T_{M_t})$ defined by any nonzero vector of Ω of type (1,0) at t is nonzero. Because of (2.3) we have

$$(\frac{\partial}{\partial \overline{t}_k} h_{i\overline{j}})(0) = \int_{M_0} (L_{\overline{v}_k} L_{v_i} J)^p_a \overline{(L_{v_j}J)^q_b}\, g_{p\overline{q}} g^{a\overline{b}} dV$$

$$+ \int_{M_0} (L_{v_i}J)^p_a \overline{(L_{v_k} L_{v_j}J)^q_b}\, g_{p\overline{q}} g^{a\overline{b}} dV$$

$$+ \int_{M_0} (L_{v_i}J)^p_a \overline{(L_{v_j}J)^q_b}\, (L_{\overline{v}_k} g)_{p\overline{q}} g^{a\overline{b}} dV$$

$$+ \int_{M_0} (L_{v_i}J)^p_a \overline{(L_{v_j}J)^q_b}\, g_{p\overline{q}} (L_{\overline{v}_k} g)^{a\overline{b}} dV$$

The last two terms vanish, because $(L_v g)_{\alpha\bar\beta}$, $(L_v g)^{\alpha\bar\beta}$, $(L_v J)^{\beta}_{\alpha}$, $(L_v J)^{\bar\beta}_{\bar\alpha}$ vanish for all canonical liftings v of real vector fields on Ω and because of type considerations. We have also used the fact that the Lie derivative of the volume form is zero.

Since $d\pi[\bar v_k, v_i] = [\frac{\partial}{\partial \bar t_k}, \frac{\partial}{\partial t_i}] = 0$, it follows that $[\bar v_k, v_i]$ is tangential to M_0 at every point. Hence the (1,0)-component of $[\bar v_k, v_i]$ is globally well-defined on M_0 and direct computation of $L_{[\bar v_k, v_i]} = L_{\bar v_k} L_{v_i} - L_{v_i} L_{\bar v_k}$ yields

$$(L_{[\bar v_k, v_i]} J)^{\beta}_{\bar\alpha} = 2\sqrt{-1}\,\partial_{\bar\alpha}[\bar v_k, v_i]^{\beta}$$

(cf. (2.1)). Hence $(L_{[\bar v_k, v_i]} J)^{\beta}_{\bar\alpha}$ is $\bar\partial$-exact. Since $(L_{v_j} J)^{\beta}_{\bar\alpha}$ is harmonic, it follows that

$$\int_{M_0} (L_{[\bar v_k, v_i]} J)^p_{\bar a}\, \overline{(L_{v_j} J)^q_{\bar b}}\; g_{p\bar q}\, g^{a\bar b} dV = 0.$$

Hence

$$\int_{M_0} (L_{\bar v_k} L_{v_i} J)^p_{\bar a}\, \overline{(L_{v_j} J)^q_{\bar b}}\; g_{p\bar q}\, g^{a\bar b} dV = \int_{M_0} (L_{v_i} L_{\bar v_k} J)^p_{\bar a}\, \overline{(L_{v_j} J)^q_{\bar b}}\; g_{p\bar q}\, g^{a\bar b} dV$$

We claim that this last integral vanishes. Since the only nonzero components of $L_{v_j} J$ are of the form $(L_{v_j} J)^{\beta}_{\bar\alpha}$, because of type considerations it suffices to show that $(L_{v_i} L_{\bar v_k} J)^{\beta}_{\bar\alpha} = 0$. Now

$$(L_{v_i} L_{\bar v_k} J)^{\beta}_{\bar\alpha} = ((L_{\bar v_k} J)^{\beta}_{\bar\alpha})^{\bullet} + (L_{\bar v_k} J)^{\beta}_{a}\partial_{\bar\alpha} v^{a} + (L_{\bar v_k} J)^{b}_{\bar\alpha}\partial_{b} v^{\beta},$$

but the nonzero components of $L_{\bar v_k} J$ must be of the type $(L_{\bar v_k} J)^{\bar\beta}_{\alpha}$. Thus $(L_{v_i} L_{\bar v_k} J)^{\beta}_{\bar\alpha} = 0$ and

$$(\frac{\partial}{\partial \bar t_k} h_{i\bar j})(0) = \int_{M_0} (L_{v_i} J)^p_{\bar a}\, \overline{(L_{v_k} L_{v_j} J)^q_{\bar b}}\; g_{a\bar b}\, g^{p\bar q} dV$$

Since M_0 is no different from other M_t, the formula continues to hold when 0 is replaced by t and M_0 is replaced by M_t.

The Kähler property of the Weil-Petersson matric follows from the vanishing of

$$(\frac{\partial}{\partial \overline{t_k}} h_{i\overline{j}})(0) - (\frac{\partial}{\partial \overline{t_j}} h_{i\overline{k}})(0) = \int_{M_0} (L_{v_i} J)^p_{\overline{a}} \overline{(L_{[v_k, v_j]} J)^q_b} g_{\overline{pq}}\, g^{a\overline{b}} dV$$

which is a consequence of the $\overline{\partial}$-exactness of $(L_{[v_k,v_j]} J)^{\beta}_{\overline{\alpha}}$ and the harmonicity of $(L_{v_i} J)^{\beta}_{\overline{\alpha}}$.

§4. First Part of the Curvature Computation.

(4.1) In this section we compute the curvature tensor of the Weil-Petersson metric and reduce it to such a form that in the one-dimensional case it agrees with the expression obtained in Wolpert [12]. Further and more complicated reduction will be done in the next section on the second part of the curvature computation.

Because of (2.3) the curvature tensor $R^{(WP)}_{\ell \overline{k} i \overline{j}}$ of the Weil-Petersson metric at 0 is given by

$$R^{(WP)}_{\ell \overline{k} i \overline{j}} = \frac{\partial}{\partial t_\ell} \frac{\partial}{\partial \overline{t_k}} h_{i\overline{j}}$$

$$= \frac{\partial}{\partial t_\ell} \int_{M_t} (L_{v_i} J)^p_{\overline{a}} \overline{(L_{v_k} L_{v_j} J)^q_b}\, g^{a\overline{b}}\, g_{p\overline{q}}\, dV$$

$$= I + II + III + IV,$$

where

$$I = \int_{M_0} (L_{v_\ell} L_{v_i} J)^p_{\overline{a}} \overline{(L_{v_k} L_{v_j} J)^q_b}\, g^{a\overline{b}}\, g_{p\overline{q}}\, dV$$

$$II = \int_{M_0} (L_{v_i} J)^p_a \overline{(L_{\bar{v}_\ell} L_{v_k} L_{v_j} J)^q_b} \, g^{a\bar{b}} \, g_{p\bar{q}} \, dV$$

$$III = \int_{M_0} (L_{v_i} J)^p_a \overline{(L_{v_k} L_{v_j} J)^q_b} \, (L_{v_\ell} g)^{a\bar{b}} \, g_{p\bar{q}} \, dV$$

$$IV = \int_{M_0} (L_{v_i} J)^p_a \overline{(L_{v_k} L_{v_j} J)^q_b} \, g^{a\bar{b}} \, (L_{v_\ell} g)_{p\bar{q}} \, dV.$$

Our goal is to show that the holomorphic bisectional curvature of the Weil-Petersson metric is negative in certain directions. For this purpose the term I is good. The term II is the most difficult one to handle. We first look at term III.

(4.2) From our computations of $L_v g$ in (2.2) and of $L_{v_i} J$ in (3.1), we have

$$(L_{v_i} J)^p_a \overline{(L_{v_k} L_{v_j} J)^q_b} \, (L_{v_\ell} g)^{a\bar{b}} \, g_{p\bar{q}} = (L_{v_i} J)^{\lambda}_{\bar{\alpha}} \overline{(L_{v_k} L_{v_j} J)^{\mu}_{\beta}} \, (L_{v_\ell} g)^{\bar{\alpha}\bar{\beta}} \, g_{\lambda\bar{\mu}}$$

$$= 2\sqrt{-1}\, \phi\left(\frac{\partial}{\partial t_i}\right)_{\bar{\alpha}\bar{\mu}} \overline{(L_{v_k} L_{v_j} J)^{\mu}_{\beta}} \; \overline{\left(-2\phi\left(\frac{\partial}{\partial t_\ell}\right)^{\alpha\beta}\right)}.$$

By applying $L_{v_k} L_{v_j}$ to both sides of $J^b_a J^c_b = -\delta^c_a$ and using the fact that the only nonzero components of $L_{v_j} J$ are of the type $(L_{v_j} J)^{\beta}_{\bar{\alpha}}$ and the fact that $J^{\beta}_{\alpha} = \sqrt{-1}\, \delta^{\beta}_{\alpha}$ and $J^{\bar{\beta}}_{\bar{\alpha}} = -\sqrt{-1}\, \delta^{\beta}_{\alpha}$, we conclude that $(L_{v_k} L_{v_j} J)^{\gamma}_{\alpha} = 0$. Hence the term III vanishes. In exactly the same way we show that the term IV vanishes.

(4.3) We now deal with the term II. Since

$$\begin{aligned}
L_{\bar{v}_\ell} L_{v_k} L_{v_j} &= L_{[\bar{v}_\ell, v_k]} L_{v_j} + L_{v_k} L_{\bar{v}_\ell} L_{v_j} \\
&= L_{[\bar{v}_\ell, v_k]} L_{v_j} + L_{v_k} L_{[\bar{v}_\ell, v_j]} + L_{v_k} L_{v_j} L_{\bar{v}_\ell} \\
&= L_{[[\bar{v}_\ell, v_k], v_j]} + L_{v_j} L_{[\bar{v}_\ell, v_k]} + L_{v_k} L_{[\bar{v}_\ell, v_j]} + L_{v_k} L_{v_j} L_{\bar{v}_\ell},
\end{aligned}$$

it follows that II = A + B + C + D, where

$$A = \int_{M_0} (L_{v_i} J)^p_a \overline{(L_{[[\bar v_\ell, v_k], v_j]} J)^q_b} g^{a\bar b} g_{p\bar q}\, dV$$

$$B = \int_{M_0} (L_{v_i} J)^p_a \overline{(L_{v_j} L_{[\bar v_\ell, v_k]} J)^q_b} g^{a\bar b} g_{p\bar q}\, dV$$

$$C = \int_{M_0} (L_{v_i} J)^p_a \overline{(L_{v_k} L_{[\bar v_\ell, v_j]} J)^q_b} g^{a\bar b} g_{p\bar q}\, dV$$

$$D = \int_{M_0} (L_{v_i} J)^p_a \overline{(L_{v_k} L_{v_j} L_{\bar v_\ell} J)^q_b} g^{a\bar b} g_{p\bar q}\, dV.$$

We immediately have the vanishing of A, because $[[\bar v_\ell, v_k], v_j]$ is tangential to M_0 and $(L_{[[\bar v_\ell, v_k], v_j]} J)^{\mu}_{\bar\beta}$ and $(L_{v_i} J)^{\lambda}_{\bar\alpha}$ are respectively $\bar\partial$-exact and harmonic $T^{1,0}_{M_0}$-valued (0,1)-forms on M_0. The second easiest term to handle is D. In (3.2) we computed $(L_{v_i} L_{\bar v_k} J)^{\beta}_{\bar\alpha}$ to be zero. Hence

$$(L_{v_k} L_{v_j} L_{\bar v_\ell} J)^{\beta}_{\bar\alpha} = (L_{v_j} L_{\bar v_\ell} J)^{\beta}_{\lambda}\, \phi(\frac{\partial}{\partial t_k})^{\lambda}_{\bar\alpha} - (L_{v_j} L_{\bar v_\ell} J)^{\bar\mu}_{\bar\alpha}\, \phi(\frac{\partial}{\partial t_k})^{\beta}_{\bar\mu}.$$

Applying $L_{v_j} L_{\bar v_\ell}$ to both sides of $J^b_a J^c_b = -\delta^c_a$ and using the computations for $L_{\bar v_\ell} J$ and $L_{v_j} J$ from (3.1), we obtain

$$(L_{v_j} L_{\bar v_\ell} J)^{\gamma}_{\alpha} = 2\sqrt{-1}\, \phi(\frac{\partial}{\partial \bar t_\ell})^{\bar\beta}_{\alpha}\, \phi(\frac{\partial}{\partial t_j})^{\gamma}_{\bar\beta}$$

$$(L_{v_j} L_{\bar v_\ell} J)^{\bar\gamma}_{\bar\alpha} = -2\sqrt{-1}\, \phi(\frac{\partial}{\partial t_j})^{\beta}_{\bar\alpha}\, \phi(\frac{\partial}{\partial \bar t_\ell})^{\bar\gamma}_{\beta}$$

Thus

$$(L_{v_k} L_{v_j} L_{\bar v_\ell} J)^{\beta}_{\bar\alpha} = 4\sqrt{-1}\, \phi(\frac{\partial}{\partial t_k})^{\lambda}_{\bar\alpha}\, \phi(\frac{\partial}{\partial \bar t_\ell})^{\bar\mu}_{\lambda}\, \phi(\frac{\partial}{\partial t_j})^{\beta}_{\bar\mu}$$

and

$$D = 8\int_{M_0} \phi(\frac{\partial}{\partial t_i})^{\beta}_{\bar\alpha}\, \phi(\frac{\partial}{\partial \bar t_k})^{\bar\gamma}_{\beta}\, \phi(\frac{\partial}{\partial t_\ell})^{\delta}_{\bar\gamma}\, \phi(\frac{\partial}{\partial \bar t_j})^{\bar\alpha}_{\delta}\, dV$$

(4.4) The terms B and C are similar and can be obtained by switching j and k. So we will handle only the term B. Since

$$\int_{M_t} (L_{v_i} J)^p_a\, \overline{(L_{[\bar v_\ell, v_k]} J)^q_b}\, g^{a\bar b} g_{p\bar q}\, dV = \int_{M_t} (L_{v_i} J)^{\lambda}_{\bar\alpha} \overline{(L_{[\bar v_\ell, v_k]} J)^{\mu}_{\bar\beta}}\, g^{\bar\alpha\beta} g_{\lambda\bar\mu}\, dV$$

vanishes due to the harmonicity of $(L_{v_i} J)^{\lambda}_{\bar\alpha}$ and the $\bar\partial$-exactness of $(L_{[\bar v_\ell, v_k]} J)^{\mu}_{\bar\beta}$, it follows by applying $\frac{\partial}{\partial \bar t_j}$ to it that

$$B = -\int_{M_0} (L_{\bar v_j} L_{v_i} J)^p_a\, \overline{(L_{[\bar v_\ell, v_k]} J)^q_b}\, g^{a\bar b}\, g_{p\bar q}\, dV$$

We write $B = B_1 + B_2 + B_3$, where

$$B_1 = -\int_{M_0} (L_{\bar v_j} L_{v_i} J)^{\lambda}_{\alpha}\, \overline{(L_{[\bar v_\ell, v_k]} J)^{\mu}_{\beta}}\, g^{\alpha\bar\beta}\, g_{\lambda\bar\mu}\, dV$$

$$B_2 = -\int_{M_0} (L_{\bar v_j} L_{v_i} J)^{\lambda}_{\bar\alpha}\, \overline{(L_{[\bar v_\ell, v_k]} J)^{\mu}_{\bar\beta}}\, g^{\bar\alpha\beta}\, g_{\lambda\bar\mu}\, dV$$

$$B_3 = -\int_{M_0} (L_{\bar v_j} L_{v_i} J)^{\bar\lambda}_{\bar\alpha}\, \overline{(L_{[\bar v_\ell, v_k]} J)^{\bar\mu}_{\bar\beta}}\, g^{\bar\alpha\beta}\, g_{\bar\lambda\mu}\, dV.$$

Since from the computations of $(L_{v_j} L_{\bar v_\ell} J)^{\gamma}_{\alpha}$ and $(L_{v_j} L_{\bar v_\ell} J)^{\bar\gamma}_{\bar\alpha}$ in (4.3)

$$(L_{[\bar v_\ell, v_k]} J)^{\mu}_{\beta} = (L_{\bar v_\ell} L_{v_k} J)^{\mu}_{\beta} - (L_{v_k} L_{\bar v_\ell} J)^{\mu}_{\beta}$$

$$= \overline{(L_{v_\ell} L_{\overline{v_k}} J)_{\overline{\beta}}^{\overline{\mu}}} - (L_{v_k} L_{\overline{v_\ell}} J)_{\beta}^{\mu} = 0,$$

it follows that $B_1 = 0$. Likewise $B_3 = 0$. We now handle the term B_2. This is the most difficult step in the first part of the computation of the curvature tensor of the Weil-Petersson metric. It will be achieved by computing $\overline{\partial}^*$ of the $T^{1,0}_{M_0}$-valued (0,1)-form $(L_{\overline{v_j}} L_{v_i} J)_{\overline{\alpha}}^{\lambda}$.

(4.5) For notational simplicity we let $T_{\overline{\alpha}}^{\beta} = (L_{v_i} J)_{\overline{\alpha}}^{\beta}$ and $v = \overline{v_j}$. Choose a smooth trivialization $w^\alpha = w^\alpha(z,t)$ for the family $\pi\colon\ M \to \Omega$ such that $\frac{\partial}{\partial \overline{t_j}} = \overline{v_j}$ at M_0 and $w^\alpha(z,0) = z^\alpha$. We use the dot $^\bullet$ in the superscript position to denote $\frac{\partial}{\partial \overline{t_j}}$ of a <u>scalar</u> function or a <u>component</u> of a tensor. As before, ∂_α, $\partial_{\overline{\alpha}}$ mean respectively $\frac{\partial}{\partial z^\alpha}$, $\frac{\partial}{\partial z^{\overline{\alpha}}}$.

Since on each M_t the $T^{1,0}_{M_0}$-valued (0,1)-form $T_{\overline{\alpha}}^{\beta}$ is harmonic, we have $g^{\alpha\overline{\beta}} \frac{\partial}{\partial w^\alpha}(g_{\overline{\delta}\gamma} T_{\overline{\beta}}^{\gamma}) \equiv 0$ on M_t for every t, that is

$$g^{\alpha\overline{\beta}}\left(\frac{\partial z^\lambda}{\partial w^\alpha}\partial_\lambda + \frac{\partial \overline{z^\lambda}}{\partial w^\alpha}\partial_{\overline{\lambda}}\right)(g_{\overline{\delta}\gamma} T_{\overline{\beta}}^{\delta}) \equiv 0.$$

Applying $\frac{\partial}{\partial \overline{t_j}}$ and setting $t = 0$, we get on M_0

$$(g^{\alpha\overline{\beta}})^\bullet\, \partial_\alpha(g_{\overline{\delta}\gamma} T_{\overline{\beta}}^{\gamma}) + g^{\alpha\overline{\beta}}\left(\left(\frac{\partial z^\lambda}{\partial w^\alpha}\right)^\bullet \partial_\lambda + \left(\frac{\partial \overline{z}^\lambda}{\partial w^\alpha}\right)^\bullet \partial_{\overline{\lambda}}\right)(g_{\overline{\delta}\gamma} T_{\overline{\beta}}^{\gamma})$$
$$+ g^{\alpha\overline{\beta}}\partial_\alpha((g_{\overline{\delta}\gamma})^\bullet\, T_{\overline{\beta}}^{\gamma}) + g^{\alpha\overline{\beta}}\partial_\alpha(g_{\overline{\delta}\gamma}(T_{\overline{\beta}}^{\gamma})^\bullet) \equiv 0.$$

Using

$$0 = (L_v g)^{\alpha\overline{\beta}} = (g^{\alpha\overline{\beta}})^\bullet - g^{\alpha\overline{\sigma}}\left(\overline{\partial_\sigma w^\beta}\right)^\bullet - g^{\tau\overline{\beta}}(\partial_\tau w^\alpha)^\bullet$$

$$0 = (\mathbf{L}_v g)_{\bar\delta\gamma} = (g_{\bar\delta\gamma})^\bullet + g_{\bar\delta\tau}(\partial_\gamma w^\tau)^\bullet + g_{\bar\sigma\gamma}(\overline{\partial_\delta w^\sigma})^\bullet$$

$$(\mathbf{L}_v T)^\gamma_{\bar\beta} = (T^\gamma_{\bar\beta})^\bullet + T^\gamma_{\bar\sigma}(\overline{\partial_\beta w^\sigma})^\bullet - T^\tau_{\bar\beta}(\partial_\tau w^\gamma)^\bullet,$$

we get

$$(g^{\alpha\bar\sigma}(\overline{\partial_\sigma w^\beta})^\bullet + g^{\tau\bar\beta}(\partial_\tau w^\alpha)^\bullet)\partial_\alpha(g_{\bar\delta\gamma}T^\gamma_{\bar\beta}) + g^{\alpha\bar\beta}((\frac{\partial z^\lambda}{\partial w^\alpha})^\bullet\partial_\lambda + (\frac{\partial \bar z^\lambda}{\partial w^\alpha})^\bullet\partial_{\bar\lambda})(g_{\bar\delta\gamma}T^\gamma_{\bar\beta})$$

$$+ g^{\alpha\bar\beta}\partial_\alpha((-g_{\bar\delta\tau}(\partial_\gamma w^\tau)^\bullet - g_{\bar\sigma\gamma}(\overline{\partial_\delta w^\sigma})^\bullet)T^\gamma_{\bar\beta})$$

$$+ g^{\alpha\bar\beta}\partial_\alpha(g_{\bar\delta\gamma}((\mathbf{L}_v T)^\gamma_{\bar\beta} - T^\gamma_{\bar\sigma}(\overline{\partial_\beta w^\sigma})^\bullet + T^\tau_{\bar\beta}(\partial_\tau w^\gamma)^\bullet)) = 0$$

Multiplying out and using normal coordinates at the point under consideration and the fact that $(\frac{\partial z^\lambda}{\partial w^\sigma})^\bullet_{t=0} = -(\partial_\sigma w^\lambda)^\bullet_{t=0}$ and $(\frac{\partial z^\lambda}{\partial \bar w^\sigma})^\bullet_{t=0} = -(\partial_{\bar\sigma} w^\lambda)^\bullet_{t=0}$ (obtained by applying respectively $\frac{\partial}{\partial t_j}$ to $\delta^\lambda_\mu = \frac{\partial z^\lambda}{\partial w^\sigma}\partial_\mu w^\sigma + \frac{\partial z^\lambda}{\partial \bar w^\sigma}\partial_\mu \overline{w^\sigma}$ and $0 = \frac{\partial z^\lambda}{\partial \bar w^\sigma}\overline{\partial_\tau w^\sigma} + \frac{\partial z^\lambda}{\partial w^\sigma}\partial_{\bar\tau} w^\sigma$), we obtain

$$g^{\alpha\bar\sigma}(\overline{\partial_\sigma w^\beta})^\bullet g_{\bar\delta\gamma}\partial_\alpha T^\gamma_{\bar\beta} + g^{\tau\bar\beta}(\partial_\tau w^\alpha)^\bullet g_{\bar\delta\gamma}\partial_\alpha T^\gamma_{\bar\beta} - g^{\alpha\bar\beta}(\partial_\alpha w^\lambda)^\bullet g_{\bar\delta\gamma}\partial_\lambda T^\gamma_{\bar\beta}$$

$$- g^{\alpha\bar\beta}(\overline{\partial_\alpha w^\lambda})^\bullet g_{\bar\delta\gamma}\partial_{\bar\lambda} T^\gamma_{\bar\beta} - g^{\alpha\bar\beta}g_{\bar\delta\tau}\partial_\alpha(\partial_\gamma w^\tau)^\bullet T^\gamma_{\bar\beta} - g^{\alpha\bar\beta}g_{\bar\delta\tau}(\partial_\gamma w^\tau)^\bullet\partial_\alpha T^\gamma_{\bar\beta}$$

$$- g^{\alpha\bar\beta}g_{\bar\sigma\gamma}\partial_\alpha(\overline{\partial_\delta w^\sigma})^\bullet T^\gamma_{\bar\beta} - g^{\alpha\bar\beta}g_{\bar\sigma\gamma}(\overline{\partial_\delta w^\sigma})^\bullet\partial_\alpha T^\gamma_{\bar\beta} + g^{\alpha\bar\beta}g_{\bar\delta\gamma}\partial_\alpha(\mathbf{L}_v T)^\gamma_{\bar\beta}$$

$$- g^{\alpha\bar\beta}g_{\bar\delta\gamma}(\partial_\alpha T^\gamma_{\bar\sigma})(\overline{\partial_\beta w^\sigma})^\bullet - g^{\alpha\bar\beta}g_{\bar\delta\gamma}T^\gamma_{\bar\sigma}\partial_\alpha(\overline{\partial_\beta w^\sigma})^\bullet + g^{\alpha\bar\beta}g_{\bar\delta\gamma}(\partial_\alpha T^\gamma_{\bar\beta})(\partial_\tau w^\gamma)^\bullet$$
$$+ g^{\alpha\bar\beta}g_{\bar\delta\gamma}T^\tau_{\bar\beta}\partial_\alpha(\partial_\tau w^\gamma)^\bullet = 0.$$

The following pairs of terms on the left-hand side cancel out after a change in the dummy indices: the first and the tenth terms, the second and the third terms, the fifth and the thirteenth terms. Moreover, the sixth, the eighth, and the twelfth terms vanish because $g^{\alpha\overline{\beta}}\partial_\alpha T^\gamma_{\overline{\beta}} = 0$ due to the harmonicity of $T^\gamma_{\overline{\beta}}$ as a $T^{1,0}_{M_0}$-valued (0,1)-form on M_0. The eleventh term vanishes because

$$g^{\alpha\overline{\beta}}\partial_\alpha\overline{(\partial_\beta w^\sigma)^\bullet} = g^{\alpha\overline{\beta}}\partial_{\overline{\beta}}\overline{\partial_\alpha (w^\sigma)^\bullet} = \overline{g^{\overline{\alpha}\beta}\partial_\beta(\phi(\frac{\partial}{\partial t_j})^\sigma_{\overline{\alpha}})} = 0$$

due to the $\overline{\partial}^*$-closedness of $\phi(\frac{\partial}{\partial t_j})^\sigma_{\overline{\alpha}}$. Thus

$$g^{\alpha\overline{\beta}}g_{\overline{\delta}\gamma}\partial_\alpha(L_V T)^\gamma_{\overline{\beta}} = g^{\alpha\overline{\beta}}(\partial_\alpha \overline{w^\lambda})^\bullet g_{\overline{\delta}\gamma}\partial_{\overline{\lambda}}T^\gamma_{\overline{\beta}} + g^{\alpha\overline{\beta}}g_{\overline{\sigma}\gamma}\partial_\alpha\overline{(\partial_\delta w^\sigma)^\bullet}T^\gamma_{\overline{\beta}}.$$

For notational simplicity we let $S^\lambda_{\overline{\alpha}} = \phi(\frac{\partial}{\partial t_j})^\lambda_{\overline{\alpha}}$ which is equal to $\partial_{\overline{\alpha}}(w^\lambda)^\bullet$ at $t = 0$.

Let $S^{\overline{\lambda}}_\alpha = \overline{S^\lambda_{\overline{\alpha}}}$. Since

$$\partial_\alpha\overline{(\partial_\delta w^\sigma)^\bullet} = \partial_{\overline{\delta}}\partial_\alpha\overline{(w^\sigma)^\bullet} = \partial_{\overline{\delta}}S^{\overline{\sigma}}_\alpha,$$

it follows that in normal coordinates

$$g^{\alpha\overline{\beta}}\partial_\alpha(L_{\overline{V}_j}T)^\gamma_{\overline{\beta}} = g^{\alpha\overline{\beta}}S^{\overline{\lambda}}_\alpha\partial_{\overline{\lambda}}T^\gamma_{\overline{\beta}} + g^{\alpha\overline{\beta}}(\partial_{\overline{\gamma}}S^{\overline{\lambda}}_\alpha)T^\lambda_{\overline{\beta}}.$$

(4.6) Since $T^\beta_{\overline{\alpha}}$ is harmonic and in particular $\overline{\partial}$-closed, it follows that $\partial_{\overline{\gamma}}T^\beta_{\overline{\alpha}}$ is symmetric in α and γ. From (2.2) we know that $T_{\overline{\alpha}\overline{\beta}} = g_{\overline{\beta}\gamma}T^\gamma_{\overline{\alpha}}$ is symmetric in α and β. Hence in normal coordinates $\partial_{\overline{\alpha}}T_{\overline{\alpha}\overline{\beta}}$ is symmetric in α, β, γ. Let $(L_{\overline{V}_j}T)_{\overline{\alpha}\overline{\gamma}} = g_{\overline{\gamma}\beta}(L_{\overline{V}_j}T)^\beta_{\overline{\alpha}}$. Then in normal coordinates

$$g^{\alpha\overline{\beta}}\partial_\alpha(L_{\overline{V}_j}T)_{\overline{\beta}\overline{\gamma}} = g^{\alpha\overline{\beta}}S^{\overline{\lambda}}_\alpha\partial_{\overline{\gamma}}T_{\overline{\beta}\overline{\lambda}} + g^{\alpha\overline{\beta}}(\partial_{\overline{\gamma}}S^{\overline{\lambda}}_\alpha)T_{\overline{\beta}\overline{\lambda}}$$

$$= \partial_{\overline{\gamma}}(g^{\alpha\overline{\beta}}S^{\overline{\lambda}}_\alpha T_{\overline{\beta}\overline{\lambda}}$$

$$= \partial_{\bar\gamma}(S^{\bar\lambda}_{\alpha} T^{\alpha}_{\bar\lambda}).$$

Since $(L_{V_i} L_{\bar V_j} J)^{\beta}_{\bar\alpha} = 0$ from (3.2), it follows that $(L_{\bar V_j} T)^{\beta}_{\bar\alpha} = (L_{[\bar V_j, V_i]} J)^{\beta}_{\bar\alpha}$ is $\bar\partial$-exact. So there exists H^{β} such that $(L_{\bar V_j} T)^{\beta}_{\bar\alpha} = \partial_{\bar\alpha} H^{\beta}$.

Let $U_{ab} = g_{ac} T^{c}_{b}$. Since the Lie derivative of a symmetric tensor is again symmetric, $(L_{\bar V_j} U)_{ab}$ is symmetric in a and b. Since $(L_{\bar V_j} g)_{\bar\alpha\gamma} = 0$ and $T^{\bar\gamma}_{\bar\beta} = 0$, it follows that $(L_{\bar V_j} T)_{\bar\alpha\bar\beta} = (L_{\bar V_j} U)_{\bar\alpha\bar\beta}$ and $(L_{\bar V_j} T)_{\bar\alpha\bar\beta}$ is symmetric in α and β. Let $H_{\bar\beta} = g_{\alpha\bar\beta} H^{\alpha}$. Then $\nabla_{\bar\alpha} H_{\bar\beta} = (L_{\bar V_j} T)_{\bar\alpha\bar\beta}$ is symmetric in α and β. Hence

$$\begin{aligned}
g^{\alpha\bar\beta}\partial_{\alpha}(L_{\bar V_j} T)_{\bar\beta\bar\gamma} &= g^{\alpha\bar\beta}\nabla_{\alpha}\nabla_{\bar\beta} H_{\bar\gamma} \\
&= g^{\alpha\bar\beta}\nabla_{\alpha}\nabla_{\bar\gamma} H_{\bar\beta} \\
&= g^{\alpha\bar\beta}\nabla_{\bar\gamma}\nabla_{\alpha} H_{\bar\beta} - g^{\alpha\bar\beta} R_{\alpha\bar\gamma} H_{\bar\beta} \\
&= \nabla_{\bar\gamma}(\nabla_{\alpha} H^{\alpha}) - e\, H_{\bar\gamma}
\end{aligned}$$

and

$$\partial_{\bar\gamma}(S^{\bar\lambda}_{\alpha} T^{\alpha}_{\bar\lambda} - \nabla_{\alpha} H^{\alpha}) = -e\, H_{\bar\gamma}.$$

It follows that $H_{\bar\gamma}$ is $\bar\partial$-exact and we can write $H_{\bar\gamma} = \partial_{\bar\gamma}\tilde f$, where $\tilde f$ is a complex-valued smooth function on M_0. Let $\Box = -g^{\alpha\bar\beta}\nabla_{\alpha}\nabla_{\bar\beta}$ be the operator on smooth functions. Then

$$\partial_{\bar\gamma}(S^{\bar\lambda}_{\alpha} T^{\alpha}_{\bar\lambda} + (\Box + e)\tilde f) = 0$$

and $S^{\bar\lambda}_{\alpha} T^{\alpha}_{\bar\lambda} + (\Box + e)\tilde f$ is holomorphic on M_0 and therefore must be equal to a constant C on M_0. Let $f = \tilde f - \frac{C}{e}$. Then

$$S^{\bar\lambda}_{\alpha} T^{\alpha}_{\bar\lambda} + (\Box + e) f = 0$$

on M_0 and

$$f = -(\Box + e)^{-1}(S^{\bar\lambda}_{\alpha} T^{\alpha}_{\bar\lambda}).$$

It follows that

$$\nabla_{\lambda} H^{\lambda} = -\Box f = \Box(\Box + e)^{-1}(S^{\bar\lambda}_{\alpha} T^{\alpha}_{\bar\lambda}).$$

We are now ready to compute B_2.

$$B_2 = -\int_{M_0} (L_{\overline{v_j}} L_{v_i} J)^{\lambda}_{\overline{\alpha}} \overline{(L_{\overline{v_\ell}} L_{v_k} J)^{\mu}_{\overline{\beta}}}\, g^{\overline{\alpha}\beta} g_{\lambda\overline{\mu}}\, dV$$

$$= -\int_{M_0} (\partial_{\overline{\alpha}} H^{\lambda}) \overline{(L_{\overline{v_\ell}} L_{v_k} J)^{\mu}_{\overline{\beta}}}\, g^{\overline{\alpha}\beta} g_{\lambda\overline{\mu}}\, dV$$

$$= \int_{M_0} H^{\lambda}\, \overline{(\nabla_{\alpha}(L_{\overline{v_\ell}} L_{v_k} J)^{\mu}_{\overline{\beta}}\, g^{\alpha\overline{\beta}} g_{\overline{\lambda}\mu})}\, dV$$

$$= \int_{M_0} H^{\lambda}\, \overline{(\partial_{\overline{\lambda}}(2\sqrt{-1}\,\phi(\frac{\partial}{\partial \overline{t}_\ell})^{\overline{\beta}}_{\alpha}\phi(\frac{\partial}{\partial t_k})^{\alpha}_{\overline{\beta}}))}\, dV$$

$$= -\int_{M_0} (\nabla_{\lambda} H^{\lambda})\, \overline{(2\sqrt{-1}\,\phi(\frac{\partial}{\partial \overline{t}_\ell})^{\overline{\beta}}_{\alpha}\phi(\frac{\partial}{\partial t_k})^{\alpha}_{\overline{\beta}})}\, dV$$

$$= -\int_{M_0} (\square(\square + e)^{-1}(S^{\overline{\lambda}}_{\alpha} T^{\alpha}_{\overline{\lambda}}))\, \overline{(2\sqrt{-1}\,\phi(\frac{\partial}{\partial \overline{t}_\ell})^{\overline{\beta}}_{\alpha}\phi(\frac{\partial}{\partial t_k})^{\alpha}_{\overline{\beta}})}\, dV$$

$$= -\int_{M_0} (S^{\overline{\lambda}}_{\alpha} T^{\alpha}_{\overline{\lambda}}))\, \overline{(2\sqrt{-1}\,\phi(\frac{\partial}{\partial \overline{t}_\ell})^{\overline{\beta}}_{\alpha}\phi(\frac{\partial}{\partial t_k})^{\alpha}_{\overline{\beta}})}\, dV$$

$$+ e\int_{M_0} (\square + e)^{-1}(S^{\overline{\lambda}}_{\alpha} T^{\alpha}_{\overline{\lambda}})\, \overline{(2\sqrt{-1}\,\phi(\frac{\partial}{\partial \overline{t}_\ell})^{\overline{\beta}}_{\alpha}\phi(\frac{\partial}{\partial t_k})^{\alpha}_{\overline{\beta}})}\, dV$$

$$= -4\int_{M_0} \phi(\frac{\partial}{\partial \overline{t}_j})^{\overline{\delta}}_{\gamma}\phi(\frac{\partial}{\partial t_i})^{\gamma}_{\overline{\delta}}\phi(\frac{\partial}{\partial \overline{t}_k})^{\overline{\beta}}_{\alpha}\phi(\frac{\partial}{\partial t_\ell})^{\alpha}_{\overline{\beta}}\, dV$$

$$+ 4e\int_{M_0} ((\square + e)^{-1}(\phi(\frac{\partial}{\partial \overline{t}_j})^{\overline{\delta}}_{\gamma}\phi(\frac{\partial}{\partial t_i})^{\gamma}_{\overline{\delta}}))(\phi(\frac{\partial}{\partial \overline{t}_k})^{\overline{\beta}}_{\alpha}\phi(\frac{\partial}{\partial t_\ell})^{\alpha}_{\overline{\beta}})\, dV.$$

(4.7) Our final conclusion in the first part of the computation of the curvature tensor of the Weil-Petersson metric is

$$R^{(WP)}_{\ell\bar{k}i\bar{j}} = \int_{M_0} (L_{v_\ell} L_{v_i} J)^p_a \overline{(L_{v_k} L_{v_j} J)^q_b}\, g^{a\bar{b}} g_{p\bar{q}}\, dV$$

$$+ 4e \int_{M_0} ((\square + e)^{-1} (\phi(\tfrac{\partial}{\partial \bar{t}_j})^{\bar{\delta}}_{\gamma} \phi(\tfrac{\partial}{\partial t_i})^{\gamma}_{\bar{\delta}}))(\phi(\tfrac{\partial}{\partial \bar{t}_k})^{\bar{\beta}}_{\alpha} \phi(\tfrac{\partial}{\partial t_\ell})^{\alpha}_{\bar{\beta}})\, dV$$

$$+ 4e \int_{M_0} ((\square + e)^{-1} (\phi(\tfrac{\partial}{\partial \bar{t}_k})^{\bar{\delta}}_{\gamma} \phi(\tfrac{\partial}{\partial t_i})^{\gamma}_{\bar{\delta}}))(\phi(\tfrac{\partial}{\partial \bar{t}_j})^{\bar{\beta}}_{\alpha} \phi(\tfrac{\partial}{\partial t_\ell})^{\alpha}_{\bar{\beta}})\, dV$$

$$+ 8 \int_{M_0} \phi(\tfrac{\partial}{\partial t_\ell})^{\rho}_{\bar{\alpha}} \phi(\tfrac{\partial}{\partial \bar{t}_k})^{\bar{\lambda}}_{\rho} \phi(\tfrac{\partial}{\partial t_i})^{\tau}_{\bar{\lambda}} \phi(\tfrac{\partial}{\partial \bar{t}_j})^{\bar{\alpha}}_{\tau}\, dV$$

$$- 4 \int_{M_0} \phi(\tfrac{\partial}{\partial \bar{t}_j})^{\bar{\delta}}_{\gamma} \phi(\tfrac{\partial}{\partial t_i})^{\gamma}_{\bar{\delta}} \phi(\tfrac{\partial}{\partial \bar{t}_k})^{\bar{\beta}}_{\alpha} \phi(\tfrac{\partial}{\partial t_\ell})^{\alpha}_{\bar{\beta}}\, dV$$

$$- 4 \int_{M_0} \phi(\tfrac{\partial}{\partial \bar{t}_k})^{\bar{\delta}}_{\gamma} \phi(\tfrac{\partial}{\partial t_i})^{\gamma}_{\bar{\delta}} \phi(\tfrac{\partial}{\partial \bar{t}_j})^{\bar{\beta}}_{\alpha} \phi(\tfrac{\partial}{\partial t_\ell})^{\alpha}_{\bar{\beta}}\, dV.$$

Except in the one-dimensional case, this formula does not give negative holomorphic sectional curvature right away. This can be seen as follows. Let $i = j = k = \ell$. At any point under consideration we can assume that $g_{\alpha\bar{\beta}} = \delta_{\alpha\beta}$ (the Kronecker delta) and $\phi(\frac{\partial}{\partial t_i})_{\bar{\lambda}\bar{\rho}} = \delta_{\lambda\rho} \phi(\frac{\partial}{\partial t_i})_{\bar{\rho}\bar{\rho}}$. The integrands in the last three integrals can be expressed at that point as follows.

$$\phi(\tfrac{\partial}{\partial t_i})^{\rho}_{\bar{\alpha}} \phi(\tfrac{\partial}{\partial \bar{t}_i})^{\bar{\lambda}}_{\rho} \phi(\tfrac{\partial}{\partial t_i})^{\tau}_{\bar{\lambda}} \phi(\tfrac{\partial}{\partial \bar{t}_i})^{\bar{\alpha}}_{\tau} = \sum_{\alpha} \left| \phi(\tfrac{\partial}{\partial t_i})_{\bar{\alpha}\bar{\alpha}} \right|^4$$

$$\phi(\frac{\partial}{\partial \bar{t}_i})_{\gamma}^{\bar{\delta}} \phi(\frac{\partial}{\partial t_i})_{\bar{\delta}}^{\gamma} \phi(\frac{\partial}{\partial \bar{t}_i})_{\alpha}^{\bar{\beta}} \phi(\frac{\partial}{\partial t_i})_{\bar{\beta}}^{\alpha} = (\sum_{\alpha} \left| \phi(\frac{\partial}{\partial t_i})_{\bar{\alpha}\bar{\alpha}} \right|^2)^2.$$

The sum of the last three integrals becomes

$$8 \int_{M_0} \sum_{\alpha} \left| \phi(\frac{\partial}{\partial t_i})_{\bar{\alpha}\bar{\alpha}} \right|^4 - 8 \int_{M_0} (\sum_{\alpha} \left| \phi(\frac{\partial}{\partial t_i})_{\bar{\alpha}\bar{\alpha}} \right|^2)^2$$

which is negative unless $\{\phi(\frac{\partial}{\partial t_i})_{\bar{\alpha}\bar{\alpha}}\}_{\alpha=1,\ldots,n}$ has only one nonzero component at every point.

One way to get negative holomorphic sectional curvature is to use the integral

$$\int_{M_0} (L_{v_\ell} L_{v_i} J)_a^p \overline{(L_{v_k} L_{v_j} J)_b^q}\, g^{a\bar{b}} g_{p\bar{q}}\, dV$$

to help cancel the undesirable contribution from the last three integrals. In the case of Reimann surfaces and the Teichmüller space this integral vanishes. One can see it as follows. We choose normal coordinates $t_1,\ldots,t_N$ at the point $t = 0$ under consideration so that $\partial_i h_{j\bar{k}} = 0$ at that point. In the case of the Teichmüller space $(L_{v_1} J)_{\bar{\alpha}}^{\beta},\ldots,(L_{v_N} J)_{\bar{\alpha}}^{\beta}$ form a basis over $\mathbb{C}$ in the space of all harmonic $T_{M_0}^{1,0}$-valued (0,1)-forms on M_0. When M_0 is a Riemann surface every $T_{M_0}^{1,0}$-valued (0,1)-form is $\bar{\partial}$-closed. We will verify below that $(L_{v_i} L_{v_j} J)_{\bar{\alpha}}^{\beta}$ is always $\bar{\partial}^*$-closed. Hence $(L_{v_i} L_{v_j} J)_{\bar{\alpha}}^{\beta}$ is harmonic, but

$$0 = (\partial_{\bar{k}} h_{i\bar{j}})(0) = \int_{M_0} (L_{v_i} J)_{\bar{\alpha}}^{\lambda} \overline{(L_{v_k} L_{v_j} J)_{\bar{\beta}}^{\mu}}\, g^{\bar{\alpha}\beta} g_{\lambda\bar{\mu}}\, dV .$$

Thus, being perpendicular to a basis, $(L_{v_k} L_{v_j} J)_{\bar{\alpha}}^{\beta}$ is zero. So, for the case of the Teichmüller space the curvature tensor of the Weil-Petersson metric is simply

$$R^{(WP)}_{\ell\bar{k}i\bar{j}} = 4e\int_{M_0} ((\square + e)^{-1}(\phi(\frac{\partial}{\partial \bar{t}_j})^{\bar{\delta}}_{\gamma}\phi(\frac{\partial}{\partial t_i})^{\gamma}_{\bar{\delta}}))(\phi(\frac{\partial}{\partial \bar{t}_k})^{\bar{\beta}}_{\alpha}\phi(\frac{\partial}{\partial t_\ell})^{\alpha}_{\bar{\beta}})dV$$

$$+ 4e\int_{M_0} ((\square + e)^{-1}(\phi(\frac{\partial}{\partial \bar{t}_k})^{\bar{\delta}}_{\gamma}\phi(\frac{\partial}{\partial t_i})^{\gamma}_{\bar{\delta}}))(\phi(\frac{\partial}{\partial \bar{t}_j})^{\bar{\beta}}_{\alpha}\phi(\frac{\partial}{\partial t_\ell})^{\alpha}_{\bar{\beta}}) dV$$

This expression agrees with that obtained by Wolpert [12].

Now we return to the general case. We are going to verify that $(L_{v_i}L_{\bar{v}_j}J)^{\beta}_{\bar{\alpha}}$ is always $\bar{\partial}^*$-closed. In the computation in (4.5) if we use $v = v_j$ instead of $\bar{v}_j$, we get in normal coordinates

$$g^{\alpha\bar{\beta}} g_{\bar{\delta}\gamma}\,\partial_{\alpha}(L_v T)^{\gamma}_{\bar{\beta}} = g^{\alpha\bar{\beta}}(\partial_{\alpha}\overline{w^{\lambda}})^{\bullet}\, g_{\bar{\delta}\gamma}\,\partial_{\bar{\lambda}} T^{\gamma}_{\bar{\beta}} + g^{\alpha\bar{\beta}} g_{\bar{\sigma}\gamma}\partial_{\alpha}(\overline{\partial_{\delta} w^{\sigma}})^{\bullet}\, T^{\gamma}_{\bar{\beta}}$$

$$= g^{\alpha\bar{\beta}}\,\overline{\phi(\frac{\partial}{\partial \bar{t}_j})^{\lambda}_{\bar{\alpha}}}\, g_{\bar{\delta}\gamma}\partial_{\bar{\lambda}} T^{\gamma}_{\bar{\beta}} + g^{\alpha\bar{\beta}}\, g_{\bar{\sigma}\gamma}\,(\partial_{\bar{\delta}}\overline{\phi(\frac{\partial}{\partial \bar{t}_j})^{\sigma}_{\bar{\alpha}}})T^{\gamma}_{\bar{\beta}}$$

which vanishes because the only nonzero components of $\phi(\frac{\partial}{\partial \bar{t}_j})^{b}_{a}$ are of the form $\phi(\frac{\partial}{\partial \bar{t}_j})^{\bar{\beta}}_{\alpha}$. Here when we use $v = v_j$ instead of $\bar{v}_j$ the eleventh term $g^{\alpha\bar{\beta}}\partial_{\alpha}(\overline{\partial_{\beta} w^{\sigma}})^{\bullet}$ on the left-hand side of the equation in (4.5) vanishes not because of the $\bar{\partial}^*$-closedness of $\phi(\frac{\partial}{\partial t_j})^{\sigma}_{\bar{\alpha}}$ but because

$$g^{\alpha\bar{\beta}}\partial_{\alpha}(\overline{\partial_{\beta} w^{\sigma}})^{\bullet} = g^{\alpha\bar{\beta}}\partial_{\bar{\beta}}(\overline{\partial_{\bar{\alpha}} w^{\sigma}})^{\bullet} = g^{\alpha\bar{\beta}}\partial_{\bar{\beta}}(\overline{\phi(\frac{\partial}{\partial \bar{t}_j})^{\sigma}_{\bar{\alpha}}})$$

and $\phi(\frac{\partial}{\partial \bar{t}_j})^{\sigma}_{\bar{\alpha}} = 0$. Thus, $\bar{\partial}^*(L_{v_j}T) = 0$. That is, $(L_{v_j}L_{v_i}J)^{\beta}_{\bar{\alpha}}$ is $\bar{\partial}^*$-closed.

§5. Second Part of the Curvature Computation.

(5.1) We continue with the notations of §4 and set out to transform the first integral in the expression for $R^{(WP)}{}_{\ell\bar{k}i\bar{j}}$ in (4.7). We will do it first for the special case $i = j = k = \ell$ and then get the general case by polarization. As a first step in the transformation we compute $\bar\partial$ of $L_{v_i} L_{v_i} J$.

Since T is $\bar\partial$-closed, $\partial_{\bar\alpha} T^{\gamma}_{\bar\beta}$ is symmetric in α and β. That is,

$\frac{\partial z^\lambda}{\partial w^{\bar\alpha}}\partial_\lambda T^\gamma_{\bar\beta} + \left(\overline{\frac{\partial z^\lambda}{\partial w^\alpha}}\right)\partial_{\bar\lambda} T^\gamma_{\bar\beta}$ is symmetric in α and β. Let the dot $^\bullet$ in the superscript position denote $\frac{\partial}{\partial t_i}$ of a <u>scalar</u> function or a <u>component</u> of a tensor. Then

$$\left(\frac{\partial z^\lambda}{\partial w^{\bar\alpha}}\right)^\bullet \partial_\lambda T^\gamma_{\bar\beta} + \frac{\partial z^\lambda}{\partial w^{\bar\alpha}}\partial_\lambda (T^\gamma_{\bar\beta})^\bullet + \left(\overline{\frac{\partial z^\lambda}{\partial w^\alpha}}\right)^\bullet \partial_{\bar\lambda}T^\gamma_{\bar\beta} + \left(\overline{\frac{\partial z^\lambda}{\partial w^\alpha}}\right)\partial_{\bar\lambda}(T^\gamma_{\bar\beta})^\bullet$$

is symmetric in α and β. Using

$$(L_{v_i}T)^\gamma_{\bar\beta} = (T^\gamma_{\bar\beta})^\bullet + T^\gamma_{\bar\sigma}\overline{(\partial_\beta w^\sigma)}^\bullet - T^\tau_{\bar\beta}\partial_\tau (w^\gamma)^\bullet,$$

we conclude that at $t = 0$

$$-(\partial_{\bar\alpha} w^\lambda)^\bullet \partial_\lambda T^\gamma_{\bar\beta} - \overline{(\partial_\alpha w^\lambda)}^\bullet \partial_{\bar\lambda} T^\gamma_{\bar\beta} + \partial_{\bar\alpha}\left((L_{v_i}T)^\gamma_{\bar\beta} - T^\gamma_{\bar\sigma}\overline{(\partial_\beta w^\sigma)}^\bullet + T^\tau_{\bar\beta}\partial_\tau(w^\gamma)^\bullet\right)$$

is symmetric in α and β. Hence

$$\frac{\sqrt{-1}}{2}T^\lambda_{\bar\alpha}\partial_\lambda T^\gamma_{\bar\beta} - \overline{(\partial_\alpha w^\lambda)}^\bullet\partial_{\bar\lambda}T^\gamma_{\bar\beta} + \partial_{\bar\alpha}(L_{v_i}T)^\gamma_{\bar\beta} - (\partial_{\bar\alpha}T^\gamma_{\bar\sigma})\overline{(\partial_\beta w^\sigma)}^\bullet$$

$$- T^\gamma_{\bar\sigma}\overline{(\partial_\alpha\partial_\beta w^\sigma)}^\bullet + \partial_{\bar\alpha}T^\tau_{\bar\beta}(\partial_\tau w^\gamma)^\bullet + T^\tau_{\bar\beta}\partial_\tau\left(-\frac{\sqrt{-1}}{2}T^\gamma_{\bar\alpha}\right)$$

is symmetric in α and β. Since the sum of the second and the fourth term is symmetric in α and β and also both the fifth term and the sixth term are symmetric in α and β, by taking the skew-symmetrization we obtain

$$\partial_{\bar\alpha}(L_{v_i}T)^\gamma_{\bar\beta} - \partial_{\bar\beta}(L_{v_i}T)^\gamma_{\bar\alpha} = \sqrt{-1}(T^\tau_{\bar\beta}\partial_\tau T^\gamma_{\bar\alpha} - T^\tau_{\bar\alpha}\partial_\tau T^\gamma_{\bar\beta}).$$

(Note that one can also obtain this result by differentiating with respect to $\frac{\partial}{\partial t_i}$ the integrability condition $\bar\partial\Phi(t) = \frac{1}{2}[\Phi(t),\Phi(t)]$ and using $\Phi^* = \frac{1}{2\sqrt{-1}}T$, where $\Phi(t)$ is the $T_{M_0}^{1,0}$-valued (0,1)-form on M_0 such that a local function f is holomorphic on M_t if as a function on M_0 through the smooth trivialization f satisfies $(\bar\partial - \Phi(t)^\lambda\partial_\lambda)f = 0$ on M_0.) We now use normal coordinates. Since $\partial_\tau T^\tau_{\bar\beta} = 0$ due to the $\bar\partial^*$-closedness of T, it follows that

$$\partial_{\bar\alpha}(L_{v_i}T)^\gamma_{\bar\beta} - \partial_{\bar\beta}(L_{v_i}T)^\gamma_{\bar\alpha} = \sqrt{-1}\,\partial_\tau(T^\tau_{\bar\beta}T^\gamma_{\bar\alpha} - T^\tau_{\bar\alpha}T^\gamma_{\bar\beta}).$$

Let

$$\Theta_{\bar\beta\bar\alpha,\bar\tau\bar\gamma} = T_{\bar\beta\bar\tau}T_{\bar\alpha\bar\gamma} - T_{\bar\alpha\bar\tau}T_{\bar\beta\bar\gamma}$$

and let $\Theta^{\beta\alpha}{}_{\bar\tau\bar\gamma}$ be obtained from $\Theta_{\bar\beta\bar\alpha,\bar\tau\bar\gamma}$ by raising the first two indices. Then the last equation can be rewritten as

$$\bar D_1(L_{v_i}T) = \sqrt{-1}\,\bar D_2^*\,\Theta,$$

when

(i) $\bar D_1$ means applying $\bar\partial$ to the first index of $(L_{v_i}T)_{\bar\beta\bar\gamma}$ (which is $g_{\lambda\bar\gamma}(L_{v_i}T)^\lambda_{\bar\beta}$), i.e. regarding $(L_{v_i}T)^\gamma_{\bar\alpha}$ as a $T_{M_0}^{1,0}$-valued (0,1)-form, applying $\bar\partial$ and lowering the index γ,

(ii) $\bar D_2^*$ means applying $\bar\partial^*$ to $\Theta^{\beta\alpha}{}_{\bar\tau\bar\gamma}$ as a $\Lambda^2 T_{M_0}^{1,0}$-valued (0,2)-form and then lowering the indices β and α, or in other words $\bar\partial^*$ is applied to the second set of indices of $\Theta_{\bar\beta\bar\alpha,\bar\tau\bar\gamma}$.

(5.2) Let **X** denote the space of all tensors $\Xi_{\bar\alpha\bar\beta,\bar\gamma\bar\delta}$ satisfying the following three symmetry relations:

(i) $\Xi_{\bar\alpha\bar\beta,\bar\gamma\bar\delta} = -\Xi_{\bar\beta\bar\alpha,\bar\gamma\bar\delta}$ (skew-symmetry in the first two indices)

(ii) $\Xi_{\bar\alpha\bar\beta,\bar\gamma\bar\delta} = \Xi_{\bar\gamma\bar\delta,\bar\alpha\bar\beta}$ (symmetry in the two sets of double indices)

(iii) $\Xi_{\bar\alpha\bar\beta,\bar\gamma\bar\delta} + \Xi_{\bar\alpha\bar\gamma,\bar\delta\bar\beta} + \Xi_{\bar\alpha\bar\delta,\bar\beta\bar\gamma} = 0$ (vanishing of the sum from the cyclic permutation of the last three indices).

Simple direct verification shows that $\Theta_{\bar\alpha\bar\beta,\bar\gamma\bar\delta}$ belongs to **X**.

For $s = 1,2$, the operator $\bar D_s$ applied to a covariant tensor with two sets of skew-symmetric indices of antiholomorphic type (e.g. an element of **X**) means the operator $\bar\partial$ applied to the sth set of skew-symmetric indices. Let $\bar D^*_s$ denote the adjoint operator of $\bar D_s$ which is the same as applying $\bar\partial^*$ to the sth pair of indices. Let $\Box_s = \bar D^*_s \bar D_s + \bar D_s \bar D^*_s$ and let $\mathbf{H}_s$ denote the projection operator onto the kernel of $\Box_s$. Let $\mathbf{G}_s$ denote the Green's operator which is zero on Ker $\Box_s$ and equals the inverse of (the identity operator minus $\mathbf{H}_s$) on the orthogonal complement of Ker $\Box_s$.

For Ξ in **X** we have the following properties

(a) $\bar D_1 \bar D_2 \Xi = \bar D_2 \bar D_1 \Xi$

(b) $\bar D_2 \bar D_1^* \Xi = \bar D_1^* \bar D_2 \Xi$

(c) $\bar D_1^* \bar D_2^* \Xi = \bar D_2^* \bar D_1^* \Xi$

(d) $\bar D_1 \bar D_2^* \Xi = \bar D_2^* \bar D_1 \Xi$

(e) $\Box_1 \Xi$ belongs to **X**

(f) $\Box_1 \Xi = \Box_2 \Xi$

(g) if $\bar D_1 \Xi = 0$, then $[\bar D_2^*, \Box_1]\Xi = e\,\bar D^*_2 \Xi$, where $[\bar D_2^*, \Box_1]$ means the commutator $\bar D_2^* \Box_1 - \Box_1 \bar D_2^*$.

Properties (a) and (c) follow from simple straightforward computations. To prove property (b), by definition we have

$$([\bar D_2, \bar D_1^*]\Xi)_{\bar\alpha,\bar\beta_0\bar\beta_1\bar\beta_2} = \sum_{\nu=0}^{2} (-1)^{\nu+1} g^{\tau\bar\sigma} [\nabla_{\bar\beta_\nu}, \nabla_\tau] \Xi_{\bar\sigma\bar\alpha,\bar\beta_0\ldots\hat{\bar\beta}_\nu\ldots\bar\beta_2}$$

$$= ① + ② + ③$$

where $\hat{\beta}_\nu$ means that the index $\bar\beta_\nu$ is omitted and

$$① = \sum_{\nu=0}^{2} (-1)^{\nu+1} g^{\sigma\bar{\tau}} R_{\sigma\bar{\beta}_\nu} \Xi_{\bar{\tau}\bar{\alpha},\bar{\beta}_0\ldots\hat{\bar{\beta}}_\nu\ldots\bar{\beta}_2}$$

$$② = \sum_{\nu=0}^{2} (-1)^{\nu+1} R_{\bar{\beta}_\nu}{}^{\bar{\tau}}{}_{\alpha}{}^{\bar{\rho}} \Xi_{\bar{\tau}\bar{\rho},\bar{\beta}_0\ldots\hat{\bar{\beta}}_\nu\ldots\bar{\beta}_2}$$

$$③ = \sum_{\nu=0}^{2} (-1)^{\nu+1} \sum_{\substack{\mu=0\\ \mu\neq\nu}}^{2} R_{\bar{\beta}_\nu}{}^{\bar{\tau}}{}_{\beta_\mu}{}^{\bar{\rho}} \Xi_{\bar{\tau}\bar{\alpha},\bar{\beta}_0\ldots(\bar{\rho})_\mu\ldots\hat{\bar{\beta}}_\nu\ldots\bar{\beta}_2} ,$$

the subscript $(\bar{\rho})_\mu$ meaning that the subscript $\bar{\beta}_\mu$ is replaced by $\bar{\rho}$.

The term ① vanishes because $R_{\sigma\bar{\beta}_\nu} = e\, g_{\sigma\bar{\beta}_\nu}$ and because of the symmetry property (iii) of Ξ. The term ② vanishes because $R_{\bar{\beta}_\nu}{}^{\bar{\tau}}{}_{\alpha}{}^{\bar{\rho}}$ is symmetric in $\bar{\tau}$ and $\bar{\rho}$ whereas $\Xi_{\bar{\tau}\bar{\rho},\bar{\beta}_0\ldots\hat{\bar{\beta}}_\nu\ldots\bar{\beta}_2}$ is skew-symmetric in $\bar{\tau}$ and $\bar{\rho}$. The term ③ is the sum of six terms which can be grouped in three pairs so that the two terms in each pair cancel out because of the symmetry of $R_{\bar{\alpha}}{}^{\bar{\beta}}{}_{\bar{\gamma}}{}^{\bar{\delta}}$ in α and γ and of the skew-symmetry of $\Xi_{\bar{\alpha}\bar{\beta},\bar{\gamma}\bar{\delta}}$ in γ and δ. Property (d) is obtained from property (b) because of the symmetry property (ii) of Ξ.

To prove properties (e), (f) and (g) we compute explicitly $\square_1\Xi$ and obtain

$$\begin{aligned}(\square_1\Xi)_{\bar{\alpha}\bar{\beta},\bar{\lambda}\bar{\mu}} &= - g^{\sigma\bar{\tau}} \nabla_\sigma \nabla_{\bar{\tau}} \Xi_{\bar{\alpha}\bar{\beta},\bar{\lambda}\bar{\mu}} - g^{\sigma\bar{\tau}} [\nabla_{\bar{\alpha}}, \nabla_\sigma] \Xi_{\bar{\tau}\bar{\beta},\bar{\lambda}\bar{\mu}} \\ &\quad + g^{\sigma\bar{\tau}} [\nabla_{\bar{\beta}}, \nabla_\sigma] \Xi_{\bar{\tau}\bar{\alpha},\bar{\lambda}\bar{\mu}} \\ &= - g^{\sigma\bar{\tau}} \nabla_\sigma \nabla_{\bar{\tau}} \Xi_{\bar{\alpha}\bar{\beta},\bar{\lambda}\bar{\mu}} - 2e\, \Xi_{\bar{\alpha}\bar{\beta},\bar{\lambda}\bar{\mu}} \\ &\quad - R_{\bar{\alpha}}{}^{\bar{\tau}}{}_{\lambda}{}^{\bar{\rho}} \Xi_{\bar{\tau}\bar{\beta},\bar{\rho}\bar{\mu}} - R_{\bar{\beta}}{}^{\bar{\tau}}{}_{\mu}{}^{\bar{\rho}} \Xi_{\bar{\tau}\bar{\alpha},\bar{\rho}\bar{\lambda}} \\ &\quad + R_{\bar{\alpha}}{}^{\bar{\tau}}{}_{\mu}{}^{\bar{\rho}} \Xi_{\bar{\tau}\bar{\beta},\bar{\rho}\bar{\lambda}} + R_{\bar{\beta}}{}^{\bar{\tau}}{}_{\lambda}{}^{\bar{\rho}} \Xi_{\bar{\tau}\bar{\alpha},\bar{\rho}\bar{\mu}}\end{aligned}$$

where $R_{\alpha\bar{\beta}} = e\, g_{\alpha\bar{\beta}}$ is used. The skew-symmetry of $(\square_1\Xi)_{\bar{\alpha}\bar{\beta},\bar{\lambda}\bar{\mu}}$ in α and β and the symmetry of $(\square_1\Xi)_{\bar{\alpha}\bar{\beta},\bar{\lambda}\bar{\mu}}$ in (α,β) and (λ,μ) are clear from this expression. To get the symmetry property (iii) for $\square_1 \Xi$, we cyclically permute β,λ,μ and take the sum from the last four terms of the above expression and get

$$
\begin{aligned}
&- R_{\bar\alpha}{}^{\bar\tau}{}_{\bar\lambda}{}^{\bar\rho}\, \Xi_{\bar\tau\bar\beta,\bar\rho\bar\mu} - R_{\bar\beta}{}^{\bar\tau}{}_{\bar\mu}{}^{\bar\rho}\, \Xi_{\bar\tau\bar\alpha,\bar\rho\bar\lambda} + R_{\bar\alpha}{}^{\bar\tau}{}_{\bar\mu}{}^{\bar\rho}\, \Xi_{\bar\tau\bar\beta,\bar\rho\bar\lambda} + R_{\bar\beta}{}^{\bar\tau}{}_{\bar\lambda}{}^{\bar\rho}\, \Xi_{\bar\tau\bar\alpha,\bar\rho\bar\mu} \\
&- R_{\bar\alpha}{}^{\bar\tau}{}_{\bar\mu}{}^{\bar\rho}\, \Xi_{\bar\tau\bar\lambda,\bar\rho\bar\beta} - R_{\bar\lambda}{}^{\bar\tau}{}_{\bar\beta}{}^{\bar\rho}\, \Xi_{\bar\tau\bar\alpha,\bar\rho\bar\mu} + R_{\bar\alpha}{}^{\bar\tau}{}_{\bar\beta}{}^{\bar\rho}\, \Xi_{\bar\tau\bar\lambda,\bar\rho\bar\mu} + R_{\bar\lambda}{}^{\bar\tau}{}_{\bar\mu}{}^{\bar\rho}\, \Xi_{\bar\tau\bar\alpha,\bar\rho\bar\beta} \\
&- R_{\bar\alpha}{}^{\bar\tau}{}_{\bar\beta}{}^{\bar\rho}\, \Xi_{\bar\tau\bar\mu,\bar\rho\bar\lambda} - R_{\bar\mu}{}^{\bar\tau}{}_{\bar\lambda}{}^{\bar\rho}\, \Xi_{\bar\tau\bar\alpha,\bar\rho\bar\beta} + R_{\bar\alpha}{}^{\bar\tau}{}_{\bar\lambda}{}^{\bar\rho}\, \Xi_{\bar\tau\bar\mu,\bar\rho\bar\beta} + R_{\bar\mu}{}^{\bar\tau}{}_{\bar\beta}{}^{\bar\rho}\, \Xi_{\bar\tau\bar\alpha,\bar\rho\bar\lambda}
\end{aligned}
$$

where the two terms in each of the following six pairs cancel out: the first and the eleventh terms, the second and the twelfth terms, the third and the fifth terms, the fourth and the sixth terms, the seventh and the ninth terms, the eighth and the tenth terms.

Property (f) follows from property (e) because applying $\square_2$ to Ξ is equivalent to switching the two pairs of indices, applying $\square_1$, and then switching the two pairs of indices again.

perty (d), we have $\bar D_2{}^* \square_1 \Xi = \bar D_2{}^* \bar D_1{}^* \bar D_1 \Xi$ and $\square_1 \bar D_2{}^* \Xi = \bar D_1 \bar D_1{}^* \bar D_2{}^* \Xi$. Direct computations from definitions yield

$$(\bar D_2{}^* \bar D_1 \bar D_1{}^* \Xi)_{\bar\alpha\bar\beta,\bar\mu} = g^{\rho\bar\lambda} g^{\sigma\bar\tau} (\nabla_\rho \nabla_{\bar\alpha} \nabla_\sigma \Xi_{\bar\tau\bar\beta,\bar\lambda\bar\mu} - \nabla_\rho \nabla_{\bar\beta} \nabla_\sigma \Xi_{\bar\tau\bar\alpha,\bar\lambda\bar\mu})$$

$$(\bar D_1 \bar D_1{}^* \bar D_2{}^* \Xi)_{\bar\alpha\bar\beta,\bar\mu} = g^{\rho\bar\lambda} g^{\sigma\bar\tau} (-\nabla_{\bar\alpha} \nabla_\rho \nabla_\sigma \Xi_{\bar\tau\bar\beta,\bar\lambda\bar\mu} + \nabla_{\bar\beta} \nabla_\rho \nabla_\sigma \Xi_{\bar\tau\bar\alpha,\bar\lambda\bar\mu}).$$

Hence

$$([\bar D_2{}^*, \square_1] \Xi)_{\bar\alpha\bar\beta,\bar\mu} = g^{\rho\bar\lambda} [\nabla_{\bar\beta}, \nabla_\rho] (g^{\sigma\bar\tau} \nabla_\sigma \Xi_{\bar\tau\bar\alpha,\bar\lambda\bar\mu})$$

$$- \text{(the expression obtained by switching } \alpha \text{ and } \beta)$$

$$
\begin{aligned}
&= [R_{\bar\beta}{}^{\bar\lambda}{}_{\bar\alpha}{}^{\bar\rho} (g^{\sigma\bar\tau} \nabla_\sigma \Xi_{\bar\tau\bar\rho,\bar\lambda\bar\mu}) \\
&\quad + R_{\bar\beta}{}^{\bar\lambda}{}_{\bar\lambda}{}^{\bar\rho} (g^{\sigma\bar\tau} \nabla_\sigma \Xi_{\bar\tau\bar\alpha,\bar\rho\bar\mu}) + R_{\bar\beta}{}^{\bar\lambda}{}_{\bar\mu}{}^{\bar\rho} (g^{\sigma\bar\tau} \nabla_\sigma \Xi_{\bar\tau\bar\alpha,\bar\lambda\bar\rho})]
\end{aligned}
$$

$$- [\text{the expression obtained by switching } \alpha \text{ and } \beta]$$

$$= e\, g^{\sigma\bar{\tau}} \nabla_{\sigma}\, \Xi_{\bar{\tau}\bar{\alpha},\bar{\beta}\bar{\mu}}$$

- (the expression obtained by switching α and β),

because $R_{\bar{\beta}}{}^{\bar{\lambda}}{}_{\bar{\alpha}}{}^{\bar{\rho}} (g^{\sigma\bar{\tau}} \nabla_{\sigma} \Xi_{\bar{\tau}\bar{\rho},\bar{\lambda}\bar{\mu}})$ is symmetric in α and β and $R_{\bar{\beta}}{}^{\bar{\lambda}}{}_{\bar{\mu}}{}^{\bar{\rho}}(g^{\sigma\bar{\tau}} \nabla_{\sigma} \Xi_{\bar{\tau}\bar{\alpha},\bar{\lambda}\bar{\rho}})$ vanishes due to the symmetry of $R_{\bar{\beta}}{}^{\bar{\lambda}}{}_{\bar{\mu}}{}^{\bar{\rho}}$ in λ,ρ and the skew-symmetry of $\Xi_{\bar{\tau}\bar{\alpha},\bar{\rho}\bar{\lambda}}$ in λ,ρ. From symmetry property (iii) of Ξ we obtain

$$([\bar{D}_2^*, \Box_1]\, \Xi)_{\bar{\alpha}\bar{\beta},\bar{\mu}} = - e\, g^{\sigma\bar{\tau}} \nabla_{\sigma} \Xi_{\bar{\alpha}\bar{\beta},\bar{\tau}\bar{\mu}} = e(\bar{D}_2^* \Xi)_{\bar{\alpha}\bar{\beta},\bar{\mu}}.$$

Let $\mathbf{X}_0$ be the set of all Ξ in $\mathbf{X}$ such that $\bar{D}_1 \Xi = 0$. Since $\bar{D}_2^* \Box_1 - \Box_1 \bar{D}_2^* = e\, \bar{D}_2^*$ on $\mathbf{X}_0$ by property (g), it follows that $\bar{D}_2^* \Box_1 = (\Box_1 + e)\bar{D}_2^*$ on $\mathbf{X}_0$ and $(\Box_1+e)^{-1} \bar{D}_2^* \Box_1 = \bar{D}_2^*$ on $\mathbf{X}_0$.

Since $\Box_1 = \Box_2$ on $\mathbf{X}$, it follows that $(\Box_1+e)^{-1}\bar{D}_2^*\Box_2 = \bar{D}_2^*$ on $\mathbf{X}_0$. Moreover, we have $\mathbf{G}_1 = \mathbf{G}_2$ on $\mathbf{X}$ and hence $\bar{D}_1 \mathbf{G}_2 = \bar{D}_1 \mathbf{G}_1 = \mathbf{G}_1 \bar{D}_1 = 0$ on $\mathbf{X}_0$. Thus $\mathbf{G}_2$ maps $\mathbf{X}_0$ to itself and $(\Box_1+e)^{-1}\bar{D}_2^*\Box_2 \mathbf{G}_2 = \bar{D}_2^* \mathbf{G}_2$ on $\mathbf{X}_0$.

Since $\Box_2 \mathbf{G}_2 \Xi = \Xi - \mathbf{H}_2 \Xi$ and $\bar{D}_2^* \mathbf{H}_2 \Xi = 0$ for Ξ in $\mathbf{X}$, it follows that $\bar{D}_2^* \Box_2 \mathbf{G}_2 \Xi = \bar{D}_2^* \Xi$ and $(\Box_1+e)^{-1}\bar{D}_2^* \Xi = \bar{D}_2^* \mathbf{G}_2 \Xi$ for Ξ in $\mathbf{X}_0$.

(5.3) We now return to our tensor $\theta_{\bar{\beta}\bar{\alpha},\bar{\tau}\bar{\gamma}} = T_{\bar{\beta}\bar{\tau}} T_{\bar{\alpha}\bar{\gamma}} - T_{\bar{\alpha}\bar{\tau}} T_{\bar{\beta}\bar{\gamma}}$ arising from the formula $\bar{D}_1(\mathbf{L}_{V_i} T) = \sqrt{-1}\bar{D}_2^* \theta$. From the harmonicity of $T_{\bar{\alpha}}^{\beta}$ we conclude from straightforward computations that θ belongs to $\mathbf{X}_0$.

Let $\tilde{T}$ denote $\mathbf{L}_{V_i} T - \mathbf{H}_1(\mathbf{L}_{V_i} T)$. Since $\bar{D}_1(\mathbf{L}_{V_i} T) = \sqrt{-1}\bar{D}_2^* \theta$ and $\bar{D}_1^* \mathbf{L}_{V_i} T = 0$ (from the end of (4.7)), it follows that $\bar{D}_1 \tilde{\mathbf{T}} = \sqrt{-1}\bar{D}_2^* \theta$ and $\bar{D}_1^* \tilde{\mathbf{T}} = 0$. From $\mathbf{H}_1 \tilde{\mathbf{T}} = 0$ it follows that

$$\begin{aligned}\tilde{\mathbf{T}} &= (\bar{D}_1^* \bar{D}_1 + \bar{D}_1 \bar{D}_1^*)\mathbf{G}_1 \tilde{\mathbf{T}} \\ &= \bar{D}_1^* \mathbf{G}_1 \bar{D}_1 \tilde{\mathbf{T}} \\ &= \bar{D}_1^* \mathbf{G}_1(\sqrt{-1}\bar{D}_2^* \theta).\end{aligned}$$

We use the notation $(\cdot,\cdot)_{M_0}$ to denote the global inner product, for example,

$$(L_{v_i}T,L_{v_i}T)_{M_0} = \int_{M_0} (L_{v_i}T)^p_a \overline{(L_{v_i}T)^q_b} g^{a\bar{b}} g_{p\bar{q}} \, dV.$$

Thus

$$(L_{v_i}T,L_{v_i}T)_{M_0} = (HL_{v_i}T,HL_{v_i}T)_{M_0} + (\tilde{T},\tilde{T})_{M_0}.$$

We now use $\tilde{T} = \bar{D}_1{}^*G_1(\sqrt{-1}\bar{D}_2{}^*\theta)$ and $\bar{D}_1\tilde{T} = \sqrt{-1}\bar{D}_2{}^*\theta$ to transform $(\tilde{T},\tilde{T})_{M_0}$.

$$\begin{aligned}(\tilde{T},\tilde{T})_{M_0} &= (\bar{D}_1{}^*G_1(\sqrt{-1}\bar{D}_2{}^*\theta),\tilde{T})_{M_0}\\ &= (G_1(\sqrt{-1}\bar{D}_2{}^*\theta),\bar{D}_1\tilde{T})_{M_0}\\ &= (G_1\bar{D}_2{}^*\theta,\bar{D}_2{}^*\theta)_{M_0}\\ &= ((G_1 - (\Box_1+e)^{-1}\bar{D}_2{}^*\theta,\bar{D}_2{}^*\theta)_{M_0} + ((\Box_1+e)^{-1}\bar{D}_2{}^*\theta,\bar{D}_2{}^*\theta)_{M_0}\end{aligned}$$

From property (g) we obtain

$$\begin{aligned}(\tilde{T},\tilde{T})_{M_0} &= ((G_1 - (\Box_1+e)^{-1})\bar{D}_2{}^*\theta,\bar{D}_2{}^*\theta)_{M_0} + (\bar{D}_2{}^*G_2\theta,\bar{D}_2{}^*\theta)_{M_0}\\ &= ((G_1 - (\Box_1+e)^{-1})\bar{D}_2{}^*\theta,\bar{D}_2{}^*\theta)_{M_0} + (\Box_2G_2\theta,\theta)_{M_0}\\ &= ((G_1 - (\Box_1+e)^{-1})\bar{D}_2{}^*\theta,\bar{D}_2{}^*\theta)_{M_0} + (\theta,\theta)_{M_0} - (H_2\theta,H_2\theta)_{M_0},\end{aligned}$$

where to get the second last expression we have used $\bar{D}_2\theta = 0$ which follows from $\bar{D}_1\theta = 0$ and the symmetry property (ii) of θ. We now compute $(\theta,\theta)_{M_0}$.

$$(\theta,\theta)_{M_0} = \frac{1}{4}\int_{M_0} (T^{\lambda}_{\bar{\alpha}}T^{\mu}_{\bar{\beta}} - T^{\lambda}_{\bar{\beta}}T^{\mu}_{\bar{\alpha}}) \overline{(T^{\bar{\alpha}}_{\lambda}T^{\bar{\beta}}_{\mu} - T^{\bar{\beta}}_{\lambda}T^{\bar{\alpha}}_{\mu})} \, dV$$

The reason for the factor $\frac{1}{4}$ is that for

$$A = \frac{1}{p!q!}\sum A_{\bar{\alpha}_1\ldots\bar{\alpha}_p,\bar{\beta}_1\ldots\bar{\beta}_q}(dz^{\bar{\alpha}_1}\wedge\ldots\wedge dz^{\bar{\alpha}_p}) \otimes (dz^{\bar{\beta}_1}\wedge\ldots\wedge dz^{\bar{\beta}_q}),$$

one has

$$(A,A)_{M_0} = \frac{1}{p!q!}\int_{M_0} A_{\bar{\alpha}_1\ldots\bar{\alpha}_p,\bar{\beta}_1\ldots\bar{\beta}_q} \overline{A^{\bar{\alpha}_1\ldots\bar{\alpha}_p,\bar{\beta}_1\ldots\bar{\beta}_q}} \, dV$$

and in the formulae we have been using for $\bar{D}_1, \bar{D}_1^*, \bar{D}_2, \bar{D}_2^*$, the components $\theta_{\overline{\alpha\beta},\overline{\lambda\mu}}$ correspond to $A_{\bar{\alpha}_1\ldots\bar{\alpha}_p,\bar{\beta}_1\ldots\bar{\beta}_q}$ with $p = q = 2$. Thus, when we choose local coordinates so that

$$T_{\overline{\alpha\beta}} = 2\sqrt{-1}\phi(\frac{\partial}{\partial t_i})_{\overline{\alpha\beta}} = 2\sqrt{-1}\delta_{\alpha\beta}\phi(\frac{\partial}{\partial t_i})_{\overline{\alpha\alpha}}$$

at the point under consideration, we have

$$(\theta,\theta)_{M_0} = 8\int_{M_0}(\sum_{\alpha}\left|\phi(\frac{\partial}{\partial t_i})_{\overline{\alpha\alpha}}\right|^2)^2 - 8\int_{M_0}\sum_{\alpha}\left|\phi(\frac{\partial}{\partial t_i})_{\overline{\alpha\alpha}}\right|^4.$$

Thus $R^{(WP)}_{i\bar{i}i\bar{i}}$ at $t = 0$ becomes

$$R^{(WP)}_{i\bar{i}i\bar{i}} = 8e\int_{M_0}(((\Box_1 + e)^{-1}(\phi(\frac{\partial}{\partial t_i})^{\beta}_{\bar{\alpha}}\,\overline{\phi(\frac{\partial}{\partial \bar{t}_i})^{\bar{\alpha}}_{\beta}}))(\phi(\frac{\partial}{\partial t_i})^{\beta}_{\bar{\alpha}}\,\overline{\phi(\frac{\partial}{\partial \bar{t}_i})^{\bar{\alpha}}_{\beta}})\,dV$$

$$+ ((G_1 - (\Box_1 + e)^{-1})\bar{D}_2^*\theta, \bar{D}_2^*\theta)_{M_0}$$

$$+ (HL_{v_i}L_{v_i}J, HL_{v_i}L_{v_i}J)_{M_0} - (H_2\theta, H_2\theta)_{M_0}.$$

Here we drop the subscript 1 for H in $HL_{v_i}L_{v_i}J$ because we regard $L_{v_i}L_{v_i}J$ as a $T_{M_0}^{1,0}$-valued (0,1)-form on M_0 and there is no need to distinguish between two different kinds of covariant indices. We will drop subscripts in similar circumstances later without further explicit mention.

We now further simplify the second and the third terms on the right-hand side of the above expression for $R^{(WP)}_{i\bar{i}i\bar{i}}$.

$$((G_1 - (\Box_1 + e)^{-1})\bar{D}_2^*\theta, \bar{D}_2^*\theta)_{M_0} = ((G_1 - (\Box_1 + e)^{-1})\bar{D}_1 L_{v_i} T, \bar{D}_1 L_{v_i} T)_{M_0}$$

$$= ((G - (\Box_1 + e)^{-1})\bar{\partial} L_{v_i}L_{v_i}J, \bar{\partial} L_{v_i}L_{v_i}J)_{M_0}$$

(with $L_{v_i}L_{v_i}J$ regarded as $T_{M_0}^{1,0}$-valued (0,1)-form)

$$= (\bar{\partial}(G - (\Box_1 + e)^{-1})L_{v_i}L_{v_i}J, \bar{\partial} L_{v_i}L_{v_i}J)_{M_0}$$

$$= (\overline{\partial}^{*}\overline{\partial}(G - (\Box_1 + e)^{-1})L_{v_i}L_{v_i}J, L_{v_i}L_{v_i}J)_{M_0}$$

$$= (\Box\,(G - (\Box_1 + e)^{-1})L_{v_i}L_{v_i}J, L_{v_i}L_{v_i}J)_{M_0}$$

(because $L_{v_i}L_{v_i}J$ is $\overline{\partial}^{*}$-closed from the end of (4.7))

$$= (\Box G L_{v_i}L_{v_i}J, L_{v_i}L_{v_i}J)_{M_0}$$

$$- (\Box(\Box_1 + e)^{-1}L_{v_i}L_{v_i}J, L_{v_i}L_{v_i}J)_{M_0}$$

$$= (L_{v_i}L_{v_i}J, L_{v_i}L_{v_i}J)_{M_0} - (HL_{v_i}L_{v_i}J, HL_{v_i}L_{v_i}J)_{M_0}$$

$$- (\Box(\Box_1 + e)^{-1}L_{v_i}L_{v_i}J, L_{v_i}L_{v_i}J)_{M_0}$$

(because $HL_{v_i}L_{v_i}J$ has the same inner product with $L_{v_i}L_{v_i}J$ as with $HL_{v_i}L_{v_i}J$)

$$= e((\Box_1 + e)^{-1}L_{v_i}L_{v_i}J, L_{v_i}L_{v_i}J)_{M_0}$$

$$- (HL_{v_i}L_{v_i}J, HL_{v_i}L_{v_i}J)_{M_0}.$$

The final expression for $R^{(WP)}{}_{i\overline{i}i\overline{i}}$ at $t = 0$ becomes

$$R^{(WP)}{}_{i\overline{i}i\overline{i}} = 8e \int_{M_0} ((\Box_1 + e)^{-1}(\phi(\tfrac{\partial}{\partial t_i})^{\beta}_{\alpha}\phi(\tfrac{\partial}{\partial \overline{t}_i})^{\overline{\alpha}}_{\beta}))(\phi(\tfrac{\partial}{\partial t_i})^{\beta}_{\alpha}\phi(\tfrac{\partial}{\partial \overline{t}_i})^{\overline{\alpha}}_{\beta})\, dV$$

$$+ e\,((\Box_1 + e)^{-1}L_{v_i}L_{v_i}J, L_{v_i}L_{v_i}J)_{M_0}$$

$$- (H_2\theta, H_2\theta)_{M_0}$$

(5.4) We now polarize the expression for $R^{(WP)}{}_{i\overline{i}i\overline{i}}$ to get the expression for $R^{(WP)}{}_{i\overline{j}k\overline{\ell}}$. We introduce the following notations to make the expression simpler. Let $<\phi(\frac{\partial}{\partial t_i}), \phi(\frac{\partial}{\partial t_j})>$ denote the function on M_0 which is the <u>pointwise</u> inner product of the $T_{M_0}^{1,0}$-valued (0,1)-forms $\phi(\frac{\partial}{\partial t_i}), \phi(\frac{\partial}{\partial t_j})$. Let $\phi(\frac{\partial}{\partial t_i}) \wedge \phi(\frac{\partial}{\partial t_k})$ denote the $\Lambda^2 T_{M_0}^{1,0}$-valued (0,2)-form with components

$$(\phi(\frac{\partial}{\partial t_i})) \wedge \phi(\frac{\partial}{\partial t_k})^{\lambda\mu}_{\bar{\alpha}\bar{\beta}} = \frac{1}{2}\,[\phi(\frac{\partial}{\partial t_i})^{\lambda}_{\bar{\alpha}}\phi(\frac{\partial}{\partial t_k})^{\mu}_{\bar{\beta}} - \phi(\frac{\partial}{\partial t_i})^{\lambda}_{\bar{\beta}}\phi(\frac{\partial}{\partial t_k})^{\mu}_{\bar{\alpha}}$$

$$+ \phi(\frac{\partial}{\partial t_k})^{\lambda}_{\bar{\alpha}}\phi(\frac{\partial}{\partial t_i})^{\mu}_{\bar{\beta}} - \phi(\frac{\partial}{\partial t_k})^{\lambda}_{\bar{\beta}}\phi(\frac{\partial}{\partial t_i})^{\mu}_{\bar{\alpha}}].$$

Then at $t = 0$

$$R^{(WP)}_{i\bar{j}k\bar{\ell}} = 4e\,((\square_1 + e)^{-1} \langle\phi(\frac{\partial}{\partial t_i}),\phi(\frac{\partial}{\partial t_j})\rangle, \langle\phi(\frac{\partial}{\partial t_k}),\phi(\frac{\partial}{\partial t_\ell})\rangle)_{M_0}$$

$$+ 4e\,((\square_1 + e)^{-1}\langle\phi(\frac{\partial}{\partial t_k}),\phi(\frac{\partial}{\partial t_j})\rangle, \langle\phi(\frac{\partial}{\partial t_i}),\phi(\frac{\partial}{\partial t_\ell})\rangle)_{M_0}$$

$$+ e\,((\square_1 + e)^{-1} L_{v_i} L_{v_k} J, L_{v_j} L_{v_\ell} J)_{M_0}$$

$$- (H(\phi(\frac{\partial}{\partial t_i}) \wedge \phi(\frac{\partial}{\partial t_k})), H(\phi(\frac{\partial}{\partial t_j}) \wedge \phi(\frac{\partial}{\partial t_\ell})))_{M_0}.$$

Here we have used the fact that $L_{[v_j, v_\ell]} J$ is a $\bar{\partial}$-exact and $L_{v_i} L_{v_k} J$ is $\bar{\partial}^*$-closed and the fact that $(\square_1 + e)^{-1}$ is self-adjoint.

Let Ψ: $H^1(M_0, T_{M_0}) \times H^1(M_0, T_{M_0}) \to H^2(M_0, \Lambda^2 T_{M_0})$ be defined by taking the skew-symmmetric part of the tensor product. That is, if ξ, η are $\bar{\partial}$-closed $T_{M_0}^{1,0}$-valued (0,1)-forms on M_0 defining the classes $[\xi], [\eta]$ in $H^1(M_0, T_{M_0})$, then $\Psi([\xi], [\eta])$ is defined by the $\bar{\partial}$-closed $\Lambda^2 T_{M_0}^{1,0}$-valued (0,2)-form $\xi \wedge \eta$ with components

$$(\xi \wedge \eta)^{\lambda\mu}_{\bar{\alpha}\bar{\beta}} = \frac{1}{2}\,(\xi^{\lambda}_{\bar{\alpha}}\eta^{\mu}_{\bar{\beta}} - \xi^{\lambda}_{\bar{\beta}}\eta^{\mu}_{\bar{\alpha}} + \eta^{\lambda}_{\bar{\alpha}}\xi^{\mu}_{\bar{\beta}} - \eta^{\lambda}_{\bar{\beta}}\xi^{\mu}_{\bar{\alpha}}).$$

(Straightforward verification shows that $\xi \wedge \eta$ is always $\bar{\partial}$-closed.)

We can now state our result on the negativity of the bisectional curvature of the Weil-Petersson metric.

(5.5) <u>Theorem.</u> <u>Let</u> π: $M \to \Omega$ <u>be a holomorphic family of compact complex manifolds parameterized by an open neighborhood</u> Ω <u>of</u> O <u>in</u> $\mathbb{C}^N$. <u>Assume that each fiber</u>

$M_t = \pi^{-1}(t)$ carries a Kähler-Einstein metric whose Ricci curvature is equal to the metric times a fixed negative constant. Assume that the element of $H^1(M_t, T_{M_t})$ defined by any nonzero tangent vector of Ω of type (0,1) at t is nonzero. Let X_1, X_2 be two tangent vectors of type (0,1) on Ω at 0 and ψ_1, ψ_2 be the elements of $H^1(M_0, T_{M_0})$ defined respectively by X_1, X_2. If the element $\Psi(\psi_1, \psi_2)$ of $H^2(M_0, \Lambda^2 T_{M_0})$ defined by taking the skew-symmetrization of the product of ψ_1 and ψ_2 is zero, then the holomorphic bisectional curvature of the Weil-Petersson metric in the directions of X_1, X_2 is negative.

To prove this theorem, without loss of generality we can assume that $X_1 = \frac{\partial}{\partial t_i}$, $X_2 = \frac{\partial}{\partial t_j}$, where $t_1, \ldots, t_N$ are the coordinates of $\mathbb{C}^N$. Then the vanishing of $\Psi(\psi_1, \psi_2)$ implies that $H(\phi(\frac{\partial}{\partial t_i}) \wedge \phi(\frac{\partial}{\partial t_j})) = 0$ and

$$R^{(WP)}_{i\bar{i}j\bar{j}} = 8e\,(\Box_1 + e)^{-1} \langle \phi(\frac{\partial}{\partial t_i}), \phi(\frac{\partial}{\partial t_i})\rangle, \langle \phi(\frac{\partial}{\partial t_j}), \phi(\frac{\partial}{\partial t_j})\rangle)_{M_0}$$

$$+ e\,((\Box_1 + e)^{-1} L_{v_i} L_{v_j} J, L_{v_i} L_{v_j} J)_{M_0}.$$

Since the operator $(\Box_1 + e)^{-1}$ is positive, it suffices to prove that the first term on the right-hand side is positive. First we show the following: for any nonnegative-valued smooth function f on M_0, $(\Box_1 + e)^{-1} f$ is also a nonnegative-valued function. Suppose the contrary. Let $g = (\Box_1 + e)^{-1} f$. Then the infimum $-A$ of g is a negative number and it is achieved at some point P of M_0 (note that $\Box$ is a real operator so that g is real-valued). From $(\Box g)(P) \leq 0$ we conclude that

$$-eA = eg(P) \geq ((\Box_1 + e)g)(P) = f(P),$$

contradicting the fact that $f(P) \geq 0$.

We now apply it to the function $f = \langle \phi(\frac{\partial}{\partial t_i}), \phi(\frac{\partial}{\partial t_i})\rangle$. Since a Kähler-Einstein metric is real-analytic, it follows from the harmonicity of $\phi(\frac{\partial}{\partial t_i})$ that f is real-analytic and $(\Box_1 + e)^{-1} f$ is also real-analytic. From the nonnegativity of the two non-identically-zero real-analytic functions $(\square_1 + e)^{-1}f$ and $\langle\phi(\frac{\partial}{\partial t_j}),\phi(\frac{\partial}{\partial t_j})\rangle$, we conclude that $((\square_1 + e)^{-1}f, \langle\phi(\frac{\partial}{\partial t_j}),\phi(\frac{\partial}{\partial t_j})\rangle)_{M_0}$ is positive.

REFERENCES

1. L. Ahlfors, Some remarks on Teichmüller's space of Riemann surfaces, Ann. of Math. 74 (1961), 171-191.

2. L. Ahlfors, Curvature properties of Teichmüller space, J. Analyse Math. 9 (1961), 161-176.

3. N. Hitchin, Compact four-dimensional Einstein manifolds, J. Diff. Geom. 9 (1974), 435-441.

4. K. Kodaira and J. Morrow, Complex Manifolds, New York: Holt, Reinhardt and Winston, 1971.

5. K. Kodaira, L. Nirenberg, and D.C. Spencer, On the existence of deformation of complex analytic structures, Ann. of Math. 68 (1958), 450-459.

6. N. Koiso, Einstein metrics and complex structures, Invent. Math. 73 (1983), 71-106.

7. H. Maass, Über eine neue Art von nichtanalytischen automorphen Funktionen, Math. Ann. 121 (1949), 141-183.

8. H. Royden, Intrinsic metrics on Teichmüller spaces, Proceedings International Congress Math. 2 (1974), 217-221.

9. H. Royden, Oral communication, detailed paper in preparation.

10. G. Schumacher, On the geometry of moduli spaces. Preprint 1984.

11. Y.-T. Siu, Complex-analyticity of harmonic maps, vanishing and Lefschetz theorems, J. Diff. Geom. 17 (1982), 55-138.

12. S. Wolpert, Chern forms and the Riemann tensor for the moduli space of curves. Preprint 1984.

13. S.-T. Yau, On the Ricci curvature of a compact Kähler manifold and the complex Monge-Ampère equation, I. Comm. Pure Applied Math. 31 (1978), 339-411.

Extension Problems and Positive Currents in Complex Analysis

H. SKODA

Université de Paris VI
4 Place Jussieu 75230 PARIS CEDEX 05

SUMMARY

This paper is a survey of recent developments in the theory of the extension of analytic sets and closed, positive currents.

INTRODUCTION

Since the fundamental works of P. Lelong ([17] and [19]) and Y.T. Siu [22], it is well known that positive closed currents are a very good generalization of the notion of analytic set because the analytic sets are exactly the sets of density of positive closed currents and because many properties of the analytic set X are in fact properties of the associated current of integration [X] and also because the Lelong-Poincaré equation of currents $\frac{i}{\pi} \partial\bar{\partial} \operatorname{Log}|f| = [X]$ gives a simple fundamental relation between the function F and the analytic hypersurface $X = f^{-1}(0)$ in the case of codimension one [18]. It is natural to ask if the basic theorems about the analytic sets are really theorems of analytic geometry involving the hard structure of analytic set or if these theorems are in fact theorems of complex analysis only involving the soft structure of positive closed current.

Since the Remmert-Stein extension theorem for currents of Y.T. Siu [22] and R. Harvey [12] it became clear that the extension of closed, positive current was a natural generalization of the extension of analytic set. In this paper, we shall make a survey of recent theorems of extension of analytic objects, essentialy of closed, positive currents. Our main reference will be the very interesting paper of N. Sibony [21].

1. EXTENSION OF ANALYTIC SETS AND OF CLOSED POSITIVE CURRENTS.

Perhaps the first motivation of these problems of extension of closed, positive currents across an exceptional set was to obtain a better understanding of the following classical result of E. Bishop (1964, [4]) :

Theorem 1 : Let A be a subvariety of the complex hermitian manifold Ω and X be a subvariety of $\Omega \setminus A$, of pure dimension p, such that for all compact set K of Ω :

$$\mathrm{vol}_{2p}(X \cap K) < +\infty$$

(where vol_{2p} means the euclidean volume of real dimension $2p$) then $\bar{X}$ is a subvariety of Ω (of pure dimension p).

Of course the main interest of the theorem is to give a necessary and sufficient condition of extension of X across A in the case $\dim X \leqslant \dim A$ which is not covered by the Remmert Stein theorem. The special case where $\dim X = \dim A$, is due to W. Stoll [25]. In 1964, W. Stoll [26] gave the following beautiful characterization of algebraic sets which is closely connected with the Bishop's theorem.

Theorem 2 : A subvariety X in $\mathbb{C}^n$, of pure dimension p, is algebraic if and only if there exists a constant $C > 0$ such that for all $r \gg 0$:

$$\mathrm{vol}_{2p}(X \cap B(0,r)) \leq Cr^{2p}$$

(where $B(0,r)$ is the euclidean ball of center 0, radius r).

The proof of W. Stoll was independent of Bishop's theorem. He used the value distribution theory of holomorphic maps. One can now give the following short proof : $\mathbb{C}^n$ is isomorphic to $\mathbb{P}_n \setminus \mathbb{P}_{n-1}$ (where $\mathbb{P}_{n-1}$ is the hyperplane to the infinite), the assumption about the growth of the volume of X in $\mathbb{C}^n$ means that X has finite volume for the Fubini metric on $\mathbb{P}_n$, therefore the Bishop's theorem (with $\Omega = \mathbb{P}_n$, $A = \mathbb{P}_{n-1}$) claims that the closure $\bar{X}$ of X in $\mathbb{P}_n$ is a subvariety of $\mathbb{P}_n$. Then Chow's theorem proves that $\bar{X}$ and X are algebraic.

The Bishop's theorem has a nice generalization to the case of closed, positive currents.

Theorem 3 : Let A be a closed complete pluripolar subset of the complex hermitian manifold Ω. If T is a closed, positive current on $\Omega \setminus A$ and if T has locally finite mass near A, then the trivial extension $\tilde{T}$ of T to Ω is a closed, positive current on Ω.

(A is complete pluripolar in Ω means that $A = \{z \in \Omega, u(z) = -\infty\}$ where $u \in PSH(\Omega)$, i.e. u is plurisubharmonic in Ω). R. Harvey and J. Polking proved the theorem in 1974 [15] when T is of bidimension (p,p) and A is an analytic subset of dimension p. I proved the theorem in 1981 [24] when A is an analytic subset of Ω and T is an arbitrary positive, closed current. My proof was quite different from that of R. Harvey and J. Polking. I directly worked with the current T instead of the plurisubharmonic potential associated to T by R. Harvey and J. Polking. In 1982, H. El Mir [8][9] generalized my method to the case where A is only closed, pluripolar. Nevertheless, these two proofs used technical estimates of the distribution of the mass of T in some neighborhood of A. Quite recently, N. Sibony [21] found a very nice formalization of these proofs. I shall now explain the Sibony's proof.
For the sake of simplicity, we suppose, T is of bidimension (1,1) (it is the decisive case) and T has finite mass in $\Omega \setminus A$ (we shrink Ω if necessary, the problem is local relatively to A). The two following lemmas give the basic information about the distribution of the mass of the current T near A.

<u>Lemma 1</u> : (Essentially a variation of a Chern-Levine-Nirenberg estimate) (One only needs to suppose that A is closed). Let K be a compact subset of Ω. There exists a constante $C(K,\Omega)$ such that for all $u \in \mathcal{C}^\infty(\Omega) \cap PSH(\Omega)$ such that u is ≥ 0, is bounded and vanishes in some neighborhood of A (depending on u), we have :

$$\int_K T \wedge i\,\partial\bar{\partial}u \leq C(K,\Omega).\|T\|\,(\Omega \setminus A).\|u\|_\infty$$

where $\|T\|$ is the mass-measure of the current T.

Proof :

Let be $\varphi \in \mathcal{E}^{\infty}(\Omega)$ a test-function with compact support in Ω such that : $\varphi \equiv 1$ on K and $0 \leq \varphi \leq 1$. Because $u \equiv 0$ is some neighborhood of A, the current $T \wedge i\,\partial\bar\partial u$ is well defined and is positive (because $u \in PSH(\Omega)$ and T is positive). Therefore, we have the inequality :

$$\int_K T \wedge i\,\partial\bar\partial u \leq \int_\Omega \varphi\, T \wedge i\,\partial\bar\partial u$$

Because T is closed, an integration by parts proves that :

$$\int_\Omega \varphi\, T \wedge i\partial\bar\partial u = \int_\Omega i\partial\bar\partial(\varphi\, T)u = \int_\Omega u\, T \wedge i\,\partial\bar\partial\,\varphi.$$

Then we have obviously :

$$\int_K T \wedge i\,\partial\bar\partial u \leq \left| \int_\Omega u T \wedge i\,\partial\bar\partial\varphi \right| \leq \|i\,\partial\bar\partial u\|_\infty\ \|T\|\ (\Omega \setminus A)\ \|u\|_\infty$$

Lemma 2 :

With the same hypothesis as in lemma 1, we have :

$$\int_{K \cap \{a \leq u(z) \leq b\}} T \wedge i\,\partial u \wedge \bar\partial u \leq C(K,\Omega)\ (b-a)\ \|T\|\ (\Omega \setminus A)\ \ \|u\|_\infty$$

for all a and b, $0 < a < b < 1$ and for all $u \in \mathcal{E}^\infty(\Omega) \cap PSH(\Omega)$ such that u is positive bounded and vanishes in some neighborhood of A.

<u>Proof</u> : For fixed a and b let $h : \mathbb{R} \to \mathbb{R}$ be a convex increasing function of class C^∞, vanishing in some neighborhood of o, such that :

$$h''(t) \geq \frac{1}{b-a} \mathbb{1}_{[a,b]} ,$$

$$\int_{\mathbb{R}} h''(t) \leq 2 \quad , \quad h(t) \leq 2t ,$$

the graph of h" beeing the following :

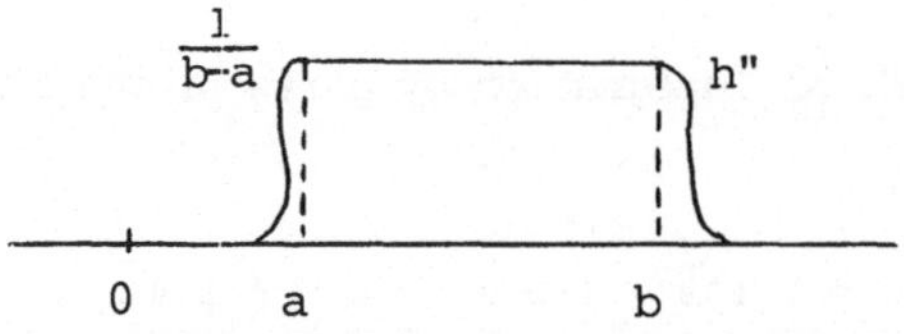

($\mathbb{1}_{[a,b]}$ is the characteristic function of [a,b]). Then $h \circ u \in \mathcal{E}^\infty(\Omega) \cap PSH(\Omega)$ and $h \circ u \equiv 0$ in some neighborhood of A. We can apply the lemma 1 to h o u :

$$\int_K T \wedge i\partial\bar\partial (h \circ u) \leq C(K,\Omega) \ \|T\| \ (\Omega \setminus A) \ \|h \circ u\|_\infty$$

But we have :

$$i\,\partial\bar\partial(h \circ u) = h' \circ u \ \ i\partial\bar\partial u + h'' \circ u \ \ i\,\partial u \wedge \bar\partial u$$

Because each term in this last equality is positive, we obtain :

$$\int_K h'' \circ u \, T \wedge i\partial u \wedge \bar\partial u \leq C(K,\Omega) \ \|T\| \ (\Omega \setminus A) \ 2\|u\|_\infty$$

As we have $h'' \circ u \geq \frac{1}{b-a} \mathbb{1}_{a \leq u \leq b}$, we obtain the announced inequality :

$$\int_{K \cap \{a \leq u(z) \leq b\}} T \wedge i\partial u \wedge \bar{\partial} u \leq 2C(K,\Omega) \quad \|T\|(\Omega \setminus A) \, \|u\|$$

<u>Lemma 3</u> :

Let A be a closed complete pluripolar subset of Ω. Shrinking Ω if necessary, there exists a sequence $u_\nu \in PSH(\Omega) \cap \mathcal{C}^\infty$ such that $0 \leq u_\nu \leq 2$, u_ν vanishes is some neighborhood of A (depending on ν) and such that pointwise $\lim_{\nu \to \infty} u_\nu = 1\!\!1_{\Omega \setminus A}$.

<u>Proof</u> :

$A = \{z \in \Omega , v(z) = -\infty\}$ for some $v \in PSH(\Omega)$. Shrinking Ω, we can suppose $v < 0$ on Ω. Let be $w_\nu := \exp(\frac{v}{\nu})$, then $w_\nu = 0$ on A, $w_\nu \to 1\!\!1_{\Omega \setminus A}$ pointwise and $w_\nu \leq 1$.

Let now χ be a convex increasing function of T whose graph is the following :

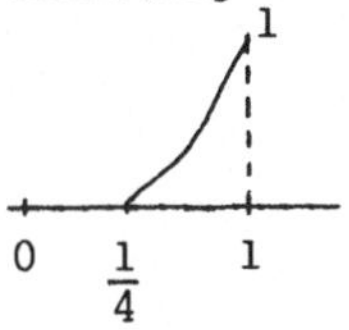

Then $\chi \circ w_\nu \in PSH(\Omega)$ and now $\chi \circ w_\nu$ vanishes in some neighborhood of Ω and $\chi \circ w_\nu \to 1\!\!1_{\Omega \setminus A}$ pointwise. We only need to regularize $\chi \circ w_\nu$ in order to archieve the construction of u_ν.

The following lemma is the Cauchy-Schwarz inequality.

<u>Lemma 4</u> :

If T is positive of bidimension $(1,1)$ and if φ and $\psi \in \mathcal{D}_{1,0}(\Omega)$ then we have :

$$|\langle T, i\,\varphi \wedge \bar{\psi}\rangle| \leqslant (\langle T, i\,\varphi \wedge \varphi\rangle)^{\frac{1}{2}} \; (T, i\,\psi \wedge \bar{\psi}\rangle)^{\frac{1}{2}} .$$

<u>Proof of theorem 3</u> :

Let $\theta \in \mathcal{E}^{\infty}(\mathbb{R})$ be a cut-off function whose graph is the following :

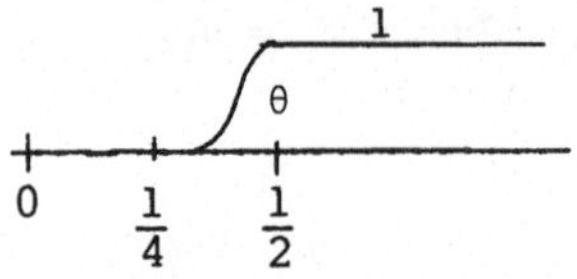

Let u_ν be the sequence in $PSH(\Omega) \cap \mathcal{E}^{\infty}(\Omega)$ given by the lemma 6, $\theta \circ u_\nu \equiv 0$ in some neighborhood of A and :

$$\lim_{\nu \to \infty} \theta \circ u_\nu = \mathbb{1}_{\Omega \setminus A} ,$$

By the definition of the trivial extension $\tilde{T}$, we have :

$$\tilde{T} = \lim_{\nu \to \infty} \theta \circ u_\nu \, T \quad ,$$

and therefore :

$$\bar{\partial}\,\tilde{T} = \lim_{\nu \to \infty} \theta' \circ u_\nu \; T \wedge \bar{\partial} u_\nu \; ,$$

in the weak sense. $\tilde{T}$ is obviously positive. The difficulty is to prove that $\tilde{T}$ is closed. Let $\varphi \in \mathcal{D}_{1,0}(\Omega)$ be a form of bidegree $(1,0)$. We have :

$$\langle\bar{\partial}\,\tilde{T}, \varphi\rangle = \lim_{\nu \to \infty} \langle T, \theta' \circ u_\nu \; \bar{\partial} u_\nu \wedge \varphi\rangle$$

and we need to prove that the limit is 0. But the Lemma 4 gives the following inequality :

$$|\langle T, \theta' o\, u_\nu\, \varphi \wedge \bar\partial u_\nu\rangle| \leq (\langle T, 1_K \theta' o\, u_\nu\; i\, \partial u_\nu \wedge \bar\partial u_\nu\rangle)^{\frac{1}{2}} \times$$

$$(\langle T, \theta' o\, u_\nu\;\; i\, \varphi \wedge \bar\varphi\rangle)^{\frac{1}{2}}.$$

But $\theta' o\, u_\nu$ is bounded and his support is a subset of $\{\frac{1}{4} \leq u_\nu \leq \frac{1}{2}\}$. The application of Lemma 2 to u_ν with $a = \frac{1}{4}$ $b = \frac{1}{2}$ and $K = \operatorname{supp} \varphi$ proves that the second term is bounded. Because $\theta' o\, u_\nu \to 0$ pointwise, $\theta' o\, u_\nu$ is bounded and T has finite mass, the third term goes to 0 when $\nu \to \infty$. We have proved :

$$\langle \bar\partial\, \tilde{T}, \varphi\rangle = 0$$

$$\bar\partial\, \tilde{T} = 0$$

<u>Remark</u> :

Of course the theorem 3 implies the Bishop's theorem because if $T = [X]$, it is easy to see that $\tilde{T} = [\bar{X}]$ where $\bar{X}$ is analytic, using the structure theorem of R. Harvey and J. King [14] or the more elementar structure theorem of J.King [16].

N. Sibony also proved a more general result than theorem 3.

In this result, the structure of the exceptionnal set A depends on the bidimension of the current T :

<u>Definition 1</u> :

Let be $p \in \mathbb{N}$. The subset A of Ω is said to be p complete pluripolar if for all $a \in \Omega$ there exists a neighborhood V of a, a local system of coordinates in V $(z_1, z_2, \ldots, z_n)$ and a finite number

of projection $\Pi_j : \mathbb{C}^n \to \mathbb{C}^p$, $1 \leq j \leq N$ such that for almost all $Z°$ in $\Pi_j(V)$, $A \cap V \cap \Pi_j^{-1}(Z°)$ is complete pluripolar in the complex linear manifold $V \cap \Pi_j^{-1}(Z°)$ and such that the differential forms $\Pi_j^\star \beta_p$ generate the space of complex forms of bidegree (p,p) where $1 \leq j \leq N$ and where β_p is the volum form on $\mathbb{C}^p$,
$\beta_p = (\frac{i}{2})^p \, dz_1 \wedge d\bar{z}_1 \wedge \ldots \wedge dz_p \wedge d\bar{z}_p$.

Notice that we need enough projections Π_j on subspaces of dimension p of $\mathbb{C}^n$, not only projections on the coordinates subspaces of dimension p $(z_1 \ldots z_n) \to (z_{i_1}, z_{i_2}, \ldots, z_{i_p})$ for a given coordinates system on $\mathbb{C}^n$.

We have the following simple examples :

1) 0-pluripolar complete sets are the usual pluripolar complete set.

2) If A is a closed subset of $\mathbb{C}^n$ of Hausdorff 2p-measure σ-finite, then A is p-pluripolar complete, because $\pi^{-1}(a)$ is countable for almost all a in $\mathbb{C}^p$ if π is a projection of $\mathbb{C}^n$ in $\mathbb{C}^p$.

Theorem 4 : ([21])

Let A be a closed p-pluripolar subset of Ω. If T is a closed, positive current on $\Omega \setminus A$ of bidimension $(p+1 ; p+1)$ and if the mass of T is locally finite in a neighborhood of A, then the trivial extension $\tilde{T}$ of T is a closed positive current on Ω.

Proof :

The problem is local. We suppose that $\Omega = V = \Delta_1 \times \Delta_2$ where Δ_1 is open in $\mathbb{C}^p$ and Δ_2 is open in $\mathbb{C}^{n-p}$. If π is the projection of V on Δ_1, we also suppose that for almost all $y \in \Delta_1$, $\pi^{-1}(y) \cap A$ is pluripolar complete in $\{y\} \times \Delta_2$. The currents $\tilde{T}$ and $d\tilde{T}$ are locally flat current on V. If ω is a form of bidegree (p,p) on Δ_1 and if φ is a form of degree 1 with compact support in V, we have the fundamental formula of the slicing of $d\tilde{T}$ (cf.[10])

$$\int_V d\tilde{T} \wedge \varphi \wedge \pi^* \omega = \int_{\Delta_1} \Big[\int_{\pi^{-1}(y)} \langle d\tilde{T}, \pi, y\rangle \wedge i_y^* \varphi \Big] \omega ,$$

where i_y is the injection of $\{y\} \times \Delta_2$ in V and where $\langle d\tilde{T}, \pi, y\rangle$ is the slice of $d\tilde{T}$ which is defined for almost all y and is a locally flat current on $\pi^{-1}(y)$.

We have [10] : $\langle d\tilde{T}, \pi, y\rangle = \langle \tilde{T}, \pi, y\rangle$ for almost all y.

$\langle \tilde{T}, \pi, y\rangle = \langle T, \tilde{\pi}, y\rangle$ is the trivial extension of the current $\langle T, \pi, y\rangle$ on $\pi^{-1}(y) \setminus A$. Because $\langle T, \pi, y\rangle$ is of bidimension $(1,1)$, closed and because $\pi^{-1}(y) \cap A$ is pluripolar complete in $\pi^{-1}(y)$, the theorem 3 proves that $\langle \tilde{T}, \pi, y\rangle = \langle T, \tilde{\pi}, y\rangle$ is closed. Therefore we have for all φ

$$\int_V d\tilde{T} \wedge \varphi \wedge \pi^* \omega = 0$$

Then we have $d\tilde{T} \wedge \pi^* \omega = 0$ for several projections π and then $d\tilde{T} = 0$.

2) Pluripositive currents.

In [24], we observe that a positive current T such that $dd^c T$

is also positive, has a well defined Lelong's number. Recently, N. Sibony [21] gaves a generalization of the preceedings results to pluripositive currents.

Definition 2 :

A current T on Ω of bidimension (p,p) is said to be pluripositive if T is normal (i.e. the coefficients of T and dT are measures), if $dd^c T$ is positive and if T is positive or negative.

Examples :

1) $T = u$ where $u \in PSH(\Omega)$ is positive or negative.

2) $T = u_o \, dd^c u_1 \wedge \ldots \wedge dd^c u_k$ where $u_j \in PSH(\Omega) \cap L^\infty(\Omega, loc)$ and where u_o is ≥ 0 or ≤ 0 (one needs the definition of E. Bedford and B. A. Taylor [3] for the product $dd^c u_1 \wedge \ldots \wedge dd^c u_k$).

3) If T is a closed, positive current and if $u \in PSH(\Omega)$ is positive, then one can prove that uT is pluripositive (cf. [21]).

We hope, pluripositive currents will be a good generalization of plurisubharmonic functions. N. Sibony proved in [21] the following result.

Theorem 5 :

Let A be a closed, p-pluripolar complete set in Ω and let T be a pluripositive current in $\Omega \setminus A$, of bidimension $(p+1,p+1)$. We suppose that T, dT and $dd^c T$ have locally finite mass in some neighborhood of A. Then we have :

i) $d\tilde{T} = \widetilde{dT}$

ii) $\widetilde{dd^cT} = dd^c\tilde{T} + S$

where S is a closed current, carried by A. Moreover, S is positive if T is positive, negative if T is negative.

The proof uses integration by parts and a Chern-Levine-Nirenberg inequality (as in Lemmas 1 and 2) :

$$\int_{K\cap\{a\leq u\leq b\}} T \wedge du \wedge d^c u \leq C(K,\Omega,T)\,(b-a)\,\|u\|_\infty$$

where T is positive, of bidimension (1,1), pluripositive in $\Omega \setminus A$ and where $u \in PSH(\Omega) \cap \mathcal{E}^\infty(\Omega)$ is positive and vanishes in some neighborhood of A ($K \subset\subset \Omega$).

Corollary 6 :

Let A be a closed p-pluripolar complete set in Ω and T a pluripositive current , of bidimension (p+1,p+1) in Ω. Then we have :

i) $d(\mathbb{1}_A T) = \mathbb{1}_A\, dT$

ii) $dd^c(\mathbb{1}_A T) \geq 0$ if T is positive

Proof : Let be $T' := T_{|\Omega \setminus A}$, then we have :

$$\mathbb{1}_A T = T - \mathbb{1}_{\Omega\setminus A} T = T - \tilde{T}'$$

we apply theorem 5 to the current T'.

3) Extension across C.R. submanifold

N. Sibony also proves in [21] extension theorems across C.R. submanifold. The main interest of these results in the fact that they don't need an assumption of finite mass.

Theorem 7 :

Let A be a totaly real submanifold of class C^2 of Ω (i.e. the real tangent space to A does not contain any complex line) and let T be a positive current on $\Omega \setminus A$ of bidimension (p,p).

1) If T is closed in $\Omega \setminus A$ and if $p \geqslant 1$, then the mass of T is locally finite in a neighborhood of A.

2) If T is pluripositif and $p \geqslant 2$, then the mass of T is also locally finite in a neighborhood of A.

Proof :

We only prove the 1) with $p = 1$ (the important case). It is a local result, therefore we suppose that the euclidean ball $B(0,1)$ is contained in Ω and that in B we have $A = \rho^{-1}(0)$ where ρ is a positive function of class C^2 in Ω such that

$$dd^c \rho \geqslant C\beta ,$$

in $\bar{B}(0,1)$, where C is a constant > 0 and β is the Kähler form on $\mathbb{C}^n$. We choose any function $h \in \mathcal{E}^\infty(\bar{B})$ such that :

$$h \equiv 0 \text{ in } \bar{B}(0,\tfrac{1}{4}) ,$$

$$h < 0 \text{ in the set } \{z \,;\, \tfrac{1}{2} \leqslant |z| \leqslant \tfrac{3}{4}\}$$

$$h \leqslant 0 \text{ everywhere.}$$

If $\delta > 0$ is small enough, we have :

$$dd^c\ (\rho+\delta h) \geqslant \frac{C}{2}\,\beta$$

in $\bar{B}(0,1)$. We fix such a δ.

We define $\rho_\varepsilon \in PSH(B(0,1)) \cap \mathcal{E}^0(B(0,1))$ by setting :

$$\rho_\varepsilon = \sup(\rho+\delta h-\varepsilon, 0).$$

Then the current $dd^c\, u_\varepsilon \wedge T$ is defined, positive and closed in $B(0,1)$. Let $g \in \mathcal{E}^\infty(B(0,1))$ a positive function such that $g \equiv 1$ on $\bar{B}(0,\frac{1}{2})$ and $g \equiv 0$ for $|z| \geqslant \frac{3}{4}$.

For $|z| \leqslant \frac{1}{4}$, we have $\rho_\varepsilon = \sup(\rho-\varepsilon,0)$ (because $h = 0$). Therefore for $|z| \leqslant \frac{1}{4}$ and $\rho \geqslant \varepsilon$, $\rho_\varepsilon = \rho-\varepsilon$ and $dd^c\,\rho_\varepsilon = dd^c\,\rho$. Because of the choice of g, we have :

$$< dd^c\,\rho_\varepsilon \wedge T, g> \geqslant C \int_{B(0,\frac{1}{4})\, \cap\, \{\varepsilon < \rho\}} T \wedge \beta$$

By Stokes formula we have :

$$< T, \rho_\varepsilon\, dd^c g > \geqslant C \int_{B(0,\frac{1}{4})\, \cap\, \{\varepsilon < \rho\}} T \wedge \beta$$

When $\rho_\varepsilon > 0$, we have $\rho+\delta h - \varepsilon > 0$, that is :

$$\rho > \varepsilon-\delta h > -\delta h .$$

But the support of $dd^c g$ is contained in $\{\frac{1}{2} \leqslant |z| \leqslant \frac{2}{4}\}$ and h is < 0 on this compact set. Therefore, there exists a constante c_o such that $-\delta h \geqslant c_o > 0$ on the support of $dd^c g$. Then on the support of $\rho_\varepsilon\, dd^c g$, we have :

$$\rho > -\delta h \geqslant c_o > 0$$

This imply that :

$$< T \, , \, \rho_\varepsilon \, dd^c g > \; \leq \; \|dd^c g\|_\infty \int_{\{z \, ; \, \rho(z) \geq c_0 \, , \, \frac{1}{2} \leq |z| \leq \frac{3}{4}\}} \|T\|$$

Let $K = \{z \, ; \, \frac{1}{2} \leq |z| \leq \frac{3}{4} \, , \, \rho(z) \geq c_0\}$. For all ε, we have the estimate :

$$\int_{B(0,\frac{1}{4}) \cap \{\varepsilon < \rho\}} T \wedge \beta \; \leq \; \frac{1}{C} \, \|dd^c g\|_\infty \; \|T\| \, (K) \; < + \infty$$

T has finite mass on $B(0,\frac{1}{4}) \setminus A$.

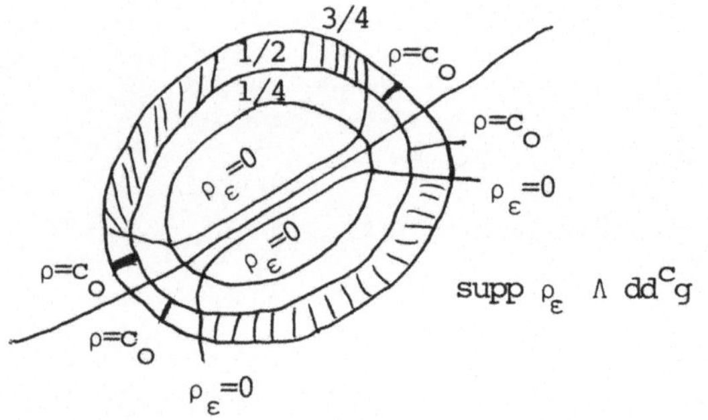

<u>Theorem 7'</u> :

Let A be a C.R. submanifold of Ω of C.R. dimension k (i.e. $\dim_{\mathbb{C}} HT_z = \dim_{\mathbb{C}}(T_z \cap i\, T_z) = k$ for all $z \in A$) and let T be a positive current of bidimension (p,p) on $\Omega \setminus A$.

1) If T is closed and $p \geq k+1$, T has locally finite mass in a neighborhood of A.

2) If T is pluripositive and $p \geq k+2$, T has locally finite mass in a neighborhood of A.

Sketch of the proof :

We choose a system of coordinates in $\mathbb{C}^n$ such that $0 \in A$ and such that :

$$HT_o = T_o A \cap i\, T_o A = \{z \; ; \; z_{k+1} = \ldots = z_n = 0\}.$$

We set $z' : (z_1,\ldots,z_k)$ $z'' = (z_{k+1},\ldots,z_n)$. There exists neighborhood V' and V'' of 0 in $\mathbb{C}^k$ and $\mathbb{C}^{n-k}$ respectively such that for all $z'_o \in V'$ the set

$$A(z'_o) = \{z \in A \; ; \; z' = z'_o\}$$

is a totally real submanifold of $\{z'_o\} \times V''$.
We apply theorem 7 to the slice $<T,\pi,z'_o>$ of T (where $\pi : \mathbb{C}^n \to \mathbb{C}^k$ is the canonical projection). One obtains a uniform estimate of the mass of the slice $<T,\pi,z'_o>$ and therefore the finiteness of the mass of T using classical estimates of slicing theory [10].

The first proof of theorem 7' assertion 1 given by H. Elmire in [8] was false. The first correct proof was given by N. Sibony in [21]. After H. Elmire gave a new proof in [9]. His proof is the same as Sibony's proof when $k=0$ but is different when $k \geqslant 1$ and don't use slicing theory.

Theorem 7" (N. Sibony)

Let A be a C.R. submanifold of Ω of C.R. dimension k and T be a positive current in $\Omega \setminus A$ of bidimension (p,p)

1) If T is closed and $p = k+1$, then $\tilde{T}$ is normal

2) If T is closed and $p \geq k+2$, T has a unique closed positive extension to Ω (this was also proved by H. Elmire).

3) If T is pluripositive and $p \geq k+2$, then $d^c T$ has a locally flat extension to Ω and one has :

$$d\tilde{T} = \widetilde{dT} \quad , \quad d^c\tilde{T} = \widetilde{d^cT} \quad , \quad dd^c\tilde{T} = \widetilde{dd^cT}.$$

Notice that dd^cT has bidimension $(p-1,p-1)$ and that $p-1 \geq k+1$, therefore dd^cT has finite mass and $\widetilde{dd^cT}$ is a normal current using assertion 1).

Easy counterexamples prove that the results are sharp.

The fundamental tool of the Sibony's proof is still a Chern-Levine-Nirenberg estimate.

$$\int_{\{a \leq |\rho| \leq b\} \cap K} T \wedge d\rho \wedge d^c\rho \leq C(K,\Omega)\,(b-a)\,[\|T\|(\Omega\setminus F) + \|dd^cT\|(\Omega \setminus F)]$$

where ρ is a C^2 function in Ω, $|\rho| \leq 1$, $F := \rho^{-1}(0)$, and T is positive, pluripositive current in $\Omega \setminus F$ of bidimension $(1,1)$, such that the mass of T and dd^cT are finite. Now the important trick is the fact that the inequality is valid for all a and b, $0 < a < b < 1$. A similar inequality is true for currents of bidimension $(2,2)$. The main idea of the proof is similar to that of the theorem 3.

N. Sibony also applied these last results to the following theorem which extends results of H. Alexander [1], J. Becker [2] and S. Shiffman [20].

Theorem 8 :

Let A be the set $\mathbb{R}^n \times \{0\} \subset \mathbb{C}^n$ and π the mapping : $(z_1,\ldots,z_n) \to (\bar{z}_1,\ldots,\bar{z}_n)$. If Ω is an open set in $\mathbb{C}^n$ such that $\pi(\Omega) = \Omega$ and if T is a positive closed current of bidimension

(1,1) in $\Omega \setminus A$, then there exists a positive, closed, current S of bidimension (1,1) on Ω such that : $S \geqslant T$, on $\Omega \setminus A$.

Proof :

Because T is positive of bidimension (1,1) , $-\pi_\star T$ is positive. We consider the closed, positive currents on $\Omega \setminus A$:

$$S := T - \pi_\star T$$

Because $\pi \circ \pi = Id$, we have :

$$\pi_\star S = -S$$

As A is a totally real submanifold, S has locally finite mass in a neighborhood of A (theorem 7). The trivial extension $\tilde{S}$ is well defined on Ω and is locally flat. Therefore $d\tilde{S}$ is locally flat on Ω and verify :

$$\pi_\star \tilde{S} = -\tilde{S}$$

$$\pi_\star d\tilde{S} = - d\tilde{S}$$

But $d\tilde{S}$ is carried by A and $\pi_{\star|A} = Id$. Because $d\tilde{S}$ is locally flat, a classical result of H. Federer [10] implies that :

$$\pi_\star d\tilde{S} = d\tilde{S}$$

We have : $\pi_\star d\tilde{S} = d\tilde{S} = -d\tilde{S}$,

$$d\tilde{S} = 0$$

The current $\tilde{S}$ is positive and closed and :

$$\tilde{S} = S \geq T$$

on $\Omega \setminus A$.

4) The complex Monge-Ampère operator and the extension of meromorphic map.

Let M be a closed subset of Ω and $u \in PSH(\Omega)$ a function of class C^∞ in $\Omega \setminus M$ or in $L^\infty(\Omega \setminus M, loc)$, we want to prove that in some cases the operator $(dd^c u)^n$ has locally finite mass in $\Omega \setminus M$. N. Sibony [21] introduces the following definition.

Definition 3 :

One says, the closed subset M of $\Omega \subset \mathbb{C}^n$ verify the condition (C) if for all $x \in M$ there exists a strictly pseudoconvex neighborhood ω of x such that : $\bar{\omega} \subset \Omega$ and $x \notin \widehat{M \cap \partial\omega}$ i.e. there exists $\varphi \in PSH(\Omega) \cap \mathcal{E}^\infty(\Omega)$ such that $\varphi(x) > 1$ but $\varphi < 0$ on $M \cap \partial\omega$.

Examples :

1) M is a closed subset of a totally real submanifold Σ.

2) M is a compact subset of the pseudoconvex open set Ω. In this case, $\partial\omega \cap M = \phi$.

3) M is a strictly pseudoconvexe real hypersurface in Ω.

4) $\Omega = B_2$ the euclidean ball in $\mathbb{C}^2$ and

$M = \{z \in B_2 \; ; \; z_2 = 0 \; , \; |z_1| \geq r\}$ where r is > 0.

(Thullen's situation).

Theorem 9 : (N. Sibony)

Let M be a subset of Ω which verifies the condition (C). Then for all $u \in PSH(\Omega) \cap \mathcal{E}^{\infty}(\Omega \setminus M)$ and all compact set $K \subset \Omega$, we have :

$$\int_{K \setminus M} (dd^c u)^u < + \infty$$

The case where M is a compact subset of the ball is due to P. Griffiths [11]. The proof is in the same vein as the proof of theorem 7, using ingenious integration by parts. Using the slicing theory, it is easy to deduce the following corollary.

Corollary 10 :

If M is a C.R. submanifold of Ω of C.R. dimension k, then for all $u \in \mathcal{E}^{\infty}(\Omega \setminus M) \cap PSH(\Omega)$ and for all compact $K \subset \Omega$, we have :

$$\int_K (dd^c u)^{n-k} < + \infty$$

The final aim of the theory is the following result in the same vein as P. Griffiths [11] and Y.T. Siu [23].

Theorem 11 :(N. Sibony)

Let M be a closed subset of the open set Ω in $\mathbb{C}^n$ and N be a compact, kähler manifold. Then a meromorphic map $f : \Omega \setminus M \to N$ extends to a meromorphic map from Ω to N when we make one of the following assumptions about M :

1) M is a C.R. submanifold of Ω of C.R. dimension $k \leqslant n-2$ and there exists a (n-2) pluripolar complete closed subset A of Ω which contains M.

2) M is compact and there exists a closed complete pluripolar set A which contains M.

Sketch of the proof :

For the sake of simplicity, we suppose f is holomorphic. Let ω be the kähler form of N. $T := f^{\star}\omega$ is a current on $\Omega \setminus M$ of bidimension $(n-1,n-1)$. Because $k \leqslant n-2$, theorem 7' implies that T has locally finite mass. The restriction of T to $\Omega \setminus A$ has also locally finite mass. Because A is $(n-2)$-pluripolar, theorem 4 proved that T has an extension $\tilde{T}$ to Ω which is of class C^{∞} in $\Omega \setminus A$. Restricting Ω if necessary, there exists $u \in PSH(\Omega) \cap \mathcal{C}^{\infty}(\Omega \setminus M)$ such that :

$$i\,\partial\bar{\partial}u = \tilde{T}$$

We can suppose that $\Omega = \Delta^n = \Delta^{n-2} \times \Delta^2$ is the product of two polydiscs in $\mathbb{C}^{n-2}$ and $\mathbb{C}^2$ and that for all $z^o = (z_1^o,\dots,z_{n-2}^o) \in \Delta^{n-2}$,

$$M(z_o) := M \cap \{z_1 = z_1^o,\dots,z_{n-2} = z_{n-2}^o\}$$

is a totally real submanifold.

Let be $\Gamma = \{(z,f(z))\ ;\ z \in \Delta^n \setminus M\}$ the graph of f.

For every $z^o \in \Delta^{n-2}$, let $\Gamma(z_o)$ be the graph of the restriction of f to $(\{z^o\} \times \Delta^2) \setminus M(z_o)$.

Then we have :

$$\mathrm{vol}(\Gamma(z^o)) = \int_{\Delta^2 \setminus M(z_o)} (\beta + dd^c u)^2 = \int_{\Delta^2 \times M(z_o)} (\beta\wedge\beta + 2\beta \wedge dd^c u + (dd^c u)^2)$$

Application of theorem 9 (with $n = 2$) proves that the last member is bounded by a constante $C(z_o)$. The details of the proof of this theorem proves that $C(z_o)$ is bounded by a constante C for all $z^o \in \Delta^{n-2}$.

Fubini's theorem allow to estimate the volume of Γ :

$$\mathrm{vol}(\Gamma) = \int_{\Delta^{n-2}} \mathrm{vol}(\Gamma(z_o))\, d\lambda(z_o) \leq C\int_{\Delta^{n-2}} d\lambda(z^o) < +\infty$$

Because Γ is an analytic subvariety of dimension n of $(\Omega \times N) \setminus (M \times N)$ of locally finite volume and because $A \times N$ is $(n-2)$-pluripolar complete, theorem 4 proves that $\bar{\Gamma}$ is a subvariety of $\Omega \times N$, i.e. f has a meromorphic extension to Ω.

5) Extention from an open set to $\mathbb{C}^n$.

J.P. Demailly in [6] considers the following problem : let T be a positive, closed, current of bidimension (p,p) on an open set Ω of $\mathbb{C}^n$. When has T a closed, positive extension to $\mathbb{C}^n$? Of course, the sets of density $E_c = \{z \in \Omega , \nu(T,z) \geq c\}$ of the current T (where c is > 0 and $\nu(T,z)$ is the Lelong's number of T at the point z) are analytic because of the Siu's theorem and are obstructions to the extension of T to $\mathbb{C}^n$. Therefore, one needs to suppose that the density $\nu(T,z)$ of T is small enough on Ω in order to extend T to $\mathbb{C}^n$.

Let $\sigma = \frac{1}{p!} T \wedge \beta^p$ be the trace measure of T, and $\sigma(z,r) = \sigma(B(z,r))$, where $B(z,r)$ is the euclidean open ball of center z, radius r and where $z \in \Omega_r$ with $\Omega_r := \{\zeta \in \mathbb{C}^n , d(\zeta, \complement\Omega) > r\}$.

Theorem 12 : J.P. Demailly [6]

Let $\Omega \subset\subset \mathbb{C}^n$ be a Runge domain and T be a closed, positive current on Ω whose cohomology class vanishes. We suppose, the

following conditions about the mass $\sigma(z,r)$ of T are verified :
For all $\varepsilon > 0$ and all compact set $K \subset \Omega_\varepsilon$,

$$(1) \qquad \sup_{z \in K} \int_0^\varepsilon \frac{\sigma(z,r)}{r^{2n-1}}\, dr < +\infty \quad \text{if } T \text{ is of bidegree } (1,1),$$

$$(2) \qquad \int_0^\varepsilon \sup_{z \in K} \frac{\sigma(z,r)^{\frac{1}{2}}}{r^n}\, dr < +\infty \quad \text{if } T \text{ is of bidegree } (q,q) \quad 1 < q < n$$

Then for all real number $\delta > \eta > 0$, there exists a current θ closed and positive on $\mathbb{C}^n$ such that $T = \theta$ on Ω_δ and such that θ is of class C^∞ outside $\bar{\Omega}_\eta$.

It is easy to extend the result to Stein manifolds. The main difficulty of the proof is to build a good potential V such that $i\,\partial\bar{\partial}\, V = T$ modulo $C^\infty(\bar{\Omega}_\delta)$, using an appropriate kernel when $1 < q < n$.
The construction is easy when T is of bidegree $(1,1)$ because the condition (1) means that the function V is locally bounded.
J.P. Demailly build ingenious counterexamples proving that the condition (1) is sharp and the reasonable sufficient condition for $1 < q < n$ is the following :

$$\sup_{z \in K} \int_0^\varepsilon \frac{\sigma(z,r)}{r^{n+p}}\, dr < +\infty .$$

He also gives a more constructive proof of the existence of totally real C^∞ submanifold of $\mathbb{C}^n$ of real dimension $n-1$ which are pluripolar complete (the existence of such manifolds is due to Diederich-Fornaess [7]).

6) Open problems

1° - There is still one open interesting question about extension of positive, closed, current.

Let $\Delta^n = \Delta^k \times \Delta^{n-k}$ be a polydisc in $\mathbb{C}^n$.

$z' = (z_1,\dots,z_k) \qquad z'' = (z_{k+1},\dots,z_n)$

$\pi : \Delta^n \to \Delta^k \quad , z \to z'$

Let T ne a positive, closed surrent in $\Delta^k \times \Delta^{n-k} \setminus (\Delta^k \times \{0\})$ of bidimension (k,k).

We consider the set :

$$E = \{z' \in \Delta^k \,,\ \| \langle T,\pi,z'\rangle \| < +\infty\}$$

where $\langle T,\pi,z'\rangle$ is the slice of the current T by π. Suppose that E has $\gtrless 0$ Lebesgue measure or more generally that E is not pluripolar in Δ^k . Has T a closed positive extension across $\Delta^k \times \{0\}$ to Δ^n (R.Harvey and J.Polking's problem) ?

It is true when $k = n-1$ (R. Harvey and Polking [15]). For $k < n-1$, I don't know the answer. When $T = [X]$ is the current of integration on an analytic subset X of $\Delta^n \setminus (\Delta^k \times \{0\})$, $\bar{X}$ is an analytic subset of Δ^n (W. Stoll [25]).

2° - If T is positive and pluripositive current on Ω, I have proved that the Lelong number $\nu(T,x)$ of T at the point x is still defined (cf. [24]). For $c > 0$, it is easy to see that the set :

$$E_c = \{x \in \Omega \ ; \ \nu(T,x) \geqslant c\}$$

is closed and of Hausdorff 2p-measure locally finite. Is E_c pluripolar in Ω ? it is also easy to see that E_c is not analytic in general.

3°- The tangent cone

We denote by $\frac{1}{r}$ the mapping : $\mathbb{C}^n \to \mathbb{C}^n$, $z \to \frac{z}{r}$ for $r > 0$. If T is a closed, positive current on a neighborhood of 0, the problem is to prove the existence of $\lim\limits_{r \to 0} \frac{1}{r}\star T$ in the weak sense. If the limit T_o exists, then T_o is conic $\frac{1}{r}\star T_o = T_o$ and is a current on $\mathbb{C}^n$.

If $T = [X]$, H. Federer [10] has proved that the limit exists and is an algebraic conic set on $\mathbb{C}^n$.

4° - The Demailly's conjecture

Let X be a projective manifold. Following J.P. Demailly [5], we denote by $J_p(X)$ the set of current $[Z] \in \mathcal{D}'_{p,p}(X)$ where Z is an irreducible analytic set of dimension p in X. $S\,P\,C^p_{\mathbb{Z}}(X)$ is the set of strongly positive, closed current of bidimension (p,p) on X which are "rational". T is "rational" if the cohomology class $cl(T)$ is in the $\mathbb{R}$-vectorspace generated by the classes of bidegree (k,k) in $H^{2k}(X,\mathbb{Z})$ $(p+k=n)$.

J.P. Demailly asked in [5] if one has the equality :

$$S\,P\,C^p_{\mathbb{Z}}(X) = \widehat{J_p(X)}$$

where $\wedge$ means the closed convex hull. It is not very difficult to see (cf. [5]) that this conjecture implies the Hodge conjecture : every rationnal cohomology class of bidegree (k,k) is the cohomology class of a holomorphic cycle $[X]$ of dimension p. But probably this conjecture is much stronger than the Hodge conjecture because roughly speaking this conjecture means that each $T \in S\,P\,C^p_{\mathbb{Z}}(X)$ can be represented as an integral $\int_\Lambda Z_\lambda \, d\mu(\lambda)$ of a family of cycle $(Z_\lambda)_{\lambda \in \Lambda}$ where μ is a probability measure on Λ.

Hodge conjecture only says that :

$$T = [Z] + dS$$

where Z is a holomorphic cycle and S a current.

J.P. Demailly [5] proved his conjecture for $p = n-1$.

BIBLIOGRAPHIE

[1] H.ALEXANDER. - Continuing 1-dimensional analytic sets. Math. Annalen, 191, 1970, p. 14 .

[2] J.BECKER. - Continuing analytic sets across $\mathbb{R}^n$. Math. Annalen, 195, 1972, 103-106.

[3] E.BEDFORD and B.A.TAYLOR. - A new capacity for plurisubharmonic functions. Acta Math., 149, 1982, 1-39.

[4] E.BISHOP. - Conditions for the analyticity of certain sets. Michigan Math. Jour., 11, 1964, p. 289-304.

[5] J.-P.DEMAILLY. - Courants positifs extrêmaux et conjecture de Hodge. Inventiones Mathematicae, 69, 1982, 347-374.

[6] J.-P.DEMAILLY. - Propagation des singularités des courants, positifs, fermés. Arkiv für Mathematik, 1983.

[7] K.DIEDERICH and J.E.FORNAESS. - Smooth but not complex-analytic pluripolar sets. Manuscripta Mathematica, 37, 1982, p. 121-125.

[8] H.EL MIR. - Sur le prolongement des courants,positifs,fermés. Thèse d'Etat soutenue à l'Université de Paris VI en Novembre 1982.

[9] H.EL MIR. - Sur le prolongement des courants,positifs,fermés. Acta Mathematica, 153, 1984, p. 1-45.

[10] H.FEDERER. - Geometric Measure Theory. Springer Berlin, Heidelberg, New-York, 1969.

[11] P.GRIFFITHS. - Two theorems on extension of holomorphic mappings. Inventiones Math., 14, 1971, 27-62.

[12] R.HARVEY. - Removable singularities for positive currents. Amer. Jour. of Math., 96, 1974, 67-68.

[13] R.HARVEY. - Holomorphic chains and their boundaries. Proceeding of Symposia in Pure Mathematics. Several complex variables, Williamstown,1975, Vol. 30, 1977, 309-382.

[14] R.HARVEY and J.R.KING. - On the structure of positive currents. Invent. Math. 15, 1972, p. 47-52.

[15] R.HARVEY and J.POLKING. - Extending analytic objects. Comm. pure and applied Math., 28, 1975, 701-727.

[16] J.KING. - The currents defined by analytic varieties. Acta Math. 127, 1971, p. 185-220.

[17] P.LELONG. - Intégration sur un ensemble analytique complexe. Bull. Soc. Math. France, 85, 1957, p. 239-262.

[18] P.LELONG. - Fonctions entières (n variables) et fonctions plurisousharmoniques d'ordre fini dans $\mathbb{C}^n$. Jour. Anal. Math. Jérusalem, 12, 1964, p. 365-407.

[19] P.LELONG..- Fonctions plurisousharmoniques et formes différentielles positives. Gordon and Breach, Dunod, Paris, Londres, New-York, 1968.

[20] B.SHIFFMAN. - On the continuation of analytic sets. Math. Ann. 185, 1970, 1-12.

[21] N.SIBONY. - Quelques problèmes de prolongement de courants en analyse complexe. Préprint Université d'Orsay, 1984, à paraître au Duke Math. Jour.,1984.

[22] Y.T.SIU. - Analyticity of Sets associated to Lelong Numbers and the extension of closed positive currents. Inventiones Math., 27, 1974, p. 53-156.

[23] H.SIU . - Extension of meromorphic maps into Kähler manifolds. Annals of Mathematics, 102, 1975, 421-462.

[24] H.SKODA. - Prolongement des courants, positifs, fermés, de masse finie, Inventiones Mathematicae, 66, 1982, p. 361-376.

[25] W.STOLL. - Über die Fortsetzbarkeit analytischer Mengen endlichen Oberflächeninhaltes. Archiv der Mathematik, 9, 1958, 167-175.

[26] W.STOLL. - The growth of the area of a transcendental analytic set I and II. Math. Annalen, t. 156, 1964, p. 47-48 et 144-170.

On the Uniformization of Parabolic Manifolds

Pit-Mann Wong
Department of Mathematics
University of Notre Dame

Introduction

The concept of an open parabolic Riemann surface is classical. It is well-known that there are many different but equivalent characterizations, each of which can be generalized in some way to complex manifolds of higher dimension. However, as is to be expected, these generalized concepts are in general not equivalent. In this article parabolicity is defined via the existence of an (unbounded) exhaustion satisfying the homogeneous complex Monge-Ampere equation which, in the case of Riemann surfaces is simply the Laplace equation. There are several reasons for choosing this as the definition of parabolic manifolds. First of all it includes affine algebraic manifolds and with this definition, the classical value distribution theory over $\mathbb{C}^n$ can be extended to parabolic manifolds in such a way that the defect relation is still intrinsic (cf. section 1 below). Secondly, this definition is indeed quite intrinsic as it is possible to obtain (under appropriate assumptions on the regularity of the exhaustion function) uniformization theorems for parabolic manifolds. Thirdly, by allowing bounded exhaustions (so that the manifolds under consideration are no longer parabolic but rather, Kobayashi hyperbolic), it is discovered that the Monge-Ampère condition is intrinsically related to the condition of certain complex curves being geodesics(or extremal disc) of the Kobayashi metric (cf. sections 4 and 5).

This article is a (partial) survey of known results. However, some of the results and proofs are new. The new gredient here is the observation that global arguments which work very well in the unbounded case can be replaced by local arguments, which work also in the bounded case, using the fact that a solution of the complex homogeneous Monge-Ampère equation is real analytic when restricted to the leaves of the associated foliation.

The definition of parabolic manifold described above, as well as many results in this area are due to Professor Wilhelm Stoll. We dedicate this paper in his honor, on the occasion of his sixtieth birthday.

The author would like to thank the National Science Foundation and the Sloan Foundation for their support.

§1 Value Distribution on Parabolic Manifolds

The fundamental paper of R. Nevanlinna in 1925 on the theory of meromorphic functions was, in the words of H. Weyl, one of the few great mathematical events of our century (cf. Weyl [34], p. 8). This theory was brought to a new height by L. Ahlfor's work [1], a magnificent demonstration of the interplay between topology, geometry and analysis, on holomorphic curves in 1941. Ahlfor's theory was extended to parabolic Riemann surfaces by H. Weyl and J. Weyl in the famous monograph [34] published in 1943. Modern treatment of this theory can also be found in Chern [8] and Wu [38]. Extension of the theory to higher dimension was first achieved by Stoll in a series of papers ([26], [27]) in the late forties and early fifties. Among other things the following defect relation was essentially obtained by Stoll.

Theorem 1.1 . Let $f : \mathbb{C}^m \longrightarrow \mathbb{CP}^n$ be a linearly non-degenerate meromorphic map and let $\mathfrak{a}$ be a set of hyperplanes in general positions in $\mathbb{CP}^n$. The sum of defects satisfy the following estimate,

$$\sum_{\alpha \in \mathfrak{a}} \delta_f(a) \leq n + 1 \ .$$

For an alternate proof of the above theorem see Vitter [33]. In the sixties and early seventies, value distribution theory attracted the attention of many mathematicians, notably the works of Carlson-Griffiths [7], Griffiths-King [12] extend the theory to algebraic manifolds in the equi-dimensional case. However in this article we shall restrict ourselves to the ideas of Stoll in extending the defect relation to a larger class of manifolds other than $\mathbb{C}^m$. This class of manifolds should be large enough so as to include all affine algebraic manifolds and on the other hand, manifolds in this class should not admit any non-constant bounded holomorphic functions (otherwise we cannot expect to have a nice simple defect relation).

Such manifolds can be produced by the following procedure (cf. Stoll [26].). For simplicity we consider only the case of Kähler manifolds. Let M^m be a complex manifold with Kähler metric g and ω the associated Kähler form. Choose a sequence of relatively compact domains with smooth boundaries $\{G_j\}$, $G_j \subset\subset G_{j+1}$ and $\cup G_j = M$. Solve the Dirichlet problem for the Laplace-Beltrami operator Δ on $G_j - G_0$,

$$u_j = 0 \quad \text{on } G_j - G_o \tag{1.1}$$

$$u_j = \begin{cases} 0 & \text{on } M - G_j \\ 1 & \text{on } \overline{G}_o \end{cases} \tag{1.2}$$

The harmonic condition (1.1) is equivalent to

$$dd^c u_j \wedge \omega^{m-1} = 0 \tag{1.3}$$

when $d^c = i(\overline{\partial} - \partial)$. Define constants

$$c_j = - \int_{\partial G_j} d^c u_j \wedge \omega^{m-1} \tag{1.4}$$

These constants are strictly positive (by Hopf's lemma) and satisfies $C_j \geq C_k$ if $k \geq j$ (by maximum principle). Thus the limit

$$C_\infty(M) = \lim_{j\to\infty} C_j \geq 0 \tag{1.5}$$

exists and is called the <u>capacity</u> of M at infinity.

Remark. In many applications we often encounter the situation where instead of a Kähler form, we have only a closed non-negative form ω. We can simply replace (1.1) by (1.3), which still makes sense, then the concept of capacity can still be defined, provided that the Dirichlet problem is solvable.

Proposition 1.1 . *A connected Kähler manifold M with zero capacity at infinity does not admit any non-constant bounded holomorphic function.*

Proof. Let ω be the fundamental Kähler form and suppose there is a non-constant bounded holomorphic function f, say $|f| \leq 1$. Then by (1.2) we have,

$$0 < \lambda = \int_{G_o} dd^c \log(1+|f|^2) \wedge \omega^{m-1} = \int_{G_o} u_j \, dd^c \log(1+|f|^2) \wedge \omega^{m-1}$$

From (1.1), (1.2) and Stoke's theorem, the last term above is equal to

$$\int_{\partial G_0 - G_j} \log(1+|f|^2) d^c u_j \wedge \omega^{m-1} \leq -\int_{\partial G_j} \log(1+|f|^2) d^c u_j \wedge \omega^{m-1} \leq c_j \log 2$$

Thus we have $0 < \lambda \leq c_j \log 2$, for any j. This is a contradiction since $c_j \longrightarrow 0$ as $j \longrightarrow \infty$, QED.

For related results on capacity see Sibony-Wong [23] and [24], also Alexander [2].

The concept of zero capacity introduced above depends on the metric, thus it is desirable to have a more intrinsic notion of "parabolicity". The following definition defining parabolicity via the homegenous complex Monge-Ampere equation is due to Stoll (cf. [29]).

Definition 1.1 . A continuous non-negative function τ on a complex manifold M^m is parabolic if

(i) τ is C^∞ on $M_* = M - \{\tau=0\}$;

(ii) τ is plurisubharmonic on M and is strictly plurisubharmonic at some point of M_*;

(iii) the function $u = \log \tau$ satisfies the complex homogeneous Monge-Ampere equation on M_*, i.e.

$$\det (u_{\alpha\bar{\beta}}) \equiv \quad \text{on } M_* \qquad (1.6)$$

where $u_{\alpha\bar{\beta}} = \partial^2 u/\partial z^\alpha \partial \bar{z}^\beta$.

A complex manifold M is parabolic if there exists a continuous exhaustion $\tau: M \longrightarrow [0,\infty)$ which is parabolic on the complement of some compact set K.

Examples. (1) Open parabolic Riemann surfaces are parabolic in the above sense; (2) Euclidean space $\mathbb{C}^m$ with $\tau(z) = |z|^2$

is parabolic; (3) an affine algebraic manifold M can be realized as a finite branched cover $\pi: M \longrightarrow \mathbb{C}^m$ over $\mathbb{C}^m$, then M with $\tau(z) = |\pi(z)|^2$ is parabolic; (4) Let L be a negative holomorphic line bundle over a projective manifold then there exists a hermitian metric h with negative chern form. This is equivalent to the parabolicity of h. Thus (L,h) is parabolic; (5) Product $M_1 \times M_2$ of parabolic manifolds (M_1,τ_1) and (M_2,τ_2) is parabolic with exhaustion $\tau_1 + \tau_2$; (6) A manifold obtained by blowing up a finite number of points of a parabolic manifold is parabolic.

On a parabolic manifold (M,τ) take r_o so that the set $M_{r_o} = \{\tau < r_o\}$ contains the exceptional compact set K. For any $r > r_o$, the function

$$u_r = \begin{cases} 0 & \text{on } M - M_r \\ \dfrac{\log r - \log \tau}{\log r - \log r_0} & \text{on } M_r - M_{r_o} \\ 1 & \text{on } M_{r_o} \end{cases}$$

solves the Dirichlet problem (1.2) and (1.3) with $\omega = dd^c \log \tau$. Capacity can be defined as before. It is easily verified that since τ is <u>unbounded</u>, the capacity at infinity $c_\infty(M)$ vanishes. We obtain from proposition 1.1 the following:

<u>Corollary</u> 1.1 . <u>Let M be a parabolic manifold then</u> $C_\infty(M) = 0$ <u>and so it does not admit any non-trivial bounded holomorphic function.</u>

The following defect relation generalize theorem 1.1 to the case of parabolic manifolds (cf. Wong [35], Stoll [32]).

<u>Theorem</u> 1.2 . <u>Let</u> (M,τ) <u>be a parabolic manifold and let</u> $f: M \longrightarrow \mathbb{CP}^n$ <u>be a linearly non-degenerate meromorphic map, then for a set</u> $\mathfrak{a}$ <u>of hyperplanes in general position in</u> $\mathbb{CP}^n$, <u>we have</u>

$$\sum_{\alpha \in \mathfrak{a}} \delta_f(a) \leq n+1 + \frac{n(n+1)}{2} R_f + 3n(n+1)Y_f .$$

The term R_f, which vanishes in the case of $\mathbb{C}^m$, reflects the topology of the manifold M. Denote by $\eta = dd^c \log \det(\tau_{\alpha\bar\beta})$

then R_f is defined as $\limsup_{r \to \infty} S(r)/T_f(r)$ where

$$S(r) = \int_{r_o}^{r} \frac{dt}{t^{2m-1}} \int_{\tau \leq t} \eta \wedge (dd^c\tau)^{m-1}$$

In the classical theory over Riemann surfaces, the term $S(r)$ is replaced by the Euler characteristic via the Gauss-Bonnet theorem, namely if $X(t)$ = Euler characteristic of the open set $\{\tau < t\}$ then

$$R_f = \limsup \{-\int_{r_o}^{r} X(t) \frac{dt}{t}\} / T_f(r)$$

If M is affine algebraic then it can be shown that $S(r) = O(\log r)$, so that $R_f = 0$ if f is transcendental and $R_f = \deg M/_{\deg f}$ if f is rational (cf. Wong [35]).

The appearance of the term Y_f is due mainly to technicality. The linearly non-degenerate condition on the map f can be expressed as a condition on the derivatives, namely $f \wedge f' \wedge \ldots \wedge f^{(n)} \not\equiv 0$. Due to the global nature of the theorem, it is necessary to define these derivaties by a global (meromorphic) vector field Z in such a way that $f \wedge Zf \wedge \ldots \wedge Z^n f \not\equiv 0$. The term f occurs in the estimate of the growth of Z, for which one does not have good control in general.

A natural way of constructing a vector field with "good" growth condition is to take the (complex) gradient vector field Z of τ, namely

$$Z = \tau^{\alpha\overline{\beta}} \tau_{\overline{\beta}} \, \partial/\partial z^{\alpha}$$

where $(\tau^{\alpha\overline{\beta}})$ is the inverse metrix of $(\tau_{\alpha\overline{\beta}})$ and where we have used the summation convention. The Monge-Ampère condition (1.6) is equivalent to $\|Z\|^2 = \tau$ where $\| \ \|$ denotes the norm of the metric $(\tau_{\alpha\overline{\beta}})$. The problem here is that Z is in general not holomorphic. However, if Z is holomorphic, the term Y_f vanishes, this is the case for parabolic Riemann surfaces and for affine algebraic manifolds.

Corollary 1.2 . Let M be as in theorem 1.2 and assume in addition that the complex gradient vector field Z of τ is holomorphic, then the term Y_f in the defect relation vanishes.

Corollary 1.3 . If M is affine algebraic, theorem 1.2 reduces to

(i) $\sum \delta_f(a) \leq n + 1$, if f is transcendental

(ii) $\sum \delta_f(a) \leq n + 1 + \frac{n(n+1)}{2} \frac{\deg M}{\deg f}$,if f is rational .

Critertion for the holomorphicity of the vector field Z will be given in the next section.

§2 Geometry of the Complex Homogeneous Monge-Ampère Equation

Let M be a complex manifold of dimension m with a positive strictly plurisubharmonic function (not necessarily an exhaustion) $\tau > 0$ of class C^2 on M, with the property,

$$(\partial\bar{\partial} \log \tau)^m \equiv 0 \tag{2.1}$$

Let $u = \log \tau$ then (2.1) together with the assumption that τ is strictly plurisubharmonic imply that the hermitian form $(u_{\alpha\bar{\beta}})$ is positive semi-definite and is of constant rank m-1. Thus it has exactly one zero eigenvalue everywhere.

By a direct calculation it is easily seen that (2.1) is equivalent to the identity

$$\tau^{\alpha\bar{\beta}} \tau_\alpha \tau_{\bar{\beta}} = \tau \tag{2.2}$$

The complex gradient vector field Z is given by

$$Z = \tau^\alpha \, \partial/\partial z^\alpha = \tau^{\alpha\bar{\beta}} \tau_{\bar{\beta}} \, \partial/\partial z^\alpha \tag{2.3}$$

then (2.2) is simply the identity $\|Z\|^2 = \tau$ when $\| \; \|$ is the norm defined by the Kähler netric $i\partial\bar{\partial}\tau$. An immediate consequence of (2.2) is that the vector field Z is the unique zero eigenvector of the hermitian form $(u_{\alpha\bar{\beta}})$,

$$\tau^\alpha \, u_{\alpha\bar{\beta}} = \tau^{\alpha\bar{\gamma}} \tau_{\bar{\gamma}} \, (\tau\tau_{\alpha\bar{\beta}} - \tau_\alpha \tau_{\bar{\beta}})/\tau^2 = 0 \tag{2.4}$$

Since $\partial\bar{\partial}u$ is a closed form, it follows that Z in integrable and M is foliate by the (complex) integral curves of M. Condition (2.5) implies that u is harmonic (first observed by Bedford- Kalka [3]) along each leaf of this foliation (henceforth will be referred to as the Monge-Ampère foliation). The following easy consequence is quite useful.

Lemma 2.1 . The function τ of class C^2 satisfying (2.1) is real analytic when restricted to a leaf of the Monge-Ampère foliation.

From now on we assume that τ is of class C^4. We shall examine the higher order derivatives of τ. We shall use the metric $\tau_{\alpha\bar{\beta}}$ to raise or lower indices, for instance $\tau^{\alpha} = \tau^{\alpha\bar{\beta}}\tau_{\bar{\beta}}$, $\tau^{\alpha}_{\beta\gamma} = \tau^{\alpha\bar{\nu}}\tau_{\beta\bar{\nu}\gamma}$ etc. The Christoffel symbols are given by,

$$\Gamma^{\alpha}_{\beta\gamma} = \tau^{\alpha}_{\beta\gamma} \tag{2.5}$$

and the components of the curvature by,

$$R^{\alpha}_{\beta\gamma\bar{\lambda}} = -\Gamma^{\alpha}_{\beta\gamma,\bar{\lambda}} = -\tau^{\alpha}_{\beta\gamma,\bar{\lambda}} \tag{2.6}$$

$$R_{\alpha\bar{\beta}\gamma\bar{\lambda}} = -\tau_{\alpha\bar{\beta}\gamma,\bar{\lambda}} \tag{2.7}$$

where, $\bar{\lambda}$ denotes covariant differentiation w.r.t. $\partial/\partial\bar{z}^{\lambda}$.

We begin by differentiating (2.2),

$$\tau_{\alpha} = \tau_{,\alpha} = \tau^{\mu\bar{\nu}}\tau_{\mu,\alpha}\tau_{\bar{\nu}} + \tau^{\mu\bar{\nu}}\tau_{\mu}\tau_{\bar{\nu}\alpha} = \tau^{\mu}\tau_{\mu,\alpha} + \tau_{\alpha}$$

from which we get the following identity involving Christoffel symbols

$$\tau^{\mu}\tau_{\alpha,\mu} = \tau^{\mu}\tau_{\mu,\alpha} = 0 \tag{2.8}$$

Note that $\tau_{\alpha,\mu} = \tau_{\mu,\alpha}$ follows from the Kähler condition.

Curvature informations are obtained by differentiating (2.8),

$$\tau^{\mu}_{,\bar{\beta}}\tau_{\alpha,\mu} + \tau^{\mu}\tau_{\alpha,\mu\bar{\beta}} = 0 \tag{2.9}$$

which implies the following theorem (cf. Wong [36] for a different derivation).

Theorem 2.1 . (i) $\|\bar{\partial}Z\|^2 + S(Z,\bar{Z}) = 0$ where S denotes the Ricci tensor, (ii) the holomorphic sectional curvature K(Z) for the plane section spanned by Z, vanishes identically.

To prove (i) we contract (2.9) with the metric,

$$\tau^{\alpha\bar{\beta}}\tau^{\mu}{}_{,\bar{\beta}}\tau_{\alpha,\mu} + \tau^{\alpha\bar{\beta}}\tau^{\mu}\tau_{\alpha,\mu\bar{\beta}} = 0 .$$

The first term above is $\|\bar{\partial}Z\|^2$ and the second term is the Ricci tensor, as can be easily deduced via the Ricci commutation formula,

$$\tau_{\alpha,\mu\bar\beta} = \tau_{\alpha,\mu\bar\beta} - \tau_{\alpha,\bar\beta\mu} = R^{\lambda}_{\alpha\mu\bar\beta}\tau_{\lambda} = R_{\alpha\bar\nu\mu\bar\beta}\tau^{\bar\nu} \qquad (2.10)$$

where the first equality follows from $\tau_{\alpha,\bar\beta\mu} = \tau_{\alpha\bar\beta,\mu} = 0$.

For the proof of (ii) we contract (2.9) with Z,

$$\tau^{\alpha}\tau^{\mu}{}_{,\bar\beta}\,\tau_{\alpha,\mu} + \tau^{\alpha}\tau^{\mu}\tau_{\alpha,\mu\bar\beta} = 0$$

where the first term vanishes by (2.8) and so we have

$$\tau^{\alpha}\tau^{\mu}\tau^{\bar\nu}\,R_{\alpha\bar\beta\mu\bar\nu} = \tau^{\alpha}\tau^{\mu}\tau_{\alpha,\mu\bar\beta} = 0 \qquad (2.11)$$

which implies (ii).

Remarks. (i) Let $M = \mathbb{C}^m$ or more generally an affine algebraic manifold with the usual exhaustion (cf. example in section 1) and metric, is Ricci flat thus Z is holomorphic by part (i) of Theorem 2.1 .

(ii) If m = 1 then the vanishing of the holomorphic sectional curvature is equivalent to the vanishing of Ricci, thus Z is holomorphic for parabolic Riemann surfaces. Actually, this can be seen more easily as follows,

$$\tau^{\bar\nu}\,\tau^{\alpha}_{,\bar\nu} = \tau^{\bar\nu}\,\tau^{\alpha\bar\beta}\,\tau_{\bar\beta,\bar\nu} = 0 \qquad (2.12)$$

by (2.8). The above identity means that Z is holomorphic in the direction of each leaf and so is holomorphic if m = 1.

Note that even though τ is only assumed to be of class C^4, it is however real analytic in the direction of Z (cf. lemma 2.1) and thus so is $||\bar\partial Z||^2$. Differentiating in the direction of Z, we get

lemma 2.2 . If τ is of class C^5 then

(i) $\tau^{a}(||\bar\partial Z||^2)_{,a} = -\,||\bar\partial Z||^2 + R^{\mu}_{a\alpha\bar\beta}\tau^{a}\tau^{\alpha}\tau^{\bar\beta}{}_{,\mu}$

(ii) $\tau^{\bar b}\tau^{a}(||\bar\partial Z||^2)_{,a\bar b} = ||\bar\partial Z||^2 - R^{\mu}_{a\alpha\bar\beta}\,\tau^{a}\tau^{\alpha}\tau^{\bar\beta}{}_{,\mu} - R^{\bar\nu}_{\bar b\alpha\bar\beta}\,\tau^{\bar b}\tau^{\bar\beta}\tau^{\alpha}{}_{,\bar\nu}$

$$+ R^{\mu}_{a\alpha\bar\beta}\tau^{a}\tau^{\alpha}R^{\bar\beta}_{\bar\eta\mu\bar b}\,\tau^{\bar b}\tau^{\bar\eta} + 2R^{\mu}_{a\alpha\bar b}\,\tau^{\alpha}\tau^{\bar b}\,R^{a}_{\mu\gamma\bar\beta}\,\tau^{\gamma}\tau^{\bar\beta}$$

Proof. By theorem 2.1 (i) we have,

$$\tau^{a}(||\bar\partial Z||^2)_{,a} = -\tau^{a}(R_{\alpha\bar\beta}\tau^{\alpha}\tau^{\bar\beta})_{,a} = -\,\tau^{a}R_{\alpha\bar\beta,a}\tau^{\alpha}\tau^{\bar\beta} - R_{\alpha\bar\beta}\tau^{\alpha}\tau^{\bar\beta} \qquad (2.13)$$

where we have also used the identities $\tau^{\alpha}{}_{,a} = \delta^{\alpha}_{a}$ and $\tau^{a}\tau^{\bar\beta}{}_{,a}=0$

Using the following commutation formula (cf. Appendix).

$$\tau_{\mu\bar\nu\alpha,\bar\beta a} = \tau_{a\bar\nu\alpha,\bar\beta\mu}$$

(we need τ of class C^5 here) and we have

$$R_{\alpha\bar\beta,a} = \tau^{\mu\bar\nu}R_{\mu\bar\nu\alpha\bar\beta,a} = -\tau^{\mu\bar\nu}\tau_{\mu\bar\nu\alpha,\bar\beta a} = -\tau^{\mu\bar\nu}\tau_{a\bar\nu\alpha,\bar\beta\mu} = \tau^{\mu\bar\nu}R_{a\bar\nu\alpha\bar\beta,\mu}.$$

Applying (2.11) to the above, we have

$$\tau^{a}R_{\alpha\bar\beta,a}\,\tau^{\alpha}\tau^{\bar\beta} = \tau^{\alpha}\tau^{\mu\bar\nu}\,R_{a\bar\nu\alpha\bar\beta,\mu}\tau^{\alpha}\tau^{\bar\beta}$$

$$= \tau^{\mu\bar\nu}\,(\tau^{a}\tau^{\alpha}\tau^{\bar\beta}\,R_{a\bar\nu\alpha\bar\beta})_{,\mu} -2R_{\alpha\bar\beta}\tau^{\alpha}\tau^{\bar\beta} - \tau^{\mu\bar\nu}R_{a\bar\nu\alpha\bar\beta}\tau^{a}\tau^{\alpha}\tau^{\bar\beta}{}_{,\mu}$$

$$= -2R_{\alpha\bar\beta}\tau^{\alpha}\tau^{\bar\beta} - R^{\mu}_{a\bar\nu\alpha\bar\beta}\,\tau^{a}\tau^{\alpha}\tau^{\bar\beta}{}_{,\mu}$$

Substituting the expression above in (2.13) gives (i).

Now for (ii) we differentiate the identity in (i),

$$\tau^{b}\tau^{a}(\|\bar\partial z\|^{2})_{,a\bar b} = \|\bar\partial z\|^{2} - R^{\bar\nu}_{\bar b\alpha\bar\beta}\,\tau^{b}\tau^{\bar\beta}\tau^{\alpha}{}_{,\bar\nu} + \tau^{\bar b}(R^{\mu}_{a\alpha\bar\beta}\tau^{a}\tau^{\alpha}\tau^{\bar\beta}{}_{,\mu})_{,\bar b}$$

where the last term above is equal to

$$\tau^{\mu\bar\nu}\tau^{\bar b}R_{a\bar\nu\alpha\bar\beta,\bar b} + \tau^{\mu\bar\nu}\,R_{a\bar\nu\alpha\bar\beta}\,\tau^{a}\tau^{\alpha}\tau^{\bar b}\tau^{\bar\beta}{}_{,\mu\bar b}\ .$$

By (2.10) we have $\tau^{\bar\beta}{}_{,\mu\bar b} = R^{\bar\beta}{}_{\bar\nu\mu\bar b}\,\tau^{\bar\nu}$ and by (2.11)

$$\tau^{a}\tau^{\alpha}\tau^{\bar b}\,R_{a\bar\nu\alpha\bar\beta,\bar b} = \tau^{a}\tau^{\alpha}\tau^{\bar b}\,R_{a\bar\nu\alpha\bar b,\bar\beta} \qquad (\tau\in C^{5}(M))$$

$$= (\tau^{a}\tau^{\alpha}\tau^{\bar b}\,R_{a\bar\nu\alpha\bar b})_{,\bar\beta} -2\tau^{a}{}_{,\bar\beta}\,\tau^{\alpha}\tau^{\bar b}\,R_{a\bar\nu\alpha\bar b} - \tau^{a}\tau^{\alpha}R_{a\bar\nu\alpha\bar\beta}$$

$$= -2R_{a\bar\nu\alpha\bar b}\,\tau^{\alpha}\tau^{\bar b}\tau^{a}{}_{,\bar\beta} - R_{a\bar\nu\alpha\bar\beta}\,\tau^{a}\tau^{\alpha}\ .$$

Combining all of these we have,

$$\tau^{\bar b}\tau^{a}(\|\bar\partial z\|^{2})_{,ab} = \|\bar\partial z\|^{2} - R^{\bar\nu}_{\bar b\alpha\bar\beta}\,\tau^{\bar b}\tau^{\bar\beta}\tau^{\alpha}{}_{,\bar\nu} - R^{\mu}_{a\alpha\bar\beta}\,\tau^{a}\tau^{\alpha}\tau^{\bar\beta}{}_{,\mu}$$

$$-2R_{a\bar\nu\alpha\bar b}\,\tau^{\alpha}\tau^{\bar b}\tau^{a}{}_{,\bar\beta}\,\tau^{\bar\beta}{}_{,\mu} + R^{\mu}_{a\alpha\bar\beta}\,\tau^{a}\tau^{\alpha}R^{\bar\beta}_{\bar\nu\mu\bar b}\,\tau^{\bar\nu}\tau^{\bar b}$$

which is (ii) by the following identity,

$$\tau^{a}{}_{,\bar\beta}\,\tau^{\bar\beta}{}_{,\mu} = (\tau^{\bar\beta}\tau^{a}{}_{,\bar\beta})_{,\mu} - \tau^{\bar\beta}\tau^{a}{}_{,\bar\beta\mu} = -\,\tau^{\bar\beta}\tau^{\gamma}R^{a}_{\gamma\mu\bar\beta}$$

where we have used (2.10) and (2.12).

The function $||\bar{\partial}Z||^2$ is real analytic when restricted to a leaf of the foliation, thus if it isn't identically zero then we can define a pseudo-hermitian metric on each leaf,

$$\omega = \frac{2}{m} ||\bar{\partial}Z||^2 \, i \, d\bar{z} \wedge d\bar{z} \qquad (2.14)$$

Corollary 2.1 . If $||\bar{\partial}Z||^2 \not\equiv 0$ on a leaf of the Monge-Ampere foliation, then Ric $\omega \geq \omega$.

Proof. For convenience define tensors A and B by

$$A^{\mu}_{\beta} = R^{\mu}_{a\alpha\beta} \tau^{a} \tau^{\alpha}, \quad B^{\mu}_{a} = R^{\mu}_{a\alpha\bar{\beta}} \tau^{\alpha} \tau^{\bar{\beta}}$$

then lemma 2.2 (i) and (ii) can be expressed as,

$$\tau^{a}(||\bar{\partial}Z||^2)_{,a} = - ||\bar{\partial}Z||^2 + \langle A, \bar{\partial}Z \rangle$$

$$\tau^{\bar{b}}\tau^{a}(||\bar{\partial}Z||^2)_{,a\bar{b}} = ||\bar{\partial}Z||^2 - \langle A, \bar{\partial}Z \rangle - \langle \bar{\partial}Z, A \rangle + ||A||^2 + 2 \, ||B||^2 \, .$$

From these we get,

$$\tau^{\bar{b}}\tau^{a}(\log||\bar{\partial}Z||^2)_{,a\bar{b}} = (||\bar{\partial}Z||^2||A||^2 - |\langle A, \bar{\partial}Z \rangle|^2 + 2||\bar{\partial}Z||^2||B||^2) / ||\bar{\partial}Z||^4$$

$$\geq 2||B||^2/||\bar{\partial}Z||^2 \, .$$

Observe that $||\bar{\partial}Z||^2 = -R_{\alpha\bar{\beta}}\tau^{\alpha}\tau^{\bar{\beta}} = - \tau^{\alpha\bar{\nu}}B_{a\bar{\nu}} = -$ trace of B and since (trace of B)$^2 \leq m \, ||B||^2$, we conclude that

$$\tau^{\bar{b}}\tau^{a}(\log \, ||\bar{\partial}Z||^2)_{,a\bar{b}} \geq \frac{2}{m} \, ||\bar{\partial}Z||^2$$

proving Corollary 2.1.

A well-known theorem of Ahlfor's asserts that there is no pseudo-hermitian metric on $\mathbb{C}$ satisfying the conclusion of Corollary 2.1, hence we have

Theorem 2.2 . If the universal cover of each leaf of the Monge-Ampère foliation is $\mathbb{C}$ then $||\bar{\partial}Z||^2 \equiv 0$, that is the Monge-Ampère foliation is holomorphic.

The above theorem is due to Burns [6], our computation is perhaps more systematic.

§3 Uniformization Theorems

Let M be a complex manifold with a parabolic exhaustion τ on M. Namely τ: M $\longrightarrow$ [0,c) is proper and continuous (3.1) and satisfies properties (i), (ii) and (iii) of definition 1.1 on $M_* = M - \{\tau=0\}$. The exhaustion is unbounded or bounded depending on $c = \infty$ or $< \infty$, in the later case, we always normalize so that $c = 1$. In the bounded case, the manifold is no longer parabolic. (cf. Corollary 1.1).

The following Uniformization theorem is due to Stoll [30] (see also Burns [6], Wong [36] for alternative proofs).

Theorem 3.1 . Let M^m be a connected complex manifold with a parabolic exhaustion τ. Assume in addition that τ is of class C^∞ and is strictly plurisubharmonic everywhere on M. Then M with the Kähler metric $i\partial\bar{\partial}\tau$ is biholomorphically isometric to (i) $\mathbb{C}^m$ with Euclidean metric if τ is unbounded, or (ii) the unit ball $\mathbb{B}^m$ in $\mathbb{C}^m$ with Euclidean metric if τ is bounded.

We present here the basic ideas of the proof of this theorem. For a strictly plurisubharmonic function, it is well-known (cf. Harvey and Wells [13]) that the Taylor series at a minimum point (i.e., where $\tau=0$) assumes the following form,

$$\tau(z) = \sum_{j=1}^{m} \{(1+\lambda_j)z_j^2 + (1-\lambda_j)y_j^2\} + \text{higher order term}$$

where $\lambda_j \in \mathbb{R}$ and $z_j = x_j + iy_j$ is a local coordinate system. Since $\log \tau$ satisfies the homogeneous Monge-Ampère equation, this implies that $\lambda_j = 0$ for all j. Thus we have

$$\tau(z) = \sum |z_j|^2 + \text{higher order term.}$$

Consequently the set $\{\tau=0\}$ is discrete and by (2.3) we know that the vector field Z is nonvanishing on M_* so that M can be homotopically deformed onto the set $\{\tau=0\}$ along the integral curves of Z. Thus $\{\tau=0\}$ consists of exactly one point $\{\underline{o}\}$ because M is connected.

If τ is an unbounded exhaustion then it is unbounded and harmonic when restricted to each leaf of the foliation (cf. lemma 2.1), thus the universal cover is $\mathbb{C}$. By theorem 2.2 the Monge-Ampère foliation is holomorphic. From this and theorem 2.1, it is easily seen (cf. Wong [36]) that the exponential map from the point $\underline{o}$ is a holomorphic isometry from the complex tangent space $T_{\underline{o}}M$ onto M. In fact, one has $\tau(\exp(z)) = |z|^2$ for $z \in T_{\underline{o}}M = \mathbb{C}^m$.

If τ is bounded we shall show that the foliation is holomorphic under the additional assumption that τ is real analytic, the general case will be treated in the next section.

We shall prove inductively that $||\bar{\partial}Z||^2$ vanishes to infinite order at $\underline{o}$, so that it vanishes identically if τ is real analytic.

From theorem 2.1, it follows that $||\bar{\partial}Z||^2 = o(1)$ since Z vanishes at $\underline{o}$. Thus we have $\tau^{\bar{\nu}}{}_{,\mu} = 0$ at $\underline{o}$ for all ν,μ.
By differentiating (2.11) w.r.t. $\partial/\partial z^\gamma$ and $\partial/\partial \bar{z}^\lambda$ respectively, we get

$$2\tau^\alpha\tau^{\bar{\nu}}R_{\alpha\bar{\beta}\gamma\bar{\nu}} + \tau^\alpha\tau^\mu\tau^{\bar{\nu}}{}_{,\gamma}R_{\alpha\bar{\beta}\mu\bar{\nu}} + \tau^\alpha\tau^\mu\tau^{\bar{\nu}}R_{\alpha\bar{\beta}\mu\bar{\nu},\gamma} = 0 \qquad (3.2)$$

$$\tau^\alpha\tau^\mu R_{\alpha\bar{\beta}\mu\bar{\lambda}} + 2\tau^\alpha{}_{,\bar{\lambda}}\tau^\mu\tau^{\bar{\nu}}R_{\alpha\bar{\beta}\mu\bar{\nu}} + \tau^\alpha\tau^\mu\tau^{\bar{\nu}}R_{\alpha\bar{\beta}\mu\bar{\nu},\bar{\lambda}} = 0 \qquad (3.3)\ .$$

The last two terms in (3.2) obviously vanish up to third order at $\underline{o}$, hence so is the first term. Thus

$$||\bar{\partial}Z||^2 = -\tau^{\alpha\bar{\beta}}R_{\alpha\bar{\beta}\mu\bar{\nu}}\tau^\mu\tau^{\bar{\nu}}$$

is of order 3 at $\underline{0}$.

Similarly from (3.3) we see that

$$\tau^\alpha\tau^\mu R_{\alpha\bar{\beta}\mu\bar{\lambda}} = o(3) \qquad (3.4)$$

which implies that the second term in (3.2) is of order o(4). By contracting (3.2) with τ^γ and using (2.11) and (2.12), we get

$$\tau^\gamma\tau^\alpha\tau^\mu\tau^{\bar{\nu}}R_{\alpha\bar{\beta}\mu\bar{\nu},\gamma} = 0 \qquad (3.5)$$

Differentiating (3.5) w.r.t. $\partial/\partial z^\eta$ and using the commutation formula we obtain

$$3\tau^\alpha\tau^\mu\tau^{\bar{\nu}}R_{\alpha\bar{\beta}\mu\bar{\nu},\eta} + \tau^\gamma\tau^\alpha\tau^\mu\tau^{\bar{\nu}}{}_{,\eta}R_{\alpha\bar{\beta}\mu\bar{\nu},\gamma} + \tau^\gamma\tau^\alpha\tau^\mu\tau^{\bar{\nu}}R_{\alpha\bar{\beta}\mu\bar{\nu},\gamma\eta} = 0 \qquad (3.6)$$

from which we conclude that the first term (= third term in (3.2)) is o(4). Thus the first term in (3.2), hence also $||\bar{\partial}Z||^2$, is of order o(4). This of course implies that $\tau^{\bar{\nu}}{}_{,\mu} = o(2)$.

Now by (3.4) we conclude that the second term in (3.2) and (3.6) are actually o(5). By contracting (3.6) with τ^η and using (3.5) and (2.12), we obtain

$$\tau^{\eta}\tau^{\gamma}\tau^{\alpha}\tau^{\mu}\tau^{\bar{\nu}} R_{\alpha\bar{\beta}\mu\bar{\nu},\gamma\eta} = 0 .$$

Differentiating the above w.r.t. $\partial/\partial z^{\rho}$ we obtain

$$4\tau^{\gamma}\tau^{\alpha}\tau^{\mu}\tau^{\nu} R_{\alpha\bar{\beta}\mu\bar{\nu},\gamma\rho} + \tau^{\eta}\tau^{\gamma}\tau^{\alpha}\tau^{\mu}\tau^{\bar{\nu}}{}_{,\rho} R_{\alpha\bar{\beta}\mu\bar{\nu},\gamma\eta} + \tau^{\eta}\tau^{\gamma}\tau^{\alpha}\tau^{\mu}\tau^{\bar{\nu}} R_{\alpha\bar{\beta}\mu\bar{\nu},\gamma\eta\rho} = 0$$

from which we see that the first term (= third term in (3.6)) is o(5). Thus the first term in (3.6) (= third term in (3.2)) is also of order o(5). This forces the first term in (3.2), hence also $||\bar{\partial}z||^2$, vanishes at $\underline{o}$ up to fifth order.

The process described above can of course be repeated as often as desired, hence $||\bar{\partial}z||^2$ vanishes at $\underline{o}$ up to infinite order as claimed.

Let L^m be a negative holomorphic line bundle over a projective manifold N^{m-1}, then there exists a fiber metric h satisfying all the hypothesis of theorem 3.1 with the exception that it is only strictly plurisubharmonic outside of the zero section = {h=0}. The following theorem is due to Burns [6].

<u>Theorem</u> 3.2 . Let (M,τ) be a connected complex manifold with unbounded exhaustion τ satisfying all hypothesis of theorem 3.1 except that τ is strictly plurisubharmonic only on M_*. Then there exists a negative holomorphic line bundle (L,h) such that the Remmert quotients $\tilde{M}$ and $\tilde{L}$ are biholomorphic .

The Remmert quotients $\tilde{M}$ and $\tilde{L}$ are obtained by blowing down $\{\tau=0\}$ and {h=0} respectively. We shall only make one remark about the above theorem. Namely, since the exhaustion is unbounded, the leaves of the foliation are uniformized by $\mathbb{C}$, hence the foliation is holomorphic.

We shall discuss the counterpart of Theorem 3.2 for bounded exhaustion in the next section.

§4 Bounded Monge-Ampere exhaustions

We begin with an existence theorem of Lempert [14].

Theorem 4.1 . Let D be a bounded strictly convex domain in $\mathbb{C}^m$ with C^∞ boundary. Then for each point $p \in D$, there exists a unique exhaustion $\tau: \bar{D} \longrightarrow [0,1]$ with $\tau^{-1}(o) = p$, $\tau^{-1}(1) = \partial D$ and satisfies (i), (ii) and (iii) of definition (1.1) up to ∂D. In addition,

$$\log \tau(z) = \log |z-p|^2 + 0(1) \text{ near } p \tag{4.1}$$

Remark . The exhaustion τ is actually strictly convex on $\bar{D} - \{p\}$.

Lempert also gave a geometric description of τ:

Theorem 4.2 . Let D and τ be as in theorem 4.1, then there exists a homeomorphism $\Phi: \bar{D} \longrightarrow \bar{B}^m$ = unit ball in $\mathbb{C}^m$ such that (i) $\Phi(p) = 0$ and Φ is a C^∞-diffeomorphism of $\bar{D} - \{p\}$ onto $\bar{\mathbb{B}}^m - \{0\}$, (ii) Φ map each leaf of the Monge-Ampère foliation biholomorphically onto a disc through the origin, (iii) $\tau(z) = |\Phi(z)|^2$.

Furthermore there is an explicit relationship between τ and the Kobayashi metric.

Theorem 4.3 . Let D and τ be as in theorem 4.1 then $\delta_D = \log (1 + \sqrt{\tau}/1-\sqrt{\tau})$ where $\delta_D(z)$ = Kobayashi distance from p to z. Furthermore, the leaves of the foliation are extremal discs (or geodesic) of the Kobayashi metric.

Recall that a holomorphic map $f: \Delta \longrightarrow D$ from the unit disc with $f(o) = p$ and $f'(o) = v$ is said to be extremal for the Kobayashi metric if for any holomorphic map $q: \Delta \longrightarrow D$ with $q(o) = p$, $q'(o) = \lambda v$ and $\lambda > 0$ then $\lambda \leq 1$.

We shall give some motivations for the construction of τ in theorem 4.1. Let ϕ be a strictly convex defining function of D. Suppose τ exists as claimed then since $u = \log \tau = 0$ on ∂D we have $u = h\phi$ with $h > 0$ and so we have $u_\alpha = h\phi_\alpha$ on ∂D. Recall the identity (2.4),

$$\tau^{\bar{\beta}} u_{\alpha,\bar{\beta}} = \tau^{\beta} u_{\alpha\bar{\beta}} = 0 \tag{4.2}$$

which means that $u_\alpha = \tau_\alpha/\tau$ is holomorphic when restricted to a leaf (except at the point $p = \tau^{-1}(o)$) of the foliation. With these in mind we recall the following definition of Lempert [14].

Definition . A map $f: \overline{\Delta} \longrightarrow \overline{D}$ is said to be stationary if (i) f is of class $C^{1/2}$ on $\overline{\Delta}$, (ii) $f|_{\Delta}: \Delta \longrightarrow D$ is a proper holomorphic map and (iii) there exists a positive function h of class $C^{1/2}$ defined for $\zeta \varepsilon \partial\Delta$ such that for all α, the function $\tilde{f}_\alpha(\zeta) = \zeta h(\zeta)\ \phi_\alpha(f(\zeta)))$ extends holomorphically to Δ.

The existence of stationary maps will provide us with a foliation and also the u_α's. Since $\overline{D}$ is a bounded domain, it is Kobayashi hyperbolic and extremal discs exist. The existence of stationary maps are then guaranteed if one can show that they are extremal discs. To see this let $g: \Delta \longrightarrow D$ be a holomorphic map with $g(o) = f(o)$ and $g'(o) = \lambda f'(o), \lambda \geq 0$. By property (iii) of a stationary map, the function

$$\mathrm{Re}\sum_\alpha (f^\alpha(\zeta)-g^\alpha(\zeta))h(\zeta)\phi_\alpha(f(\zeta)) = \mathrm{Re}\sum_\alpha \frac{f^\alpha(\zeta)-g^\alpha(\zeta)}{\zeta}\zeta h(\zeta)\phi_\alpha(f(\zeta))$$

extends holomorphically to all $\zeta\ \varepsilon\ \Delta$. By strict convexity the left-handed side above is ≥ 0 (with equality iff $f \equiv g$). Thus we have , at the origin

$$\mathrm{Re}\ \sum (1-\lambda)\frac{\partial f^\alpha}{\partial \zeta}(o)\ \tilde{f}_\alpha(o) \geq o$$

with equality iff $f \equiv g$. In particular, taking g to be a constant map, we get $\mathrm{Re}\ \sum(1-\ )\ \frac{\partial f^\alpha}{\partial\zeta}\ (o)\ \tilde{f}_\alpha\ (o) \geq 0$, and so in general we have $\lambda \leq 1$. Thus a stationary map is extremal.

It can be shown easily that a stationary map is a biholomorphic map of τ onto its image. We conclude that the exhaustion $\tau \circ f$ is strictly subharmonic on Δ and $u \circ f = \log \tau \circ f$ is harmonic on Δ and so $\tau \circ f(\zeta) = |\zeta|^2$ by theorem 3.1 (actually for m=1, it is not difficult to see this).

In other words τ is obtained by first constructing the stationary maps (= extremal discs) f and then pulling back the absolute value function on Δ via f.

By examining the situation of the strictly convex domain carefully, one realizes that the special relationship between τ and the Kobayashi metric remains valid if the Monge-Ampère exhaustion τ satisfies,

$\tau^{-1}(o)$ consists of exactly one point $\{\underline{o}\}$ (4.3)

τ is C^∞ everywhere after blowing up the point $\{\underline{o}\}$ (4.4)

and $u = \log \tau$ satisfies (4.1) on a neighborhood of $\underline{o}$.

Theorem 4.4 . Let (M,τ) be a complex manifold with a bounded Monge-Ampère exhaustion satisfying (4.1), (4.3) and (4.4) then the leaves of the Monge-Ampère foliation of M_* extends across o and are extremal discs of the Kobayashi metric of M.

Remark. The exhaustion τ restricted to a leaf is then the pull-back of the absolute value function on Δ via an extremal disc through $\underline{o}$. In particular τ is real analytic on each leaf through $\underline{o}$.

We also remark that the conditions on τ in theorem 4.1 do not characterize strictly convex domains. However, if we know in addition that the Monge-Ampère foliation is holomorphic then we have,

Theorem 4.5 . Let (M,τ) be as in theorem 4.4 and assume in addition that the Monge-Ampère foliation is holomorphic then M is biholomorphic to a bounded strictly pseudo-convex circular domain in $\mathbb{C}^m$.

A domain D in $\mathbb{C}^m$ is circular if $z \in D$ then $\lambda z \in D$ for all complex numbers λ with $|\lambda| \leq 1$. For details of theorems 4.4 and 4.5 see [18], [19], [20], and [37].

Returning now to the proof of theorem 3.1 in the previous section where we had established the holomorphicity of the Monge-Ampère foliation in the case of bounded exhaustion under the additional assumption that τ is real analytic. However, under the C^∞ assumption, what we had shown in §3, clearly imply that the hypothesis of thereom 4.4 is satisfied, thus the leaves of the foliation extend across $\underline{o}$ as extremal discs and τ is real analytic (hence so is $||\bar{\partial} Z||^2$) when restricted to each leaf. Since $||\bar{\partial} Z||^2$ vanishes up to infinite order at $\underline{o}$, it vanishes identically on each leaf. Thus Z is holomorphic, completing the proof of theorem 3.1.

Notice that in theorem 4.5 (unlike theorem 3.1) the exhaustion τ is not assumed to be of class C^∞ at the origin, therefore holomorphicity of the foliation has to be assumed. A bounded circular domain D in $\mathbb{C}^m$ can always be defined by $D = \{z \in \mathbb{C}^m | \tau(z) < 1\}$ where $\tau(z) = e^{q} |z|^2$ and the function q is constant along each complex line through the origin, i.e., q is a function on $\mathbb{C}P^{n-1}$. Thus τ is not smooth at the origin unless D is biholomorphic to the ball.

§5 Intrinsic metrics in the bounded case

From the results of the previous section, we know that the Monge-Ampère exhaustion is intrinsically related to the Kobayashi metric, we shall further exploit this relationship in this section.

Let τ be a non-negative Monge-Ampère function (not necessarily an exhaustion) which is bounded. Without loss of generality assumed that $\sup\tau = 1$. In section 2, we have studied the geometry of the Kähler metric $h = \tau_{\alpha\overline{\beta}} dz^{\alpha} d\overline{z}^{\beta}$. Consider now also the hermitian metric

$$g = (1-\tau)^{-2}\, h \qquad (5.1)$$

The following is a consequence of theorem 2.1,

<u>Theorem</u> 5.1 . <u>Denote by S_q and S_h respectively the Ricci tenor of q and h. Let K_q be the holomorphic sectional curvature of the metric q. Then on M_* we have</u>

(i) $S_q = S_h - 2m\,(1-\tau)q = S_h - \frac{2mh}{1-\tau}$, in particular

$S_q(Z,\overline{Z}) = -\,||\overline{\partial}Z||^2 - 2m\tau/1-\tau < 0,$

(ii) $K_q(Z,\overline{Z}) \equiv -1.$

By a direct computation we have

$$\Omega_{\beta\overline{\gamma}\mu\overline{\nu}} = (1-\tau)^{-2} R_{\beta\overline{\gamma}\mu\overline{\nu}} + 2(1-\tau)^{-4}\tau_{\beta\overline{\gamma}}[-\tau_{\mu\overline{\nu}} + \tau^2 u_{\mu\overline{\nu}}]$$

where $\Omega_{\beta\overline{\gamma}\mu\overline{\nu}}$ and $R_{\beta\overline{\gamma}\mu\overline{\nu}}$ are the curvature of q and h respectively. From this we get

$$K_g(Z) = \frac{1}{2}\,\Omega_{\beta\overline{\gamma}\mu\overline{\nu}}\, \tau^{\beta}\tau^{\overline{\gamma}}\tau^{\mu}\tau^{\overline{\nu}}/|Z|^4_q$$

$$= \frac{1}{2}\, \frac{|Z|^4_h}{(1-\tau)^2 |Z|^4_g}\, K_h(Z) - \frac{\tau^2}{(1-\tau)^4\;\, |Z|^4_g} \equiv -1.$$

Specialize to the case where M is a strictly convex domain with smooth boundary in $\mathbb{C}^m$ or, more generally where M satisfies the hypothesis of theorem 4.4, then the leaves of the foliation extend through the fixed point and are in fact extremal discs of the Kobayashi metric. The metric $g = i\partial\overline{\partial}\tau/(1-\tau)^2$ pull back to the unit discs in $\mathbb{C}$ via the extremal maps is the Poincare metric on Δ. Thus it can be thought of as a generalization of the Poincare metric. We have the following theorem concerning isometries of this metric.

Theorem 5.1 . Let (M,τ) and $(\tilde{M},\tilde{\tau})$ be complex manifolds satisfying the hypothesis of theorem 4.4. Let $\phi: M \longrightarrow \tilde{M}$ be an isometry of the corresponding metrics g and $\tilde{g}$ respectively defined by (5.1). If ϕ preserves the corresponding Monge-Ampère foliations then ϕ is biholomorphic or antibiholomorphic.

The above theorem was proved in [16] for strictly convex domains with the exhaustion obtained in theorem 4.1. The proof in the more general situation above is analogous. Basically the assumptions guaranteed that ϕ maps leaves onto leaves and is biholomorphic (or anti-biholomorphic) on each leaf. Thus $\overline{X}\phi = 0$ if X is tangent to a leaf. Since for each direction at the point $\underline{o}$ there is a leaf through $\underline{o}$ in that direction, we have $\overline{\partial}\phi = o$ at $\underline{o}$. By an argument similar to the argument at the end of section 4 we see that $\overline{\partial}\phi$ actually vanishes to infinite order at $\underline{o}$. But $||\overline{\partial}\phi||^2$ is real analytic on each leaf because both ϕ and the metrics are. Hence $||\overline{\partial}\phi||^2$ vanishes identically.

For manifolds (M,τ) and $(\tilde{M},\tilde{\tau})$ in theorem 5.1, let M_r and $\tilde{M}_r$ be respectively the Kobayashi balls of radius r from $\underline{o}$ and $\underline{\tilde{o}}$. The following is an easy consequence of theorem 5.1. (cf. [16]).

Corollary 5.1 . With the notations as above, then every biholomorphic map $\phi: M_r \longrightarrow \tilde{M}_r$ is the restriction of a biholomorphic map from M onto $\tilde{M}$.

The corollary above was first observed by Bland, Duchamp and Kalka [5] where M and $\tilde{M}$ are strictly convex domains. This theorem can be thought of as a "weak form" of unique analytic continuation for the complex homogeneous Monge-Ampère equation.

For applications of this result we refer to [5] and [16].

Consider now a bounded strictly convex domain D with smooth boundary in $\mathbb{C}^m$. Denote by δ_D and C_D respectively the Kobayashi and Caratheodory distances from a fixed point $p \in D$. Let τ be the unique Monqe-Ampère exhaustion centered at p. Take an extremal disc $f: \overline{\Delta} \longrightarrow \overline{D}$ of the Kobayashi metric throuqh p. Construct a complex vector bundle E over $f(\partial D) \subset \partial D$ by assigning to each point $z_o \in f(\partial D)$ the complex tanqent space to D at z_o, i.e., $E_{z_o} = \{z| \sum_{\alpha=1}^{m} u_\alpha(z_o)(z^\alpha - z_o^\alpha) = 0\}$ where $u_\alpha = (\log \tau)_\alpha$. Since an extremal disc is a stationary map, the vector bundle E extends to a holomorphic vector bundle $\pi: E \longrightarrow f(\Delta)$ over $f(\Delta)$. By the convexity of ∂D, the domain D lies entirely on one side of each tanqent space to the boundary, thus $D \subset \tilde{E}$. The restriction $\pi|_D: D \longrightarrow f(\Delta)$ is holomorphic and surjective onto the extremal disc. Thus every bounded holomorphic function on $f(\Delta)$ extends to a holomorphic function on D with the same bound, i.e., $f(\Delta)$ is also an extremal disc for the Caratheodory metric, therefore $C_D \equiv \delta_D$. More generally for any convex domain (bounded or unbounded and not necessarily with smooth boundary), by taking an exhaustion with strictly convex domains with smooth boundaries, a limit argument shows that the same is true.

<u>Theorem</u> 5.2 . <u>Let</u> D <u>be a convex domain in</u> $\mathbb{C}^m$ <u>then the Caratheodory metric and the Kobayashi metric of</u> D <u>are identically</u>.

The above theorem was proved by Lempert [15] and independently by Royden-Wong [21]. For the more general situation of manifolds satisfying the hypothesis of theorem 4.4, one expects to have a bounded $\lambda > 0$ so that $C_D \geq \lambda \delta_D$ on D, however, it is not clear how λ depends on M. A clear understanding of this constant will be extremely helpful on the problem of existence of <u>bounded</u> holomorphic functions.

Appendix

We collect here, for the convenience of the reader, some commutation formulas that were used in the computation in §2.

(I) For 1-tensors

$$\phi_{i,k\bar j} - \phi_{i,\bar j k} = R^{s}_{ik\bar j}\,\phi_s$$

(II) For 2-tensors

$$\phi_{i\bar j,k\bar\ell} - \phi_{i\bar j,k\bar\ell} = R^{s}_{ik\ell}\,\phi_{s\bar j} - R^{\bar t}_{\bar j\bar\ell k}\,\phi_{i\bar t}$$

$$\phi_{i\bar j,k\bar\ell} = \phi_{k\bar j,i\bar\ell}$$

$$\phi_{i\bar j,k\bar\ell} - \phi_{i\bar\ell,k\bar j} = R^{s}_{ik\bar\ell}\,\phi_{s\bar j} - R^{s}_{ik\bar j}\,\phi_{s\bar\ell}$$

$$\phi_{ij,k\bar\ell} - \phi_{ij,\bar\ell k} = R^{s}_{kj\bar\ell}\,\phi_{si} + R^{s}_{ik\bar\ell}\,\phi_{sj}$$

(III) For 3-tensors

$$\phi_{i\bar j k,p\bar q} - \phi_{i\bar j k,\bar q p} = R^{s}_{pk\bar q}\,\phi_{s\bar j i} + R^{s}_{ip\bar q}\,\phi_{sjk} R^{t}_{\bar j\bar q p}\phi_{i\bar t k}$$

$$\phi_{i\bar j k,p\bar q} = \phi_{k\bar j i,p\bar q}\,,\quad \phi_{i\bar j k,\bar q p} = \phi_{i\bar q k,\bar j p}$$

$$\phi_{i\bar j k,p\bar q} - \phi_{i\bar j p,k\bar q} = R^{s}_{ip\bar q}\,\phi_{s\bar j k} - R^{s}_{ikq}\,\phi_{s\bar j p}$$
$$- \Gamma^{s}_{ip}\,\phi_{s\bar j k\bar q} + \Gamma^{s}_{ip}\,\Gamma^{\bar t}_{\bar j\bar q}\,\phi_{s\bar t k} + \Gamma^{s}_{ip}\,\phi_{s\bar j k\bar q}$$
$$+ \Gamma^{s}_{ik}\,\Gamma^{\bar t}_{\bar j\bar q}\,\phi_{s\bar t p}$$

$$\phi_{i\bar j k,\bar q p} - \phi_{i\bar j p,\bar q k} = -\Gamma^{s}_{ip}\,\phi_{s\bar j k\bar q} + \Gamma^{s}_{ik}\,\phi_{s\bar j p\bar q}$$
$$+ R^{\bar t}_{\bar j\bar q p}\,\phi_{i\bar t k} - R^{\bar t}_{\bar j\bar q k}\,\phi_{i\bar t p} + \Gamma^{\bar t}_{\bar j\bar q}\,\Gamma^{s}_{ip}\,\phi_{s\bar t k}$$
$$- \Gamma^{\bar t}_{\bar j\bar q}\,\Gamma^{s}_{ik}\,\phi_{s\bar t p}$$

$$\phi_{i\bar j k,\bar q p} = \phi_{i\bar j p,\bar q k}$$

For the special case in §2 where τ is a function and $\tau_{p\bar q}$ is the metric, we have

$$\tau_{i\bar j k,\bar q p} - \tau_{i\bar j p,\bar q k} = R^{\bar t}_{\bar j\bar q p}\,\tau_{i\bar t k} - R^{\bar t}_{\bar j\bar q k}\,\tau_{i\bar t p}$$

$$- \tau_{ip}^{s} \tau_{s\bar{j}k\bar{q}} + \tau_{ik}^{s} \tau_{s\bar{j}p\bar{q}} + \Gamma_{jq}^{\bar{t}} \tau_{ip}^{s} \tau_{s\bar{t}k}$$

$$- \tau_{\overline{jq}}^{\bar{t}} \tau_{ik}^{s} \tau_{s\bar{t}p}$$

where

$$R_{\overline{jq}p}^{\bar{t}} \tau_{i\bar{t}k} = -\tau^{s\bar{t}}(\tau_{\bar{j}s\bar{q}p} - \tau_{sp}^{\ell} \tau_{\bar{j}\ell\bar{q}}) \tau_{i\bar{t}k}$$

$$= -\tau_{\bar{j}sqp} \tau_{ik}^{s} + \tau_{sp}^{\ell} \tau_{\bar{j}\ell\bar{q}} \tau_{ik}^{s}$$

$$- R_{\overline{jq}k}^{\bar{t}} \tau_{i\bar{t}p} = \tau_{\bar{j}s\bar{q}k} \tau_{ip}^{s} - \tau_{sk}^{\ell} \tau_{\bar{j}\ell\bar{q}} \tau_{ip}^{s}$$

$$\tau_{\overline{jq}}^{\bar{t}} \tau_{ip}^{s} \tau_{s\bar{t}k} = \tau_{\bar{j}\ell\bar{q}} \tau_{ip}^{s} \tau_{sk}^{\ell}$$

$$-\tau_{\overline{jq}}^{\bar{t}} \tau_{ik}^{s} \tau_{s\bar{t}p} = -\tau_{\bar{j}\ell\bar{q}} \tau_{ik}^{s} \tau_{sp}^{\ell}.$$

from which we get the following formula,

$$\tau_{i\bar{j}k,\bar{q}p} = \tau_{i\bar{j}p,\bar{q}k} \quad .$$

References

[1] Ahlfors, L., The theory of meromorphic curves, Acta Soc. Fenn. Nova Ser. A 3 (4), (1941).

[2] Alexander, H., Projective Capacity, Recent Development in Several Complex Variables, edited by J. Fornaess, Ann. Math Studies 100 (1981).

[3] Bedford, E., and Kalka, M., Foliations and complex Monge-Ampere equations, Comm. Pure and Appl. Math, 30, 543-571 (1977).

[4] Bedford, E. and Taylor, B.A., Variational properties of the complex Monge-Ampere equation II, intrinsic norms, Am. J. Math 101, 1131-1166 (1977).

[5] Bland, J., Duchamp, T. and Kalka, M., On the automorphism group of strictly convex domains in $\mathbb{C}^n$, preprint, (1984).

[6] Burns, D., Curvature of Monge-Ampere foliations and parabolic manifolds, Ann. of Math. 115, 349-373 (1982).

[7] Carlson, J., and Griffiths, P., A defect relation for equidimensional holomorphic mappings between algebraic varieties, Ann. of Math. 95, 557-584. (1972).

[8] Chern, S.S., Complex manifolds without potential theory.

[9] Chern, S.S., Holomorphic curves in the plane, Diff. Geometry in honor of K. Yano, Kinokuniya, Tokyo, 73-94 (1972).

[10] Foote, R., Curvature estimates for Monge-Ampère foliations, Univ. of Michigan Thesis (1983).

[11] Greene, R.E., and Wu, H., Function theory on manifolds which possess a pole, Springer-Verlag Lecture note series, Vol. 699(1979).

[12] Griffiths, P. and King, J., Nevanlinna theory and holomorphic mappings between algebraic varieties, Acta Math. 130, 145-220 (1973).

[13] Harvey, R. and Wells, R.O., Zero sets of non-negative strictly plurisubharmonic functions, Math. Ann. 201, 165-170 (1973).

[14] Lempert, L., La métrique de Kobayashi et la representation des domains sur la boule,Bull. Soc. Math. France 109, 427-474 (1981).

[15] Lempert, L., Intrinsic norms, Proc. Symp, Pure Math., Vol. 41, 147-150 (1984)

[16] Leung, K.L., Patrizio, G. and Wong, P.M., Isometries of intrinsic metrics on strictly convex domains, preprint (1985).

[17] Noguchi, J., Holomorphic curves in algebraic varieties, Hiroshima Math. J. 7, 833-853 (1977); Supplement ibid 10, 229-231 (1980).

[18] Patrizio, G., Parabolic exhaustions for strictly convex domains, Manuscripts Math. 47 271-309 (1984).

[19] Patrizio, G., A characterization of complex manifolds biholomorphic to a circular domain, to appear in Math Z.

[20] Patrizio, G. and Wong, P.M., Stability of the Monge-Ampere foliation, Math. Ann. 263, 13-29 (1983).

[21] Royden, H. and Wong, P.M., Caratheodory and Kobayashi metrics on convex domains, preprint, (1984).

[22] Shiffman, B., On holomorphic curves and meromorphic maps in projective space, Indiana Univ. Math. J. 28, 627-641 (1979).

[23] Sibony, N. and Wong, P.M., Some results on global analytic sets. Sem. P. Lelong/H. Skoda, années 1978-79, Springer Lecture Notes Vol. 822, 221-237.

[24] Sibony, N. and Wong, P.M., Some remarks on the Casorati-Weierstrass theorem, Ann. Polon. Math XXXIX, (1981).

[25] Siu, Y.T. and Yau, S.T., Complete Kähler manifolds with non-positive curvature of faster than quadratic decay. Ann. of Math. 105, 225-264 (1977).

[26] Stoll, W., Die beiden Hauptsätze der Wertverteilungstheorie bei Funktionen mehrerer komplexer Veränderlichen I, Acta Math. 90, 1-115 (1950) II ibid 92, 55-169, 1954.

[27] Stoll, W., Mehrfache Integrale auf komplexan Mannigfaltigkeiten. Math. Z. 57, 116-154 (1952).

[28] Stoll, W., About the value distribution of holomorphic maps into projective space, Acta Math. 123, 83-114, (1969).

[29] Stoll, W., Value distribution on parabolic spaces, Springer Lecture Notes Vol. 600 (1977).

[30] Stoll, W., The characterization of strictly parabolic manifolds, Ann. Scuola Norm. Sup. Pisa, VII, 87-154, (1980).

[31] Stoll, W., The characterization of strictly parabolic spaces, Compositio Math. 44, 305-373 (1981).

[32] Stoll, W., The Ahlfors-Weyl theory of meromorphic maps on parabolic manifolds, Springer Lecture Notes 981, 101-219 (1983).

[33] Vitter, A., The lemma of the logarithmic derivative in several complex variables, Duke Math. J. 44, 89-104, (1977).

[34] Weyl, H. and Weyl, J., Meromorphic functions and analytic curves, Princeton University Press, (1943).

[35] Wong, P.M., Defect relations for maps on parabolic spaces and Kobayashi metrics on projective spaces omitting hyperplanes, Notre Dame Thesis (1976).

[36] Wong, P.M., Geometry of the complex homogeneous Monge-Ampere equation, Invent. Math 67, 261-274 (1982).

[37] Wong, P.M., On umbilical hypersurfaces and uniformization of circular domains, Proc. Symp., Pure Math, Vol. 41, 225-252 (1984).

[38] Wu, H., The equidistribution theory of holomorphic curves, Ann. of Math. Studies 64, Princeton Univ. Press (1970).

[39] Wu, H., Mappings of Riemann surfaces (Navanlinna theory), Proc. Symp. Pure Math, 11, 480-532 (1968).

[40] Yau, S.T., On the Ricci curvature of compact Kähler manifolds and complex Monge-Ampere equations I, Comm. Pure and Appl. Math., 31, 339-411 (1978).

Wilhelm Stoll

Value Distribution Theory for Meromorphic Maps

1985. XII, 347 pp. 16,2 X 22,9 cm. (Aspects of Mathematics, Vol. E7, ed by Klas Diederich.) Softcover

Contents: Preface – Letters – Introduction – Value Distribution Theory for Fixed Targets – Value Distribution Theory for Moving Targets – Hermitian Geometry – Meromorphic Maps on Parabolic Manifolds – The First Main Theorem – Associated Maps – Frenet Frames – The Ahlfors Estimates – General Position – The Second Main Theorem – Value Distribution over a Function Field – An Example – The Theorem of Nevanlinna-Mori – References – Index.

In this book a value distribution theory for meromorphic maps is developed. A meromorphic map from a parabolic manifold into complex projective space is considered for its intersection with a family of hyperplanes as targets. The movement of the targets is parameterized by meromorphic maps into the dual projective space. A Nevanlinna theory with several First Main Theorems, Second Main Theorems, and Defect Relations is created. The recent work of B. Shiffman and S. Mori is extended. The Ahlfors-Weyl Theory modified by the curvature method of Cowen and Griffiths is used. In part A of the introduction, the theory for fixed targets is sketched to provide the background for those who are familiar with complex analysis but who are not acquainted with value distribution theory. In part B of the introduction, the results of this book are outlined. In eleven chapters, the theory is developed. All the pertinent concepts are explained.

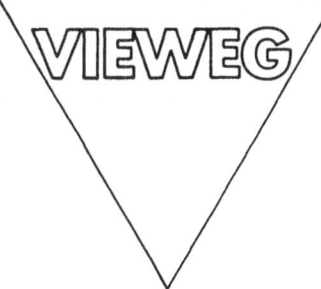

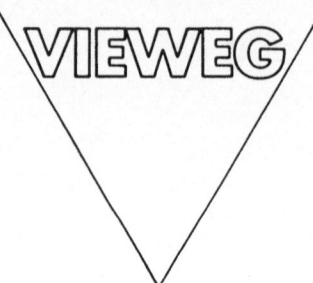